Evolutionary Games
and Population Dynamics

Evolutionary Games
and Population Dynamics

Josef Hofbauer, Karl Sigmund

CAMBRIDGE
UNIVERSITY PRESS

PUBLISHED BY THE PRESS SYNDICATE OF THE UNIVERSITY OF CAMBRIDGE
The Pitt Building, Trumpington Street, Cambridge, United Kingdom

CAMBRIDGE UNIVERSITY PRESS
The Edinburgh Building, Cambridge CB2 2RU, UK
40 West 20th Street, New York, NY 10011-4211, USA
477 Williamstown Road, Port Melbourne, VIC 3207, Australia
Ruiz de Alarcón 13, 28014 Madrid, Spain
Dock House, The Waterfront, Cape Town 8001, South Africa

http://www.cambridge.org

First published 1998
Reprinted 2002

Typeset in 10pt Monotype Times [TAG]

A catalogue record for this book is available from the British Library

ISBN 0 521 62365 0 hardback
ISBN 0 521 62570 X paperback

Transferred to digital printing 2003

Contents

Preface

This book replaces our *Theory of Evolution and Dynamical Systems*, which was published in 1988 and has been reprinted several times. It now deserves to be put to rest.

The present text, which is totally restructured and contains a lot of new material, is no longer an interdisciplinary exploration but a tightly organized mathematical textbook on replicator dynamics and Lotka–Volterra equations.

Two important developments during the last decade have made it imperative to write this new book. Within the social sciences, game theory has gained a lot of ground; and within game theory, evolutionary and dynamical aspects have exploded. In our former book, it took us 150 pages of biological motivation to tentatively introduce the notion of a replicator equation. This is no longer warranted today: replicator dynamics is a firmly established subject, and it has grown so tremendously that our old volume definitely looks dated today.

It was exciting for us to see how many mathematical results obtained within the last decade could now be added to the curriculum. Of course, we had to economize elsewhere. This meant that the chapters on ecology, genetics and sociobiology which introduced the biological ideas underlying the mathematical models had to go. No regrets! All these aspects have been covered by one of us, in more readable form and for a general audience, in the Penguin book *Games of Life*. But we have kept to the principle of introducing every new aspect of replicator dynamics by a simple example of basic importance.

The need for a dynamical approach to game theory was felt by John von Neumann and Oskar Morgenstern already. In fact, one can argue that the very term of *moves* of a game suggests *motion* already. The static 'solutions' of classical game theory obtained by analysing the behaviour of 'rational agents' are fairly unrealistic. Clearly, every form of real-life behaviour is shaped by trial and error. Such stepwise adaptation can occur through individual learning or through natural selection. It is no coincidence that the single most decisive impetus for evolutionary game theory came from a theoretical biologist, namely John Maynard Smith. This was not the only impetus, of course. With hindsight, it was understood that John Nash (a winner of the Nobel prize in 1994) had a population dynamical setting in mind when he conceived his *equilibrium* notion, and that Reinhard Selten (who shared that prize) had taken, with his principle of the *trembling hand*, an essential step away from the rationality doctrine.

Dynamical models, like the method of fictitious play by Brown and Robinson or the replicator equation introduced by Taylor and Jonker, were originally used as tools for studying equilibria. But during the last few years, it has become increasingly clear that the analysis of equilibria cannot be nearly enough.

This book approaches game theory as a branch of dynamical systems. The first part, accordingly, sets up the dynamical framework by an elementary study of Lotka–Volterra equations, which form the backbone of ecological modelling and are equivalent to replicator equations. The second part, which is the core of the book, offers a systematic introduction to non-cooperative games via replicator dynamics and many other game dynamics. Part three explores the global properties of replicator dynamics, and in particular the notion of permanence. Part four turns to genetics and investigates, in particular, the connection between the strategic approach of evolutionary game theory and the genetic mechanisms of selection, mutation and recombination.

The book is divided into chapters and sections. The theorems and exercises are referred to by their sections, and numbered consecutively. Many results are given in the form of exercises.

We are indebted to E. Akin, E. Amann, A. Arneodo, U. Berger, K. G. Binmore, I. Bomze, P. Brunovský, R. Bürger, C. Cannings, F. B. Christiansen, J. E. Cohen, R. Cressman, P. Coullet, T. Czárán, O. Diekmann, U. Dieckmann, M. Eigen, I. Eshel, H. I. Freedman, B. Garay, A. Gaunersdorfer, S. A. H. Geritz, K. P. Hadeler, W. H. Hamilton, P. Hammerstein, W. G. S. Hines, V. Hutson, V. Jansen, Y. M. Kaniovski, G. Kirlinger, M. Koth, M. Krupa, Y. A. Kuznetsov, R. Law, O. Leimar, S. Lessard, K. Lindgren, M. Lipsitch, D. O. Logofet, V. Losert, R. M. May, J. Maynard Smith, J. A. J. Metz, P. Molander, D. Monderer, T. Nagylaki, J. Nash, S. Nee, M. A. Nowak, G. Parker, M. Plank, M. Posch, D. Rand, K. Ritzberger, K. H. Schlag, P. Schuster, R. Selten, H. L. Smith, J. W. H. So, P. F. Stadler, Y. M. Svirezhev, J. Swinkels, E. Szathmáry, Y. Takeuchi, B. Thomas, J. van

Baalen, E. van Damme, P. van den Driessche, G. T. Vickers, P. Waltman, J. Weibull, F. Weissing, P. Young, E. C. Zeeman and M. L. Zeeman for helpful discussions.

M. Posch deserves our special thanks for producing the figures and helping with TEXnical questions.

Introduction for game theorists

The decline and fall of the rational player

Evolutionary game theory has been a latecomer in the evolution of game theory. The initial aim of game theorists was to find principles of rational behaviour, by means of thought experiments involving fictitious players who were assumed to know such a theory, and to know that their equally fictitious co-players would use it. At the same time, it was expected that rational behaviour would prove to be optimal against irrational behaviour too. It turned out that this was asking for too much.

The fictitious species of rational players reached a slippery slope when the so-called 'trembling hand' doctrine became common practice among game theorists. According to this eminently sensible approach, a perfect strategy would take into account that the co-player, instead of being a faultless demi-god, occasionally does the wrong thing. How often is 'occasionally', one may ask, and what does it matter whether the players lucidly conceived the right move but failed to implement it? From allowing for an infinitesimal margin of error to assuming that the faculties of the players are limited, it takes only a small step; and once the word of 'bounded rationality' went the round, the mystique of rationality collapsed. It was like observing that the Emperor was only boundedly covered by his new clothes.

For game theory, the rout of the rational players proved an unmixed blessing. It opened up a vast realm of applications in the social sciences ranging from ethics to economics, and from political affairs to animal behaviour. As soon as players were no longer constrained to be rational, they could learn, adapt, and evolve.

It became a major task of game theory to describe the dynamical outcome of model games defined by strategies, payoffs, *and* adaptive mechanisms, rather than to prescribe 'solutions' based on a priori reasoning. The simplest examples showed that it could not be taken for granted that dynamic adaptations would always lead to stationary solutions.

John von Neumann

Let us start with a highly simplified version of Poker. Suppose that there are only two players in the game, say Johnny and Oscar, and two cards, for instance an Ace and a King. At the start, both players pay one dollar into the pot. Then Johnny draws a card, and looks at it. He now has the option of folding (in which case Oscar gets the pot) or of raising the stakes by adding one dollar to the pot. Now it is Oscar's turn. Oscar, who does not know Johnny's card, can either fold (thereby losing his original dollar) or add one further dollar to the pot. In this case, Johnny has to show his card. If it is the Ace, he wins the pot; if it is the King, Oscar does.

Each player has two strategies. Johnny may as well decide in advance what he is going to do, before even looking at his card. He can chose between two strategies, which we call 'bluff' (raise the stakes no matter which card is drawn) and 'no bluff' (raise only if the card is an Ace). Oscar, who has to act only if Johnny raised the stakes, can also decide in advance what to do: he can chose between 'call' (add a dollar if Johnny did) and 'no call'.

The outcome is uncertain, since it depends on the card Johnny draws. But the expected payoff for Johnny is easy to compute in every case. If he choses 'no bluff' and Oscar choses 'call', for instance, then Johnny's expected payoff is fifty cents. Indeed, with probability 1/2, Johnny will draw the Ace, raise the stakes, and win 2 dollars. With probability 1/2, Johnny will draw the King, fold, and lose 1 dollar. The payoff for Johnny is encapsulated in the following matrix. Note that what one player wins is what the other loses: this is a zero-sum game.

	if Oscar calls	if Oscar does not call
if Johnny bluffs	0	1
if Johnny does not bluff	1/2	0

Clearly, if Johnny never bluffs, Oscar should never call. But if Johnny can count on Oscar never calling, he will always bluff. Of course, Oscar should wise up and start always calling. But then, Johnny should stop bluffing altogether, etc. The players will quickly start to use random strategies, and bluff (or call) with certain probabilities only.

Let x be the probability that Johnny bluffs, and y the probability that Oscar calls. Johnny's expected payoff is $x + y/2 - (3/2)xy$. Suppose he choses an $x > 1/3$. If Oscar can guess this, he will certainly call, and thus minimize Johnny's payoff to $(1 - x)/2$, a number less than $1/3$. If Johnny bluffs with a probabilty $x < 1/3$, however, then Oscar, if he can guess this, won't call, thereby minimising Johnny's payoff to x, which again is less than $1/3$. But if Johnny bluffs with a probability exactly equal to $1/3$, then he will have an expected gain of one-third of a dollar, no matter what Oscar does. Thus Johnny has found a way of *maximising his minimal payoff*. Oscar can do just the same. If he choses $y = 2/3$, he can guarantee that Johnny gains not more than one-third of a dollar, on average. Every other value of y would allow Johnny to get away with more — *if* he manages to guess it. The strategies given by $x = 1/3$ and $y = 2/3$ are maximin strategies — the best if one assumes the worst. But why should one always assume the worst?

Suppose that Johnny, for instance, is a timid person and does not dare to bluff with a probability as high as $x = 1/3$. He will not be penalized for deviating from the right probability, as long as Oscar keeps to his equilibrium value $y = 2/3$. And Oscar will not be penalized either. Of course he would be better off if he switched to calling less frequently. The right reply for Johnny, in that case, is not to meekly reassume his maximin strategy, but to overcome his timidity and switch all the way to always bluffing. Thus the maximin solution seems a rather spurious equilibrium. If Johnny deviates from his maximin strategy, he is not led back to it. Rather it is Oscar who is led to deviate too.

John Nash

The maximin solution is used, not just for this simplified version of Poker, but for all zero-sum games where the gain of one player is the loss of the other. But most games are not zero-sum. Consider the following game, usually called Chicken: Johnny and Oscar have the option to escalate a brawl or to give in. If both give in, they get nothing. If only one player gives in, he pays 1 dollar to the other. But if both escalate the fight, each has an expected loss of 10 dollars, say, for medical treatment. The payoff matrix for Johnny is given by the following matrix:

	if Oscar escalates	if Oscar yields
if Johnny escalates	−10	1
if Johnny yields	−1	0

Clearly, by giving in, Johnny can maximize his minimal payoff. Oscar is in the same position: he also could maximize his minimal payoff by giving in. But will both players give in? Hardly so. If they guess that the other will give in, they will certainly escalate. But if both escalate, both are worse off.

If x and y are the probabilities that Johnny and Oscar escalate, then the expected payoff for Johnny is $-10xy + x - y$. If Oscar escalates with a probability larger than $1/10$, Johnny should quit. If Oscar escalates with a probability smaller than $1/10$, Johnny should escalate. If both Johnny and Oscar escalate with a probability of exactly $1/10$, they are in a *Nash equilibrium*: neither of them has anything to gain by deviating unilaterally from his equilibrium. But neither Johnny nor Oscar has any reason *not* to deviate from $1/10$, either, as long as the other sticks to $1/10$. Oscar has no reason to care one way or the other, as long as Johnny escalates with a probability of $1/10$. But if Johnny has any reason for believing that Oscar escalates with a higher probability, he should never escalate; and if Oscar suspects that Johnny has such a reason, then he should certainly escalate. What should Oscar do, for instance, if Johnny has escalated twice in the first five rounds? Should he conclude that this was a statistical fluke? Even if it were such a fluke, Johnny could reasonably suspect that Oscar would attribute it to a higher propensity to escalate. Again, the argument for $1/10$ looks rather spurious.

John Maynard Smith

Oddly enough, it was a biologist who offered a convincing explanation. John Maynard Smith, who was studying animal contests at the time, viewed the Chicken game in a *population-dynamical* setting. There were no longer just Johnny and Oscar engaged in the game, but a large number of players meeting randomly in contests where they had to decide whether to escalate or not. It makes a lot of sense, now, to assume that the players escalate with a probability of $1/10$. Indeed, if the overall probability were higher, it would obviously pay to escalate less often, and vice versa. In this sense, *self-regulation* leads to the value of $1/10$ — self-regulation, not between two players, but within a population.

The value $1/10$, then, is an example of an evolutionarily stable strategy. It is the result of a population-dynamical approach which considers questions like: When will the frequencies of certain strategies increase? When will they reach a stable equilibrium?

Enter ecology

Such self-regulation is reminiscent of the self-regulation encountered in population ecology. Interacting species can regulate their population densities.

A scarcity of prey, for instance, will cause a population of predators to dwindle; as a result, the number of prey will increase. This, in turn, will cause the frequency of predators to increase, eventually leading to a decline in the prey, etc. This looks a lot like the up and down of the frequencies of calling and bluffing in Poker.

On second thoughts, it therefore appears considerably less surprising that a biologist would approach game theory with population dynamics in mind. It comes quite natural to naturalists to think of *self-regulation via frequency dependence*. This is a long-established theme among ecologists. Charles Darwin had already been thrilled by this dynamical aspect. He relished working out how, if 'certain insectivorous birds were to increase in Paraguay', a species of flies would decrease; how — since these flies parasitize newborn calves — this decrease would cause cattle to become abundant; which 'would certainly greatly alter the vegetation'; and 'how this again would largely affect the insects; and this again the insectivorous birds ... and so onwards in ever-increasing circles of complexity.' The mathematics underlying this complexity is the theory of dynamical systems.

Evolutionary game theory proved very popular with economists who had not felt too comfortable with the classical approach of analysing the behaviour of unboundedly rational players. It even led to a modest revival of rationality; it turned out more than once that the prescriptions for rational play agreed with the outcomes of game dynamics. And in retrospect, it was even discovered that John Nash had had a population setting in mind when he introduced his equilibrium notion. In his unpublished thesis he wrote 'it is unnecessary to assume that the participants have ... the ability to go through any complex reasoning processes. But the participants are supposed to accumulate empirical information on the various pure strategies at their disposal .... We assume that there is a population ... of participants ... and that there is a stable average frequency with which a pure strategy is employed by the "average member" of the appropriate population'.

This *mass action interpretation* foreshadows the population dynamical point of view of evolutionary game theory by more than twenty years. Why was it not pursued for such a long time? One possible reason is that evolutionarily stable equilibria do not always exist (whereas Nash equilibria do). It may well be that the adaptation process between players does not converge to a standstill. But the self-regulation of population densities in an ecosystem needn't converge either. The possibility of chaotic population oscillations was understood at about the same time as the ideas of evolutionary game theory started spreading.

One may safely conclude that ecology is the godfather of evolutionary game theory. We stress this theme throughout our book. The most common models for the dynamics of population numbers in ecosystems (the Lotka–Volterra equations) and the most common models for the dynamics of frequencies of strategies (the replicator equations) are mathematically equivalent. It pays to study them together.

Introduction for biologists

Striking a balance

If we repeated Noah's experiment — starting a new ecosystem with one couple of each species — we would certainly not expect a restoration of the *old régime*. Numbers matter. The fate of a population depends on the frequencies of other populations.

The interdependency of different species can be wonderfully intricate. Darwin relished working out 'how plants and animals, most remote in the scale of nature, are bound together by a web of complex relations', pointing out, as an instance, that bumble-bees are indispensable to the fertilization of heartsease, and that field-mice cause havoc among the nests and combs of bumble-bees. Since the number of mice is largely dependent on the number of cats, it is consequently 'quite credible that the presence of a feline animal in large numbers might determine, through the intervention first of mice and then of bees, the frequency of certain flowers!'

This self-regulation of population frequencies has been a dominant theme of mathematical ecology. It started in the 1920s with Alfred Lotka modelling the cycle of mosquitoes and humans in transmitting malaria, and Vito Volterra analysing the dynamics of predators and prey among fish in the Adriatic. They came up with differential equations describing the dynamics of such systems. But the first generations of mathematical ecologists concentrated mostly on investigating static aspects of ecological communities. Some of their main problems, like the validity of the exclusion principle (when can there be more species than niches?) and the relation between the complexity and the stability of an ecosystem (do species that are more interconnected produce assemblies that are more robust?), were phrased in terms of stability properties of equilibria. Only in the 1970s did the prevalence of irregular oscillations, which had always been known to field workers, filter down to theoreticians. Ecological models became a major impetus to chaos theory.

Survival vs. equilibrium

For a long time, the efforts of mathematical ecologists to analyze the stability of bio-communities were marred by a misunderstanding. If one simply adopts the stability concepts of physicists or engineers, then one will call an ecosystem stable if population numbers converge to an equilibrium which is promptly re-assumed after every small perturbation. But field ecologists would not expect to find, in the wild, the static, well-controlled state of affairs implied by such a stability notion. For ecologists unspoilt by physics courses, the proverbial lynx–hare cycle, whose undamped oscillations have been recorded for two hundred years, epitomizes stability. They little care whether the population numbers converge, or oscillate in a regular or chaotic fashion. For such ecologists, stability means that population numbers do not vanish; that the species making up the ecosystem do persist. Survival, not equilibrium, is what counts. This second form of stability is one of the main themes of our book.

All biological communities are transient, of course; but some are more so than others. They collapse right away, without having to wait for the construction of an interstate highway, the mutation of a parasitic strain, or a series of harsh winters. They are doomed from the start: they are *unsustainable* — that is, impossible in the long run. They do occur in nature, but it is hard to make them out before they are replaced by less fleeting configurations. Ecologists who wish to understand what happened have no time for a leisurely *post mortem*. Yet they must know why communities fail if they want to learn about those which persist. An empirical approach to this question is always difficult and often painful; mathematical models are less risky and, if they are cleverly set up, more revealing.

The history of ecological communities is chronicled in terms of invasions and extinctions. The fate of a population depends mostly on what happens when it is rare. Can an invader spread? Can a population recover after being decimated? Every new adaptation is initially rare, and every lineage has been tested in countless bottlenecks. A population must be able to gain a foothold when in a minority. It must be able to grow in a world which is essentially determined by the others. Only then will its numbers become large enough to affect the environment, the frequencies of the other species, and its own growth. If we translate this into static notions, we find that we have to investigate two types of stability for an equilibrium: *inner stability* (will a small perturbation of the prevailing distribution be offset by self-regulation?) and *outer stability* (will a new population entering as a minority be able to grow?). Obviously the same two questions apply to dynamic regimes, too: (a) what happens if one adds or removes a few heads of a population currently present in the ecosystem, and (b) what happens if one adds a few heads of a new, intruding species?

Breeding and games

Every growth will of course eventually be checked. Darwin termed this the 'struggle for life', stressing that it often had little to do with animals actually fighting each other. He compared the 'face of nature' to a yielding surface, 'with ten thousand sharp wedges packed close together and driven inwards by incessant blows.' And he emphasized that this struggle of life was almost invariably most severe between individuals and varieties of the same species.

This suggests applying ecological modelling to animal behaviour, since different types of behaviour correspond to different varieties. One of the first to think about behaviour in terms of invasion and self-regulation of frequencies was John Maynard Smith. He applied it originally to actual contests within one species, using it to explain the prevalence of conventional fighting. Stags, for instance, engage in roaring contests, a parallel walk and a pushing match with interlocked antlers. Only rarely do they escalate to an all-out fight likely to have a lethal conclusion. Their usual restraint is obviously good for the species, but this advantage cannot explain it.

John Maynard Smith and John Price couched the contests in game theoretic terms. Stags, in their thought experiment, came in two brands: the 'hawks' ruthlessly escalating every contest, and the 'doves' sticking to conventional displays and fleeing whenever their opponent gets rough. In a population of doves, hawks do well and will spread, since they win all their contests. But in a population of hawks, they have only a fifty per cent chance of winning; it is just as likely that they will end up seriously injured. Doves, who avoid this fearsome risk, will do much better and spread. Hence hawk populations can be invaded by doves and vice versa; the outcome should be a mixture where the frequency p of hawks is inversely proportional to the cost of an injury (a cost expressed in fitness, i.e. reproductive success). Among heavily armed species like stags, this cost is very high, so that escalated conflicts will be rare.

A fighting behaviour corresponds in game theoretical terms to a strategy. Hawks and doves are so-called pure strategies, and each is a best reply to the other. The structure of the struggle is exactly that of a once popular pastime of American teenagers, which went by the name *Chicken* and attracted the interest of early game theorists. The thing to avoid, in such a game, is to act like your adversary — but you do not know beforehand what he will do. It turns out that it is best to let chance decide — if you know how to weight the dice properly.

Hedging the bets

What about mixed strategies in the hawk–dove game, i.e. behavioural programs telling the player to escalate with a such and such a probability? It is easy to see that there exists one such strategy — namely escalating with the probability p given by the frequency of hawks in a hawk–dove mixture — which is a best reply to itself, a so-called Nash equilibrium. If all stags use this mixed strategy, no stag can expect to do better by escalating with a different probability. Actually, stags that do so will not do worse either. Nevertheless, they cannot invade, and this for a rather subtle reason. The offspring of the mutant (who inherit its propensity for escalation) do as well as the resident population in all contests against the residents; but they do less well than the residents in the (admittedly rare) contests against their own. Hence their type cannot spread: it checks its own increase even when rare.

A similar argument explains the prevalence of the sex ratio 1/2. If the sex ratio (the proportion of males among the offspring) were different, it would pay to produce offspring of the rarer sex. The success of one strategy (produce more sons) or the other (produce more daughters) depends on the mean sex ratio in the population, and hence on the frequencies of the strategies. Success, as always in biology, means reproductive success; hence the successful strategies spread, change the composition in the population, and therefore affect their own success. Again, one can show that investing equally in the production of the two sexes is evolutionarily stable. If everyone does it, any deviation will be self-defeating, because it affects the frequencies of the sexes in the wrong way.

The definition of an evolutionarily stable strategy (if the residents adopt it, no mutant can invade) is based on an implicit dynamics. It is easy to make this dynamics explicit, by assuming that like begets like. This yields the replicator equation describing the evolution of the frequencies of different strategies in a population. This dynamics is not merely a prop to sustain arguments from equilibrium theory. For many games, equilibria alone do not suffice to describe what happens, and a static outcome cannot be expected. The simplest example is the rock–scissors–paper game, where strategy A is beaten by B, which is beaten by C, which is beaten by A. There exists a mixed equilibrium with so much of A, B and C each. But it may happen that this equilibrium is never reached. Instead, the cyclic succession of populations which are almost entirely composed of A, B or C builds up to an increasingly jerky roundabout of upheavals.

It used to be thought that the rock–scissors–paper game was just a conundrum devised for the amusement of theoreticians, until it was found out that lizards do play it: one of their species, *Uta stansburiana*, has three types of male with different mating strategies (they are conveniently distinguished by their throat colour). Type A keeps one female and guards it closely; type B keeps several females, and necessarily guards them less closely; and C guards no female at all and looks out for sneaky matings with unguarded females. The three types can invade each other cyclically.

Similar ratchets can occur in parasitology. The immune system of the host acts as a combination lock which the parasites try to break. By trial and error, they will eventually succeed. Of course they are usually most efficient in attacking the most common immunotype in their host. For the host, it can be deadly to adopt a combination code which is currently widespread in the population. It pays to belong to a minority. But since it pays in offspring — the Darwinian currency — such a minority will yield a new majority, and come under concentrated attack. Every solution is self-defeating in the long run. This leads to arms races without a finish line.

Short vs. long term

In studying evolutionary chronicles, we are led to consider two time scales. Short-term evolution describes how the frequencies of adaptive traits regulate each other via natural selection, i.e. how the distribution of the types actually present in a population changes from generation to generation. Long-term evolution describes how new types can invade through mutation. Not surprisingly, this leads to rather different dynamics, and hence also to different stability notions. In particular, an evolutionarily stable strategy need not attract. It may well be that a population is invasion-proof in the sense that mutations cannot lead out of it, and at the same time inaccessible, so that mutations will lead any nearby populations further away from it. Such 'Garden of Eden' configurations can never occur as evolutionary outcomes. The widespread idea that a population will somehow evolve until it happily reaches the safe haven of evolutionary stability is not always valid. The dynamics of evolution can be a lot more exciting.

Replicator dynamics and adaptive dynamics describe the short-term and long-term evolution under the simplifying assumption that 'like begets like'. Most organisms do not spread by cloning, however, but mix their genes while reproducing. The connection between population genetics and replicator dynamics is a vast field which is currently under intensive study, the tentative consensus being that sexual reproduction certainly complicates the short term dynamics but often allows to the same long term predictions. In any case, it is firmly established that game-theoretical arguments provide an indispensable tool for handling the complexities of frequency-dependent genetics.

This being said, we must not forget that inheritance is just one of several ways for behavioural programs to spread in a population. The dynamics of imitation and learning, which act on a much faster scale, are budding off from the replicator equation and quickly developing into an exciting field of their own. We have tried to do justice to the recent progress in this field. Some aspects, like the intriguing best-response dynamics, can be viewed as a return to the method of fictitious play, one of the earliest dynamical concepts of game theory. It is becoming increasingly clear that transmission mechanisms are as essential for the full description of a game as are strategies and payoffs.

In spite of the diversity of game-dynamical models, replicator dynamics retains its central position, and will most likely keep its benchmark role. It emerges in various guises as an optimal imitation dynamics, and some of its asymptotic features turn up, quite unexpectedly, in adaptation mechanisms which at first glance seem very remote from 'like begets like'-inheritance. Moreover, game-theoretical replicator equations are not only motivated by ecological thinking, but actually lead to exactly the same dynamics as the time-hallowed equations of mathematical ecology. The most common models for the dynamics of population numbers in ecosystems (the Lotka–Volterra equations) and the most common models for the dynamics of frequencies of strategies (the replicator equations) are mathematically equivalent. It pays to study them together.

About this book

The first part of this book is a simple introduction to ecological modelling, aiming mostly at explaining the dynamical aspects (long-term behaviour, equilibria, basins of attraction, regular and irregular oscillations etc.).

The second part is a concise course in the dynamics of evolutionary game theory. We start out with the notion of evolutionary stability, and go on to explore the underlying field of replicator dynamics, which is based on the assumption that strategies spawn copies of themselves in proportion to their average payoff. The success of this approach should not make one forget, however, that there are other ways that strategies can spread, based on learning, for instance, or alternative mechanisms. We have emphasized the wide range of such alternative models in our chapters on adaptive dynamics, best-reply dynamics and imitation dynamics.

In the third part of the book, we concentrate on Lotka–Volterra equations and replicator dynamics and study them in parallel, stressing the special aspects of these dynamical systems, for instance the occurrence of heteroclinic attractors. But we keep the main population dynamical motivation firmly in the foreground, emphasizing notions like permanence and saturated equilibria.

In the last part, we return to the original reason for introducing game dynamics. It was intended at first as a shortcut to understanding frequency-dependent genetics. We deal with dynamical systems in classical genetics describing selection, mutation, recombination and differential fertility, and proceed to address the basic issue which motivated Maynard Smith: how will mutants introducing new behavioural strategies fare under selection?

For game theorists, we suggest skimming through part I, concentrating on part II and winding up with chapter 22 in part IV.

For mathematical ecologists, read part I, chapter 7 in part II, and part III with the exception of chapter 14.

For those whose interests lean more towards evolutionary genetics, we suggest part I, chapters 6, 7, 9 and 10 in part II, chapter 16 in part III and the whole of part IV.

Part one

Dynamical Systems and Lotka–Volterra
Equations

1

Logistic growth

Some of the simplest models for the dynamics of a single population exhibit very complicated behaviour, including bifurcations and chaos.

1.1 Population dynamics and density dependence

One of the major tasks of mathematical ecology is to investigate the dynamics of population densities. In many ecosystems, thousands of different species interact in complex patterns. Even the interactions between two species can be quite complicated, involving the effects of seasonal variations, age structure, spatial distribution and the like. In the following, we shall exclusively investigate the effects of density dependence. As a first approximation, we may distinguish here three basic situations.

(a) *Competition.* Two species may be rivals in the exploitation of a common resource. The more there is of one species, the worse for the other. Because of the importance of competition as a limiting factor in evolution, such situations have attracted considerable attention.

(b) *Mutualism.* This is the reverse situation: the species benefit from each other. The more there is of one species, the better for the other. This is the case for lichen, a coalition between algae and fungi, or for the strange partnership between hermit crabs and sea anemones.

(c) *Host–parasite relationship.* The situation, here, is asymmetrical. The parasites benefit from the host but they do it no good. In this sense, *predators* can be viewed as parasites of their *prey*. (For a parasite, however, the host is not only a resource, but also a means for spreading.)

Before turning to ecological equations modelling such interactions between two or more species, we shall consider in this first chapter the effect of density dependence upon a single species.

1.2 Exponential growth

Let R be the rate of growth of a population with discrete generations. This means that

$$x' = Rx \tag{1.1}$$

where x is the density in one generation and x' in the next. If R remains constant, then the density after t generations will be $R^t x$, which for $R > 1$ means explosive growth to infinity.

Of course such a *population explosion* — a kind of chain reaction — will be checked sooner or later. Before dealing with limits to growth, however, let us consider unrestricted growth for populations whose generations are not discrete, but blend continuously into each other. If $x(t)$ is the population number at time t, then

$$\frac{x(t + \triangle t) - x(t)}{\triangle t}$$

is the average rate of growth in the time interval $[t, t + \triangle t]$. The function $x(t)$ is integer valued and hence not differentiable, of course. Still, if the density is very large, then the jumps caused by individual births and deaths will look negligibly small on a graph of $x(t)$. So let us postulate the existence of a time derivative

$$\frac{dx(t)}{dt} = \lim_{\triangle t \to 0} \frac{x(t + \triangle t) - x(t)}{\triangle t} \tag{1.2}$$

which we shall usually denote by $\dot{x}(t)$. The quantity $\dot{x}/x$, which is equal to $(\log x)'$, may be viewed as the (per capita) *growth rate* of the population, or as the average contribution of one individual to the population growth. If this rate is constant, i.e. if

$$\dot{x} = rx \tag{1.3}$$

then

$$x(t) = x(0)e^{rt} \, . \tag{1.4}$$

For the continuous model, therefore, we have exponential growth again.

1.3 Logistic growth

There are various ways in which exponential growth will eventually be checked. A larger population means fewer resources, and this implies a smaller rate of growth. We shall start by considering the simplest case, where the rate decreases linearly as a function of x. It is then of the form $r(1 - \frac{x}{K})$, with positive constants r and K. This yields the *logistic equation*

$$\dot{x} = rx\left(1 - \frac{x}{K}\right) . \tag{1.5}$$

The behaviour of (1.5) is easy to analyse. The increase $\dot{x}$ is zero if $x = 0$ (obviously) and if $x = K$. In these two cases, the density does not change. For $0 < x < K$, it increases, and for $x > K$ it decreases. The solution of (1.5) is given by

$$x(t) = \frac{Kx(0)e^{rt}}{K + x(0)(e^{rt} - 1)} \tag{1.6}$$

as can be verified directly.

Exercise 1.3.1 Show that (1.6) can be obtained by the method of *separation of variables*. (Write (1.5) as

$$\frac{dt}{dx} = \frac{1}{r}\left(\frac{1}{x} + \frac{1}{K - x}\right),$$

integrate to obtain $t(x)$ and invert this function.) Sketch the shape of the solution.

The behaviour in the region between 0 and K is called *logistic growth*: for very small values of x, it is almost exponential; for larger values it slows down and levels off asymptotically to the constant K. K is the *carrying capacity* of the environment and r the rate of growth for small population numbers.

1.4 The recurrence relation $x' = Rx(1 - x)$

Let us return to populations with non-overlapping generations — this is the case for certain insect populations, for instance. We denote by y and y' the population numbers in one generation and the next. Obviously $(y' - y)/y$ can be viewed as per capita rate of increase. If we assume, as with the logistic equation, that this rate decreases linearly in y, we obtain

$$y' = Ry\left(1 - \frac{y}{K}\right) . \tag{1.7}$$

Equation (1.7) has the unrealistic feature that if the population number y exceeds the carrying capacity K, then $y' < 0$. Thus it cannot be used for values y larger than K. On the other hand if the parameter R is larger than 4, and y is near $K/2$, then y' is larger than K and the model becomes nonsensical again. Hence we have to restrict R to values between 0 and 4. It may well be asked whether it is worthwhile to pursue the study of an equation with such glaring defects. It is highly to be recommended, in fact, but less for biological than for mathematical reasons. (1.7) has been called a 'mathematical morality play' because it helps in understanding some types of dynamic behaviour which also occur in many models which are more realistic.

In order to simplify the discussion, we set $x = y/K$ and $x' = y'/K$. This leads to the difference equation $x' = F(x)$ with

$$F(x) = Rx(1 - x) \ .$$

This is surely the simplest kind of nonlinear recurrence relation: however, as we shall presently see, its dynamics can be of amazing complexity, quite in contrast to its continuous counterpart (1.5). We do not attempt to give a complete discussion: larger volumes than this one would be needed for such a task. But we use it to introduce two central concepts: *bifurcation* and *chaotic behaviour*.

1.5 Stable and unstable fixed points

For values of R between 0 and 4, the map

$$x \to Rx(1 - x) \tag{1.8}$$

defines a dynamical system on the interval $[0, 1]$. The point 0 is obviously a fixed point. If $R \leq 1$, one has $x' < x$ for all $x \in \,]0, 1]$. The orbit of x, then, decreases monotonically to 0. From now on we shall only consider $R > 1$. The graph of F is a parabola intersecting the x-axis at the points 0 and 1 and achieving its maximum $R/4$ at $1/2$. It intersects the diagonal $y = x$ in a unique point $\mathbf{P}$ in the interior of the unit square. The abscissa p of $\mathbf{P}$ satisfies $p' = p$ and corresponds therefore to a further fixed point $p = \frac{R-1}{R}$ (see fig. 1.1).

What about the stability of p? By the mean value theorem,

$$F(x) - p = F(x) - F(p) = (x - p)\frac{dF}{dx}(c)$$

for some suitable c between x and p. If $|\frac{dF}{dx}(p)| < 1$ then $|\frac{dF}{dx}(c)| < 1$ whenever

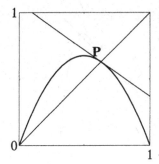 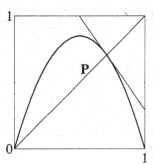

Fig. 1.1. The graph of the logistic map (1.8) for $R = 2.7$ and $R = 3.4$

x (and hence c) is sufficiently close to p. Then

$$|F(x) - p| < |x - p|$$

i.e. x' is closer than x to the fixed point p. This implies that if x is in some suitable neighbourhood of p, then the orbit of x, i.e. the sequence of its iterates under (1.8), remains close to p and even converges to p. The fixed point p is *asymptotically stable* in this sense. If $|\frac{dF}{dx}(p)| > 1$, however, then

$$|F(x) - p| > |x - p|$$

i.e. the orbit of x is driven away from the fixed point. In such a case p is *unstable*.

Now $F(x) = Rx(1 - x)$, hence $\frac{dF}{dx}(p) = 2 - R$, i.e. p is asymptotically stable for $1 < R < 3$, but not for $3 < R < 4$. What happens if p is unstable?

Exercise 1.5.1 Show that for $1 < R \leq 3$, p is even a global attractor, i.e. *all* orbits in $]0, 1[$ converge to p. For $R \leq 2$ this convergence is ultimately monotone, for $2 < R \leq 3$ the orbits ultimately oscillate around p.

Exercise 1.5.2 Study the stability of the fixed point of the difference equation $x' = xe^{r(1-x)}$.

Exercise 1.5.3 Do the same for the map $x' = Rx(1 + ax)^{-\beta}$.

1.6 Bifurcations

In two generations, x is transformed into $F(F(x)) = F^{(2)}(x)$. In our case, $F^{(2)}(x)$ is a polynomial of degree 4, with a local minimum at $1/2$ and two local maxima, symmetrically to the left and to the right of $1/2$. (See fig. 1.2.) Again, the diagonal and the graph of $F^{(2)}$ intersect at **P**. Indeed,

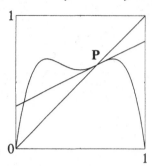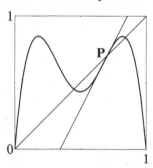

Fig. 1.2. Graphs of the second iterate of the logistic map for $R = 2.7$ and $R = 3.4$

$p = F(p) = F^{(2)}(p)$. The derivative of $F^{(2)}$ at p is given by the chain rule as $(2 - R)^2$.

If $1 < R \leq 3$, i.e. if p is asymptotically stable, then **P** is the only intersection of the diagonal with the graph of $F^{(2)}$; but for $3 < R < 4$ one obtains to the left and to the right an additional intersection, since the slope of the tangent at **P** is larger than 1. These two intersections correspond to points p_1 and p_2 with period 2: indeed, since $F^{(2)}(p_1) = p_1$ and $F^{(2)}(p_2) = p_2$, one must have $F(p_1) = p_2$ and $F(p_2) = p_1$.

Quite generally, a point **x** is said to be a *periodic point* for the dynamical system $T : X \to X$ if there exists a $k > 1$ such that $T^k \mathbf{x} = \mathbf{x}$ (but $T^j \mathbf{x} \neq \mathbf{x}$ for $j = 1, \ldots, k - 1$). The integer k is called the *period* of **x**. The occurrence of periodic oscillations in our model of population growth is not surprising. In contrast to the continuous adjustment of the rate of growth in the logistic equation (1.5), the regulation in the discrete model (1.8) operates with a delay of one generation. This delay can lead to 'overshooting', i.e. to jumping from one side of the fixed point to the other, and to oscillations which refuse to settle down. Similar effects plague many steering mechanisms and control devices.

The parameter value $R = 3$ is a *bifurcation point*: for slightly smaller values of R, the fixed point p is asymptotically stable and no periodic oscillations take place; for slightly larger values of R, p is no longer stable and periodic points appear. Thus the behaviour of the dynamical system (1.8) changes drastically if the parameter R (under some exterior influence, say) crosses the level 3.

One can show that if R is not much larger than 3, the points p_1 and p_2 are asymptotically stable as fixed points of $x \to F^{(2)}(x)$. For larger values of R, they become unstable. More bifurcations occur: the number of periodic points grows. The periods, at first, are all of the form 2^n. For

$R = 3.5700$ there are already infinitely many periodic points. For $R = 3.6786$ the first odd periods appear, from $R = 3.82\cdots$ onwards, all periods occur. The structure of the orbits becomes extremely complicated, and the system behaves 'chaotically' in a way we shall briefly describe by an example.

Exercise 1.6.1 Show a similar phenomenon for the difference equation from exercise 1.5.2. For which values of β does period doubling occur in the map from exercise 1.5.3?

Exercise 1.6.2 If x is a periodic point of period n for (1.8) then the time average $\frac{1}{n}(x + F(x) + \cdots + F^{(n-1)}(x)) < p$. For the map from exercise 1.5.2 there is equality.

1.7 Chaotic motion

Let us consider (1.8) with $R = 4$. The maximum of the parabola reaches 1. The intervals $I_0 = [0, \frac{1}{2}]$ and $I_1 = [\frac{1}{2}, 1]$ are both mapped by F bijectively onto the entire interval $[0,1]$. I_0 can be decomposed into two compact subintervals $I_{00} = [0, q]$ and $I_{01} = [q, \frac{1}{2}]$ which are mapped onto I_0 and I_1, respectively (here $q = \frac{1}{2}(1 - \frac{1}{\sqrt{2}})$). Similarly, I_1 can be decomposed into $I_{10} = [1 - q, 1]$ and $I_{11} = [\frac{1}{2}, 1 - q]$: the first interval is mapped onto I_0, the second one onto I_1.

Hence $[0, 1]$ is decomposed into two compact intervals of rank 1, namely I_0 and I_1, each of which is mapped by F bijectively onto $[0,1]$. I_0 and I_1, in turn, are decomposed into compact intervals of rank 2, namely I_{00}, I_{01}, I_{10} and I_{11}, each of which is mapped by $F^{(2)}$ onto $[0,1]$. Each of these intervals, in turn, can be decomposed into two compact subintervals of rank 3; one of them is mapped by $F^{(2)}$ onto I_0, the other one onto I_1.

We can repeat this procedure inductively, and obtain 2^n compact intervals of rank n. Each of them is mapped by $F^{(n)}$ bijectively onto $[0,1]$, and hence may be decomposed into one subinterval mapped onto I_0 and one mapped onto I_1. These are the subintervals of rank $n + 1$.

The notation for these intervals is suggestive of the *binary expansion*. If $x \in I_{0100}$, for example, then $x \in I_0$, $F(x) \in I_1$, $F^{(2)}(x) \in I_0$, $F^{(3)}(x) \in I_0$.

Hence, if we are given any finite, or even infinite, sequence of zeros and ones — obtained, for example, by repeatedly tossing a coin — we can find a point which visits I_0 and I_1 in the corresponding sequence. On the other hand, even if two points are extremely close, they will lie in different subintervals of rank n (for sufficiently large n), and hence have quite uncorrelated fates. Since we know the initial value x only up to some error, we cannot predict its orbit too far into the future. (In fact, even if we knew it precisely, our computer would introduce some round-off error.) This 'randomness' of (1.8) is not restricted to $R = 4$. It holds for many other parameter values as well.

What is the lesson, then, of this 'morality play'? Essentially that 'computable' is not 'predictable', and that deterministic motion may be indistinguishable from random motion in the long run. Even an innocent-looking recurrence relation may display bifurcations and chaotic oscillations. Another lesson is that the corresponding model for continuous generations may be quite tame, compared with its discrete counterpart. This is, to be sure, not always so, and continuous dynamics can offer all kinds of exciting behaviour. Still, if one looks for regularity, it often pays to use as model a differential rather than a difference equation.

Exercise 1.7.1 The map $S : [0, 1] \rightarrow [0, 1]$ defined by $S(y) = 2y$ (mod 1) sends y onto $2y$ for $y \in [0, \frac{1}{2}]$ and onto $2y - 1$ for $y \in [\frac{1}{2}, 1]$. Show that this map is chaotic: in fact, for almost all y, the orbit $S^n(y)$ is uniformly distributed in the sense that the frequency of its visits in every subinterval $[a, b]$ of $[0, 1]$ is proportional to the length of the subinterval, i.e.

$$\frac{1}{N} \text{card}\{n : 0 \leq n \leq N - 1 \quad \text{and} \quad S^{(n)}(y) \in [a, b]\}$$

converges to $b - a$, when $N \rightarrow \infty$.

Exercise 1.7.2 This map S is conjugate to the map $T : x \rightarrow 4x(1 - x)$ in the sense that orbits under S are mapped to orbits under T by a bijection $\theta : [0, 1] \rightarrow [0, 1]$, namely $\theta(y) = \sin^2(\frac{\pi y}{2})$. Show that for almost all x, the orbit $T^n(x)$ visits the subinterval $[a, b]$ with a frequency converging to

$$\int_{[a,b]} \frac{1}{\sqrt{\pi x(1 - x)}} dx.$$

1.8 Notes

The complex behaviour of (1.8) was displayed in a landmark article by Robert May (1976) which contributed a lot to launch chaos upon an unsuspecting world. For a superb introduction to the intricacies of one-dimensional chaos, see Devaney (1986). For the advanced reader we recommend de Melo and van Strien (1993). The relevance of chaos for evolutionary ecology is discussed in Ferrière and Fox (1995), see also Ferrière and Gatto (1995).

2

Lotka–Volterra equations for predator–prey systems

*The Lotka–Volterra equations describing the dynamics of interacting popula-
tions of predators and prey are used to introduce some of the basic notions
of the qualitative theory of dynamical systems, including constants of motion,
Lyapunov functions and ω-limits.*

2.1 A predator–prey equation

In the years after the First World War, the amount of predatory fish in the
Adriatic was found to be considerably higher than in the years before. The
hostilities between Austria and Italy had disrupted fishery to a great extent,
of course, but why was this more favourable to predators than to their prey?
When this question was posed to Vito Volterra, he denoted by x the density
of the prey fish, by y that of the predators, and came up with a differential
equation which explained the increase in predators.

Volterra assumed that the rate of growth of the prey population, in the
absence of predators, is given by some constant a, but decreases linearly as
a function of the density y of predators. This leads to $\dot{x}/x = a - by$ (with
$a, b > 0$). In the absence of prey, the predatory fish would have to die, which
means a negative rate of growth; but this rate picks up with the density x
of prey fish, hence $\dot{y}/y = -c + dx$ (with $c, d > 0$). Together, this yields

$$\begin{aligned}
\dot{x} &= x(a - by) \\
\dot{y} &= y(-c + dx) \,.
\end{aligned} \tag{2.1}$$

This is a differential equation. What can one do with it?

11

2.2 Solutions of differential equations

We shall write

$$\dot{\mathbf{x}}(t) = \mathbf{f}(t, \mathbf{x}) \tag{2.2}$$

for the *ordinary differential equation* (or ODE)

$$\dot{x}_i = f_i(t, x_1, \dots, x_n) \qquad i = 1, \dots, n . \tag{2.3}$$

Here, the functions f_i are defined on some open subset of $\mathbb{R}^{n+1}$ and continuously differentiable in all variables. A *solution* is a map

$$t \to \mathbf{x}(t) = (x_1(t), \dots, x_n(t)) \tag{2.4}$$

from some interval I into $\mathbb{R}^n$ such that the components $x_i(t)$ are differentiable and satisfy (for $i = 1, \dots, n$ and all $t \in I$):

$$\dot{x}_i(t) = f_i(t, x_1(t), \dots, x_n(t)).$$

We picture this in the following way: at every time t there is attached to every point $\mathbf{x}$ in $\mathbb{R}^n$ (or in some open subset of $\mathbb{R}^n$, the *domain of definition* of the differential equation) an n-dimensional vector $\mathbf{f}(t, \mathbf{x})$ whose components are the $f_i(t, x_1, \dots, x_n)$. This vector may be viewed as the 'wind velocity' at the point $\mathbf{x}$ and at time t. A path $t \to \mathbf{x}(t)$ is the description of the motion of a 'particle' in n-space: both position $\mathbf{x}(t) = (x_1(t), \dots, x_n(t))$ and velocity $\dot{\mathbf{x}}(t) = (\dot{x}_1(t), \dots, \dot{x}_n(t))$ at time t are n-dimensional vectors. A solution of the differential equation (2.2) is a path whose velocity $\dot{\mathbf{x}}(t)$ coincides at every instant with the wind velocity $\mathbf{f}(t, \mathbf{x}(t))$ at the point $\mathbf{x}(t)$.

An *initial condition* is the specification of the position of the particle at some given time. The *theorem on the existence and uniqueness* of solutions states that *for every initial condition, there exists a unique solution to the differential equation* (2.2). Our initial condition will always be that the position at time 0 is at some point $\mathbf{x}$: the corresponding solution will usually be denoted by $\mathbf{x}(t)$. With this convention, then, $\mathbf{x}(0) = \mathbf{x}$. For fixed t, the point $\mathbf{x}(t)$ depends continuously (and even differentiably) on the initial point $\mathbf{x}$.

Let us note that a solution need not be defined for all times. There exists a maximal open interval $]a, b[$ such that the solution $\mathbf{x}(t)$ is defined for all $t \in \ ]a, b[$, but $]a, b[$ is not necessarily $]-\infty, +\infty[$. A case in point is the equation $\dot{x} = 1 + x^2$ in $\mathbb{R}$. The function $x(t) = \tan t$ is a solution satisfying the initial condition $x(0) = 0$, but it is only defined (as a solution) for $t \in \ ]-\frac{\pi}{2}, \frac{\pi}{2}[$. In a way, the solution $x(t)$ picks up speed so quickly (i.e. travels so rapidly through the real line $\mathbb{R}$) that it reaches infinity at time $\pi/2$.

Of particular interest are the *time-independent* differential equations

$$\dot{\mathbf{x}} = \mathbf{f}(\mathbf{x}) \qquad \text{or} \qquad \dot{x}_i = f_i(x_1, \dots, x_n). \tag{2.5}$$

The right hand side, here, does not depend on t, i.e. the 'wind velocity' at $\mathbf{x}$ does not change with time. If we let two 'particles' start at point $\mathbf{x}$, one of them T time units after the other, then both will travel through the same trajectory, one of them always T time units late. In other words, $\mathbf{x}(T) = \mathbf{y}$ implies $\mathbf{x}(T + t) = \mathbf{y}(t)$ for all t for which the solutions exist.

In an important case the solution is defined for all times t: if if there is some compact set K in the domain of definition of $\mathbf{f}$ which $\mathbf{x}(t)$ never leaves for $t \in \,]a, b[$, then $]a, b[\,= \,]-\infty, +\infty[$.

If there exists a set X in the domain of definition of $\mathbf{f}$ such that for all $\mathbf{x} \in X$ and all $t \in \mathbb{R}$, the solution $\mathbf{x}(t)$ is defined and lies in X, then the differential equation (2.5) determines a *continuous time dynamical system* on X. To every $\mathbf{x} \in X$ corresponds its *orbit* $\{\mathbf{x}(t) : t \in \mathbb{R}\}$.

Three types of solution $\mathbf{x}(t)$ can occur:

(a) if $\mathbf{x}(t) \equiv \mathbf{x}$ for all $t \in \mathbb{R}$, i.e. if $\mathbf{x}(t)$ is a constant, then $\mathbf{x}$ is called a *rest point* (or fixed point, or stationary state). A rest point $\mathbf{x}$ is characterized by $\mathbf{f}(\mathbf{x}) = \mathbf{0}$. If one starts at such a point, one remains there forever.

(b) If $\mathbf{x}(T) = \mathbf{x}$ for some $T > 0$, but $\mathbf{x}(t) \neq \mathbf{x}$ for all $t \in \,]0, T[$, then $\mathbf{x}$ is called a *periodic point* and T is called the *period*. All other points on the orbit are also periodic with period T. The motion describes an endless periodic oscillation. Topologically, i.e. up to a continuous transformation, the orbit looks like a circle and the solution travels round and round.

(c) If $t \to \mathbf{x}(t)$ is injective, then the orbit never intersects itself. The orbit may be bent, knotted and twisted, but topologically it looks like a line.

2.3 Analysis of the Lotka–Volterra predator–prey equation

Let us now return to the differential equation (2.1). We may write down three solutions immediately:

(i) $x(t) = y(t) = 0$
(ii) $x(t) = 0, \quad y(t) = y(0)e^{-ct}$ (for any $y(0) > 0$)
(iii) $y(t) = 0, \quad x(t) = x(0)e^{at}$ (for any $x(0) > 0$).

This means that if the density of predators or prey is zero at some given time, then it is always zero. In the absence of prey, predators will become extinct ($y(t)$ converges to 0, for $t \to +\infty$). In the absence of predators, the prey population explodes ($x(t) \to +\infty$). This last feature is absurd, and will be amended in subsequent models.

To the three solutions (i), (ii) and (iii) correspond three orbits: (i) the origin $(0, 0)$, which is a rest point; (ii) the positive y-axis; (iii) the positive

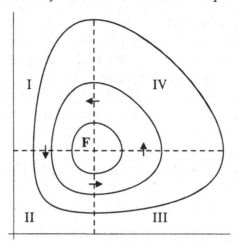

Fig. 2.1. Phase Portrait of the Lotka–Volterra predator–prey equation

x-axis. Together, these three orbits form the boundary of the nonnegative orthant

$$\mathbb{R}^2_+ = \{(x, y) \in \mathbb{R}^2 : x \geq 0, y \geq 0\} \, . \tag{2.6}$$

Since population densities have to be nonnegative, we shall only consider the restriction of (2.1) to $\mathbb{R}^2_+$. This set is *invariant* in the sense that any solution which starts in it remains there for all (positive and negative) time for which it is defined. Indeed, as we have seen, the boundary bd $\mathbb{R}^2_+$ is invariant. Since no orbit can cross another, the interior

$$\text{int} \, \mathbb{R}^2_+ = \{(x, y) \in \mathbb{R}^2 : x > 0, y > 0\}$$

is also invariant.

There is a unique rest point in int $\mathbb{R}^2_+$. Indeed, such a rest point $\mathbf{F} = (\bar{x}, \bar{y})$ must satisfy $\bar{x}(a - b\bar{y}) = 0$ and $\bar{y}(-c + d\bar{x}) = 0$. Since $\bar{x} > 0$ and $\bar{y} > 0$, this implies

$$\bar{x} = \frac{c}{d} \quad \text{and} \quad \bar{y} = \frac{a}{b} \, . \tag{2.7}$$

The signs of $\dot{x}$ and $\dot{y}$ depend on whether y is larger or smaller than $\bar{y}$, and x larger or smaller than $\bar{x}$. Thus $\mathbb{R}^2_+$ is divided into four regions I, II, III, IV (see fig. 2.1). As we shall presently see, $\mathbf{F}$ is surrounded by periodic orbits which travel from I to II, from II to III, etc. in a counterclockwise rotation.

Indeed, if we multiply the first line in (2.1) by $\frac{c-dx}{x}$, and the second line by $\frac{a-by}{y}$, and then add, we obtain

$$\left(\frac{c}{x} - d\right)\dot{x} + \left(\frac{a}{y} - b\right)\dot{y} = 0$$

or

$$\frac{d}{dt}\left[c \log x - dx + a \log y - by\right] = 0 \ . \tag{2.8}$$

We shall write this in a slightly different way. With

$$H(x) = \bar{x} \log x - x, \quad G(y) = \bar{y} \log y - y, \tag{2.9}$$

and

$$V(x, y) = dH(x) + bG(y) \ , \tag{2.10}$$

(2.8) turns into

$$\frac{d}{dt}V\left(x(t), y(t)\right) = 0 \tag{2.11}$$

or

$$V\left(x(t), y(t)\right) = \text{const.} \tag{2.12}$$

The function V, which is defined in int $\mathbb{R}_+^2$, remains constant along the orbits of (2.1): it is a so-called *constant of motion*.

Since $H(x)$ satisfies

$$\frac{dH}{dx} = \frac{\bar{x}}{x} - 1, \qquad \frac{d^2H}{dx^2} = -\frac{\bar{x}}{x^2} < 0,$$

it attains its maximum at $x = \bar{x}$; $G(y)$ takes its maximum at $y = \bar{y}$. Thus $V(x, y)$ has its unique maximum at the rest point $\mathbf{F} = (\bar{x}, \bar{y})$. It is easy to check that it decreases to $-\infty$ along every half-line issuing from $\mathbf{F}$. The constant level sets $\{(x, y) \in \text{int}\,\mathbb{R}_+^2 : V(x, y) = \text{const.}\}$ are closed curves around $\mathbf{F}$. (We may interpret $V(x, y)$ as the 'height' at the point (x, y): then $\mathbf{F}$ is the unique summit of this landscape). The solutions have to remain on the constant level sets, and thus return to their starting point. The orbits, therefore, are periodic.

2.4 Volterra's principle

The densities of predator and prey will oscillate periodically, with both the amplitude and frequency of the oscillations depending on the initial conditions. The *time averages* of the densities, however, will remain constant, and in fact equal to the corresponding values at the rest point **F**:

$$\frac{1}{T} \int_0^T x(t)dt = \bar{x}, \quad \frac{1}{T} \int_0^T y(t)dt = \bar{y}, \tag{2.13}$$

where T is the period of the solution. Indeed, from

$$\frac{d}{dt}(\log x) = \frac{\dot{x}}{x} = a - by$$

it follows by integration that

$$\int_0^T \frac{d}{dt} \log x(t)dt = \int_0^T (a - by(t))dt$$

i.e.

$$\log x(T) - \log x(0) = aT - b \int_0^T y(t)dt . \tag{2.14}$$

Since $x(T) = x(0)$, this implies that

$$\frac{1}{T} \int_0^T y(t)dt = \frac{a}{b} = \bar{y},$$

and an analogous result holds for the x-averages.

   We are ready now for Volterra's explanation of the increase of predatory fish during the war. Fishing reduces the rate of increase of the prey (instead of a, we now have some smaller value $a - k$) and it augments the rate of decrease of the predators (instead of c, we get some larger value $c + m$). However, the interaction constants b and d do not change. The average density of predators is now $\frac{a-k}{b}$ and hence smaller, that of the prey is $\frac{c+m}{d}$ and hence larger than in the unperturbed state. Ceasing to fish leads to an increase of predators and a decrease of prey.

2.5 The predator–prey equation with intraspecific competition

We have seen that the differential equation (2.1) displays the rather unrealistic feature that in the absence of predators, the prey population is subject to exponential growth: $\dot{x} = ax$. This is easily remedied by taking competition within the prey species into account and assuming logistic growth: $\dot{x} = x(a - ex)$. If we wish we may also allow for competition within the predators. (This is less crucial, however, as their population does not explode anyway.)

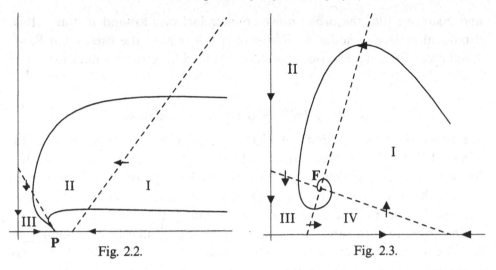

Fig. 2.2.   Fig. 2.3.

In lieu of (2.1), we obtain

$$\begin{aligned} \dot{x} &= x(a - ex - by) \\ \dot{y} &= y(-c + dx - fy) \end{aligned}$$

(2.15)

with $e > 0$ and $f \geq 0$. Again, $\mathbb{R}_+^2$ is invariant. Its boundary consists of five orbits: the two rest points $\mathbf{0} = (0,0)$ and $\mathbf{P} = \left(\frac{a}{e}, 0\right)$, the two intervals $]0, \frac{a}{c}[$ and $]\frac{a}{c}, +\infty[$ of the x-axis, and the positive y-axis.

In order to get some rough feeling about what happens in $\operatorname{int} \mathbb{R}_+^2$, i.e. in the presence of both populations, we shall have a look at the *isoclines*. The x-isocline is the set where $\dot{x} = 0$, i.e. where the vector field is vertical: in $\operatorname{int} \mathbb{R}_+^2$, this is the set where

$$ex + by = a .$$

(2.16)

Similarly, the y-isocline, where the vector field is horizontal, is the set where

$$dx - fy = c .$$

(2.17)

Depending on the parameters, these lines may or may not intersect in $\operatorname{int} \mathbb{R}_+^2$. If they don't, they divide $\operatorname{int} \mathbb{R}_+^2$ into three regions I, II, III (see fig. 2.2). In this case, every orbit has to converge towards $\mathbf{P}$. The predators will therefore vanish; the prey density converges towards the limit $\frac{a}{e}$ that corresponds to the carrying capacity of the logistic equation $\dot{x} = x(a - ex)$ which, in the absence of predators, governs their growth.

If the isoclines intersect at some point $\mathbf{F} = (\bar{x}, \bar{y})$ in $\operatorname{int} \mathbb{R}_+^2$, this point is a rest point. Its coordinates solve the linear equation (2.16–17). In this case, $\operatorname{int} \mathbb{R}_+^2$ is divided into four regions I to IV (see fig. 2.3). The signs of $\dot{x}$

and $\dot{y}$ suggest that the orbits move counterclockwise around $\mathbf{F}$. But is this rotational motion periodic, as before, or will it tend to the fixed point $\mathbf{F}$, or spiral away from it? The isoclines do not suffice to settle this question.

2.6 On ω-limits and Lyapunov functions

We know that the solutions of (2.15) exist, but we don't know how to compute them. This is the trouble with most differential equations. We are left with two complementary courses: to calculate approximate orbits or to analyse the qualitative behaviour. We shall pursue the latter line, since it is the final outcome of the dynamics which interests us most.

The asymptotic features of a solution are encapsulated in its ω-limit. Let $\dot{\mathbf{x}} = \mathbf{f}(\mathbf{x})$ be a time-independent ODE in some region of $\mathbb{R}^n$ and let $\mathbf{x}(t)$ be a solution defined for all $t \geq 0$ and satisfying the initial condition $\mathbf{x}(0) = \mathbf{x}$. The ω-*limit* of $\mathbf{x}$ is the set of all accumulation points of $\mathbf{x}(t)$, for $t \to +\infty$:

$$\omega(\mathbf{x}) = \{\mathbf{y} \in \mathbb{R}^n : \mathbf{x}(t_k) \to \mathbf{y} \quad \text{for some sequence } t_k \to +\infty\} . \tag{2.18}$$

The points in the ω-limit have the property that all of their neighbourhoods keep getting visited by the solution $\mathbf{x}(t)$, even after an arbitrarily long time (α-limits are defined in the same way, but for $t_k \to -\infty$).

The ω-limit of a point $\mathbf{x}$ may be empty. This is the case for the equation $\dot{x} = 1$ on $\mathbb{R}^1$, for instance. The solution $x(t) = x + t$ travels along the line at constant speed and never returns. But if the orbit, or at least the positive semiorbit $\{\mathbf{x}(t) : t \geq 0\}$, remains in some compact set K, then every sequence $\mathbf{x}(t_k)$ must admit accumulation points, and $\omega(\mathbf{x})$ cannot be empty. Any point $\mathbf{z}$ on the orbit of $\mathbf{x}$ has the same ω-limit: indeed, in that case $\mathbf{z} = \mathbf{x}(T)$ for some time T. If $\mathbf{x}(t_k)$ converges to $\mathbf{y}$, then so does $\mathbf{z}(t_k - T)$.

It is easy to see that $\omega(\mathbf{x})$ is *closed*, being a set of accumulation points. We can represent $\omega(\mathbf{x})$ as an intersection of closed sets:

$$\omega(\mathbf{x}) = \bigcap_{t \geq 0} \overline{\{\mathbf{x}(s) : s \geq t\}} .$$

The set $\omega(\mathbf{x})$, furthermore, is *invariant*. Indeed let $\mathbf{y}$ be in $\omega(\mathbf{x})$, and t arbitrary. Since $\mathbf{x}(t_k) \to \mathbf{y}$ for some sequence t_k, and since solutions are continuous functions of their initial data, one has $\mathbf{x}(t_k + t) \to \mathbf{y}(t)$ for $k \to \infty$. Hence $\mathbf{y}(t)$ belongs to $\omega(\mathbf{x})$.

Rest points and periodic orbits constitute their own ω-limits. If $\omega(\mathbf{x})$ is compact, it is also connected, in the sense that any pair $\mathbf{y}, \mathbf{z} \in \omega(\mathbf{x})$ can be joined by a continuous path in $\omega(\mathbf{x})$.

One can get a handle on ω-limits even if one does not know the solution. This is the gist of *Lyapunov's theorem*:

Theorem 2.6.1 *Let* $\dot{\mathbf{x}} = \mathbf{f}(\mathbf{x})$ *be a time-independent ODE defined on some subset* G *of* $\mathbb{R}^n$. *Let* $V : G \to \mathbb{R}$ *be continuously differentiable. If for some solution* $\mathbf{x}(t)$, *the derivative* $\dot{V}$ *of the map* $t \to V(\mathbf{x}(t))$ *satisfies the inequality* $\dot{V} \geq 0$ *(or* $\dot{V} \leq 0$*), then* $\omega(\mathbf{x}) \cap G$ *is contained in the set* $\{\mathbf{x} \in G : \dot{V}(\mathbf{x}) = 0\}$ *(and so is* $\alpha(\mathbf{x}) \cap G$*).*

Proof If $\mathbf{y} \in \omega(\mathbf{x}) \cap G$, there is a sequence $t_k \to +\infty$ with $\mathbf{x}(t_k) \to \mathbf{y}$. Since $\dot{V} \geq 0$ along the orbit of $\mathbf{x}$, one has $\dot{V}(\mathbf{y}) \geq 0$ by continuity. Suppose that $\dot{V}(\mathbf{y}) = 0$ does not hold. Then $\dot{V}(\mathbf{y}) > 0$. Since the value of V can never decrease along an orbit, this implies

$$V(\mathbf{y}(t)) > V(\mathbf{y}) \tag{2.19}$$

for $t > 0$. The function $V(\mathbf{x}(t))$ is also monotonically increasing. Since V is continuous, $V(\mathbf{x}(t_k))$ converges to $V(\mathbf{y})$, and hence

$$V(\mathbf{x}(t)) \leq V(\mathbf{y}) \tag{2.20}$$

for every $t \in \mathbb{R}$. From $\mathbf{x}(t_k) \to \mathbf{y}$ it follows that $\mathbf{x}(t_k + t) \to \mathbf{y}(t)$ and hence

$$V(\mathbf{x}(t_k + t)) \to V(\mathbf{y}(t)), \tag{2.21}$$

so that by (2.19)

$$V(\mathbf{x}(t_k + t)) > V(\mathbf{y})$$

for k sufficiently large. This contradicts (2.20). $\qquad\square$

This theorem does not tell how to find such a *Lyapunov function V*. There is no general recipe, in fact.

2.7 Coexistence of predators and prey

Let us now return to (2.15) and consider the case of intersecting isoclines. There exists, then, a rest point $\mathbf{F} = (\bar{x}, \bar{y})$ in $\text{int}\,\mathbb{R}_+^2$. At the corresponding densities, predators and prey can coexist. But is such a rest point stable?

Let us try the function V defined in (2.10),

$$V(x, y) = dH(x) + bG(y)$$

with

$$H(x) = \bar{x} \log x - x \quad \text{and} \quad G(y) = \bar{y} \log y - y .$$

The derivative of the function $t \to V(x(t), y(t))$ is

$$\dot{V}(x, y) = \frac{\partial V}{\partial x}\dot{x} + \frac{\partial V}{\partial y}\dot{y}$$

$$= d\left(\frac{\bar{x}}{x} - 1\right)x(a - by - ex) + b\left(\frac{\bar{y}}{y} - 1\right)y(-c + dx - fy) \quad .$$

Since $\bar{x}$ and $\bar{y}$ are the solutions of (2.16) and (2.17), we may replace a and c by $e\bar{x} + b\bar{y}$ and $d\bar{x} - f\bar{y}$, respectively. This yields

$$\dot{V}(x, y) = d(\bar{x} - x)(b\bar{y} + e\bar{x} - by - ex) + b(\bar{y} - y)(-d\bar{x} + f\bar{y} + dx - fy)$$

$$= de(\bar{x} - x)^2 + bf(\bar{y} - y)^2 \geq 0 \quad . \tag{2.22}$$

We may, therefore, apply the theorem of Lyapunov. The ω-limit of every orbit in int $\mathbb{R}^2_+$ is contained in the set $\{(x, y) : \dot{V}(x, y) = 0\}$. For $f > 0$, this consists only of the point **F**. For $f = 0$, it is the set $K = \{(x, y) \in \mathbb{R}^2 : y > 0, x = \bar{x}\}$. But the ω-limit must be an invariant subset of K, and hence as before reduces to **F**. Every solution in int $\mathbb{R}^2_+$ converges to this rest point.

Let us interpret $V(x, y)$ again as the height of a landscape at the point (x, y). While the solutions of (2.1) follow the contour lines and remain at the same altitude, those of (2.15) ascend. A *strict Lyapunov function* — one for which $\dot{V}(\mathbf{x}) > 0$ whenever $\mathbf{x}$ is not a rest point — describes a steady uphill movement. Systems admitting such a function are said to be *gradient-like*.

A rest point $\mathbf{z}$ of an ODE $\dot{\mathbf{x}} = \mathbf{f}(\mathbf{x})$ is said to be *stable* if, for any neighbourhood U of $\mathbf{z}$, there exists a neighbourhood W of $\mathbf{z}$ such that any orbit through W remains in U for ever (i.e. $\mathbf{x} \in W$ implies $\mathbf{x}(t) \in U$ for all $t \geq 0$). It is said to be *asymptotically stable* if, in addition, such orbits converge to $\mathbf{z}$ (i.e. $\mathbf{x}(t) \to \mathbf{z}$ for all $\mathbf{x} \in W$). The set of points $\mathbf{x}$ with $\mathbf{x}(t) \to \mathbf{z}$, as $t \to +\infty$, is called the *basin of attraction* of $\mathbf{z}$. It is an open invariant set. If it is the whole state space — or at least its interior — then $\mathbf{z}$ is said to be *globally stable*.

The rest point **F** is stable for (2.1), and asymptotically stable — in fact, even globally stable — for (2.15). For the neighbourhood W, we can choose the interior of a contour line of V contained in U.

Note that asymptotic stability means much more than stability alone. A small perturbation away from the rest point **F** will be promptly offset by the dynamics of (2.15), since the solution will tend back towards the rest point: it is like the action of a spring pulling the state back. In contrast to this, the response of the dynamics of (2.1) to a perturbation away from the rest point is flaccid and ineffective: the state will remain on a periodic orbit and not return to the rest point. In fact, a sequence of perturbations may send the state from periodic orbit to periodic orbit, and ultimately to bd $\mathbb{R}_+^2$, corresponding to the extinction of one population.

There is another sense in which (2.15) is more stable than (2.1). A small change in the vector field (2.15) will slightly shift the position of the rest point **F**, but not essentially alter the behaviour of the orbits, which will still spiral towards **F**. In contrast to this, the behaviour of (2.1) is radically altered by the introduction of the competition term $-ex^2$ in the equation governing the growth of the prey, no matter how small it is. Instead of the undamped periodic oscillations of (2.1), one gets damped oscillations and convergence to a steady state. The dynamical system (2.1) is not *structurally stable*. In fact, its main features — existence of a constant of motion, periodicity of all orbits, a rest point which is stable but not asymptotically stable — are all nongeneric.

2.8 Notes

We refer to Scudo and Ziegler (1978) for a re-edition of classic papers from the 'golden age' of mathematical ecology. Jansen (1995) discovered that in a spatial model of (2.1) only small-amplitude cycles remain stable. There are many excellent textbooks on differential equations, e.g. Hirsch and Smale (1974), Arrowsmith and Place (1990), Verhulst (1990).

3

The Lotka–Volterra equation for two competing species

We use the Lotka–Volterra equation describing two competing populations to introduce local stability analysis of rest points, and briefly discuss more general competitive systems.

3.1 Linear differential equations

Most of the differential equations in this book are nonlinear. So are those in nature. Still, the all-important question of the stability of rest points can in many cases be reduced to an associated linear equation. Hence we shall briefly describe some of the main features of *linear systems*.

Let A be a real $n \times n$ matrix and

$$\dot{\mathbf{x}} = A\mathbf{x} \tag{3.1}$$

the corresponding linear differential equation in $\mathbb{R}^n$. The solution $\mathbf{x}(t)$ can be written as $e^{At}\mathbf{x}(0)$ where the matrix e^{At} is given by

$$e^{At} = I + A\frac{t}{1!} + A^2\frac{t^2}{2!} + \cdots . \tag{3.2}$$

The eigenvalues of A can be real or complex: in the latter case, they occur in conjugate pairs. The components $x_i(t)$ of the solutions of (3.1) are linear combinations (with constant coefficients) of the following functions:

(i) $e^{\lambda t}$, whenever λ is a real eigenvalue of A;

(ii) $e^{at} \cos bt$ and $e^{at} \sin bt$, i.e. the real and imaginary parts of $e^{\mu t}$, whenever $\mu = a + ib$ is a complex eigenvalue of A;

(iii) $t^j e^{\lambda t}$, or $t^j e^{at} \cos bt$ and $t^j e^{at} \sin bt$, with $0 \le j < m$, if the eigenvalue λ or μ occurs with multiplicity m.

We note that the complex eigenvalues μ introduce an oscillatory part into the solutions. These oscillations will be damped if and only if $a < 0$.

The origin $\mathbf{0}$ is obviously a rest point of (3.1). It is called

(a) a *sink*, if the real parts of the eigenvalues, i.e. the λ's and the a's, are all negative. In this case, $e^{\lambda t} \to 0$ and $e^{at} \to 0$ for $t \to +\infty$, and hence $\{\mathbf{0}\}$ is the ω-limit of every orbit;

(b) a *source*, if the real parts of the eigenvalues are all positive. In this case, $\{\mathbf{0}\}$ is the α-limit of every orbit;

(c) a *saddle*, if some eigenvalues are in the left half and some in the right half of the complex plane $\mathbb{C}$, but none on the imaginary axis. The orbits whose ω-limit is $\{\mathbf{0}\}$ form a linear submanifold of $\mathbb{R}^n$ called the *stable manifold*; those whose α-limit is $\{\mathbf{0}\}$ form the *unstable manifold*; and these subspaces span $\mathbb{R}^n$.

The rest point $\mathbf{0}$ is called *hyperbolic* if it is a source, a saddle or a sink, i.e. if no eigenvalue of A has real part 0. Eigenvalues on the imaginary axis correspond to 'degenerate' solutions: an eigenvalue 0 corresponds to a linear manifold of rest points, and a pair of purely imaginary eigenvalues $\pm ib$ to a linear manifold of periodic orbits with period $2\pi/b$. If all eigenvalues are on the imaginary axis, $\mathbf{0}$ is called a *centre*. If $\mathbf{0}$ is not hyperbolic, then an arbitrarily small perturbation of the coefficients of A can change $\mathbf{0}$ into a source, a sink or a saddle and so lead to an altogether different behaviour. On the other hand, the hyperbolic case is *structurally stable*: if the perturbation of the coefficients is sufficiently small, the orbits will exhibit an essentially unchanged behaviour.

Similar results hold for discrete linear dynamics. A linear map $\mathbf{x} \to A\mathbf{x}$ from $\mathbb{R}^n$ to $\mathbb{R}^n$ has the origin $\mathbf{0}$ as fixed point. This origin is asymptotically stable if and only if all eigenvalues of A have absolute value less than 1. (The necessity is obvious by considering an eigenvector $\mathbf{x}$ of the eigenvalue λ. The sufficiency follows from the representation of A in triangular form.) If one of the eigenvalues has absolute value larger than 1, then there exists an orbit diverging exponentially fast towards infinity, and $\mathbf{0}$ is unstable.

Exercise 3.1.1 If $\mathbf{0}$ is a sink for (3.1), i.e. if all eigenvalues of A have negative real part, then A is said to be a *stable* matrix. Show that A is stable if and only if there exists a symmetric positive definite matrix Q such that $QA + A^t Q$ is negative definite. Geometrically this means that the quadratic form $V(\mathbf{x}) = \mathbf{x} \cdot Q\mathbf{x}$ is a Lyapunov function for (3.1).

Exercise 3.1.2 Find a similar characterization for maps.

Exercise 3.1.3 Show that a 2×2 matrix

$$A = \begin{pmatrix} a & b \\ c & d \end{pmatrix}$$

is stable if and only if trace $A = a + d < 0$ and $\det A = ad - bc > 0$. $\mathbf{0}$ is a saddle if and only if $\det A < 0$.

3.2 Linearization

Let us now consider the local behaviour of the solutions of

$$\dot{\mathbf{x}} = \mathbf{f}(\mathbf{x}) \tag{3.3}$$

near a point $\mathbf{z}$ in $\mathbb{R}^n$.

If $\mathbf{z}$ is not a rest point, there exists a neighbourhood U of $\mathbf{z}$ where the orbits can be 'straightened out' by a continuous transformation, so that they turn into parallel lines. This *flow box theorem* is plausible enough, and follows from the fact that the first term of the Taylor expansion of $\mathbf{f}$ at $\mathbf{z}$ is the constant $\mathbf{f}(\mathbf{z}) \neq \mathbf{0}$.

If $\mathbf{z}$ is a rest point, however, then the local behaviour is less easy to sketch. The constant term $\mathbf{f}(\mathbf{z})$ vanishes now. The next term — the linear one — is given by the *Jacobian matrix* $D_{\mathbf{z}}\mathbf{f} = A$ of first order partial derivatives:

$$A = \begin{bmatrix} \frac{\partial f_1}{\partial x_1}(\mathbf{z}) & \cdots & \frac{\partial f_1}{\partial x_n}(\mathbf{z}) \\ \vdots & & \vdots \\ \frac{\partial f_n}{\partial x_1}(\mathbf{z}) & \cdots & \frac{\partial f_n}{\partial x_n}(\mathbf{z}) \end{bmatrix}.$$

(Instead of $\partial f_i / \partial x_j$ one sometimes writes $\partial \dot{x}_i / \partial x_j$.) The linear equation

$$\dot{\mathbf{y}} = A\mathbf{y} \tag{3.4}$$

can be solved explicitly. What has it to do with (3.3)? A lot, as long as $\mathbf{z}$ is hyperbolic in the sense that all eigenvalues of A have nonvanishing real part. This is the content of the *theorem of Hartman and Grobman*:

Theorem 3.2.1 *For any hyperbolic rest point $\mathbf{z}$ of (3.3), there exists a neighbourhood U and a homeomorphism $\mathbf{h}$ from U to some neighbourhood V of the origin $\mathbf{0}$ such that $\mathbf{y} = \mathbf{h}(\mathbf{x})$ implies $\mathbf{y}(t) = \mathbf{h}(\mathbf{x}(t))$ for all $t \in \mathbb{R}$ with $\mathbf{x}(t) \in U$. (Here, $\mathbf{x}(t)$ is the solution of (3.3) with $\mathbf{x}(0) = \mathbf{x}$ and $\mathbf{y}(t)$ the solution of the linear equation (3.4) with $\mathbf{y}(t) = \mathbf{y}$. A homeomorphism is a continuous map with a continuous inverse.)*

Locally, then, the orbits of (3.3) near $\mathbf{z}$ look like those of (3.4) near $\mathbf{0}$. We may speak again of sinks, sources, saddles, and stable and unstable manifolds, just as in the linear case. In particular, if $\mathbf{z}$ is a sink, then it is asymptotically stable. If an eigenvalue of A has strictly positive real part, then $\mathbf{z}$ is unstable.

We emphasize that the theorem of Hartman and Grobman says nothing about the nonhyperbolic case and in particular about centres, where all eigenvalues have real part 0. The local behaviour, there, depends on the higher-order terms of the Taylor expansion of $\mathbf{f}$.

Exercise 3.2.2 Linearize (2.15) at the interior rest point **F**.

Again, similar results hold for discrete dynamical systems. If $\mathbf{z}$ is a hyperbolic fixed point of $\mathbf{x} \rightarrow \mathbf{f}(\mathbf{x})$, i.e. if no eigenvalue of $D_{\mathbf{z}}\mathbf{f}$ has absolute value 1 or 0, then the local behaviour of $\mathbf{f}$ is completely reflected by its linearization $\mathbf{x} \rightarrow D_{\mathbf{z}}\mathbf{f}(\mathbf{x})$. There exists a homeomorphism $\mathbf{g}$ as above such that $\mathbf{g}(\mathbf{f}(\mathbf{x})) = D_{\mathbf{z}}\mathbf{f}(\mathbf{g}(\mathbf{x}))$ for all $\mathbf{x}$. In particular, if all eigenvalues of $D_{\mathbf{z}}\mathbf{f}$ are in the interior of the unit circle, then $\mathbf{z}$ is an asymptotically stable fixed point.

It is instructive to compare these stability criteria for differential and difference equations. To a differential equation (3.3) in $\mathbb{R}^n$ corresponds in a natural way the difference equation

$$\mathbf{x}' = \mathbf{x} + h\mathbf{f}(\mathbf{x}) \tag{3.5}$$

whose increment $\mathbf{x}' - \mathbf{x}$ points into the same direction as the vector field (3.3), with h as *step length* (one often chooses $h = 1$). In numerical analysis this is called the *Euler scheme*. If h is small then the orbits of (3.5) will stay close to those of (3.3) for some finite length of time.

Suppose now that $\mathbf{z}$ is a fixed point of (3.3), and hence of (3.5), and set $A = D_{\mathbf{z}}\mathbf{f}$. Let us compare the local behaviour near $\mathbf{z}$ in both equations by considering their linearizations

$$\dot{\mathbf{y}} = A\mathbf{y} \tag{3.6}$$

and

$$\mathbf{y}' = \mathbf{y} + hA\mathbf{y} = (I + hA)\mathbf{y} \ . \tag{3.7}$$

$\mathbf{0}$ is stable for (3.6) if and only if all eigenvalues of A have negative real part. If λ is an eigenvalue of A then $1 + h\lambda$ is an eigenvalue of $I + hA$. Thus for stability of $\mathbf{0}$ in (3.7) we need

$$|1 + h\lambda| < 1,$$

i.e. λ has to lie in the interior of the circle with centre $-\frac{1}{h}$ and radius $\frac{1}{h}$. Thus the stability condition for the difference equation is more restrictive than for the differential equation. If $\mathbf{0}$ is stable for (3.6) it will be stable for (3.7) only if h is small enough. If h becomes too large then $\mathbf{0}$ loses stability for discrete time. For example, if there is a real eigenvalue $\lambda < 0$ then for $h > -\frac{2}{\lambda}$ overshooting will occur, which is often accompanied by period doubling for the nonlinear equation (3.5), as we observed in section 1.6.

On the other hand, if $\mathbf{0}$ is a centre for (3.6) then it will be unstable for (3.7) for *every* choice of h, although it could be asymptotically stable for (3.3).

Exercise 3.2.3 If $\mathbf{z}$ is a hyperbolic fixed point then for h sufficiently small, show that the local behaviours of (3.3) and (3.5) are similar.

3.3 A competition equation

Let us return to ecology, and model the interaction of two *competing species*. If x and y denote their densities, then the rates of growth $\dot{x}/x$ and $\dot{y}/y$ will be decreasing functions of both x and y, since competition will act both within and between the species. The most simple-minded assumption would be that this decrease is linear. This leads to

$$
\begin{aligned}
\dot{x} &= x(a - bx - cy) \\
\dot{y} &= y(d - ex - fy)
\end{aligned}
\tag{3.8}
$$

with positive constants a to f. Again, since the boundary of $\mathbb{R}_+^2$ is invariant, so is $\mathbb{R}_+^2$ itself. In fact, if one population is absent, the other obeys the familiar logistic growth law.

The x- and y-isoclines are given by

$$
\begin{aligned}
a - bx - cy &= 0 \\
d - ex - fy &= 0
\end{aligned}
$$

in $\operatorname{int} \mathbb{R}_+^2$. These are straight lines with negative slopes.

Exercise 3.3.1 (a) If the isoclines coincide, then show that xy^{-k} (with $k = a/d$) is a constant of motion.
(b) If the isoclines do not intersect in $\operatorname{int} \mathbb{R}_+^2$, one of the species (which one?) tends to extinction (see fig. 3.1). One species, in this sense, is *dominant*.

There remains the case of a unique intersection $\mathbf{F} = (\bar{x}, \bar{y})$ of the isoclines in $\operatorname{int} \mathbb{R}_+^2$, with

$$
\bar{x} = \frac{af - cd}{bf - ce} \qquad \bar{y} = \frac{bd - ae}{bf - ce} \quad .
\tag{3.9}
$$

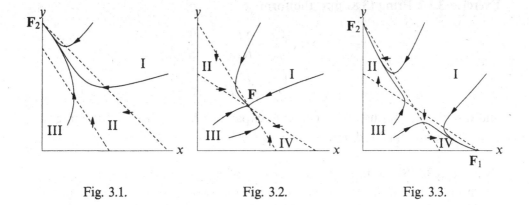

Fig. 3.1.   Fig. 3.2.   Fig. 3.3.

The Jacobian of (3.8) at **F** is

$$A = \begin{bmatrix} -b\bar{x} & -c\bar{x} \\ -e\bar{y} & -f\bar{y} \end{bmatrix}. \tag{3.10}$$

We have to distinguish two situations:

(a) If $bf > ce$, the denominator in (3.9) is positive. This implies $af - cd > 0$, $bd - ae > 0$ and hence

$$\frac{b}{e} > \frac{a}{d} > \frac{c}{f}. \tag{3.11}$$

From the signs of $\dot{x}$ and $\dot{y}$ in the regions I, II, III, IV (see fig. 3.2) we infer that every orbit in $\mathrm{int}\,\mathbb{R}_+^2$ converges to **F**. This agrees with the fact that the eigenvalues of (3.10) are negative, i.e. that **F** is a sink. This is the case of *stable coexistence*.

(b) Otherwise

$$\frac{c}{f} > \frac{a}{d} > \frac{b}{e}. \tag{3.12}$$

As seen from fig. 3.3, all orbits in region II converge to the y-axis and all those in region IV to the x-axis. Since $\det A = \bar{x}\bar{y}(bf - ce) < 0$, **F** is a saddle. Its stable manifold consists of two orbits converging to **F**. One of them must lie in region III, the other in region I. Together, they divide $\mathrm{int}\,\mathbb{R}_+^2$ into two basins of attraction. All orbits from one basin converge to $\mathbf{F}_2 = (0, d/f)$, all those from the other to $\mathbf{F}_1 = (a/e, 0)$. This means that — depending on the initial condition — one or the other species gets eliminated. This is the so-called *bistable case*.

Exercise 3.3.2 Bring (3.8) into the form

$$\dot{x}_1 = x_1 r_1 \left(1 - \frac{x_1}{K_1} - \alpha_{12} \frac{x_2}{K_2} \right)$$

$$\dot{x}_2 = x_2 r_2 \left(1 - \alpha_{21} \frac{x_1}{K_1} - \frac{x_2}{K_2} \right)$$

and discuss it in terms of the carrying capacities K_1 and K_2 and the interaction coefficients α_{12} and α_{21}.

Exercise 3.3.3 Show that in the case (a) of stable coexistence, the interior rest point $\mathbf{F}$ lies above the line connecting the two one-species rest points $\mathbf{F}_1$ and $\mathbf{F}_2$, and in the bistable case (b) below.

Exercise 3.3.4 Show that the quadratic form

$$Q(x, y) = be(x - \bar{x})^2 + 2ce(x - \bar{x})(y - \bar{y}) + cf(y - \bar{y})^2$$

is a Lyapunov function for the competition equation (3.8).

Exercise 3.3.5 Show that if two competing populations depend on one resource, one of them will vanish. (Assume that the resource R is given by $\bar{R} - c_1 x - c_2 y$ (with $\bar{R}, c_1, c_2 > 0$) and set $\dot{x}/x = b_1 R - \alpha_1$, $\dot{y}/y = b_2 R - \alpha_2$.) Show that coexistence is possible if the two species compete for two distinct resources.

Exercise 3.3.6 Set up and discuss a model for two mutualistic species. Show that it has unbounded solutions if $\det A < 0$ and a globally stable rest point if $\det A > 0$.

3.4 Cooperative systems

A differential equation $\dot{x} = f(x)$ defined on $G \subseteq \mathbb{R}^n$ is said to be *cooperative* if $\frac{\partial f_i}{\partial x_j}(x) \geq 0$ for all $x \in G$ and all $i \neq j$ (the growth of every component is enhanced by an increase in any other component).

Theorem 3.4.1 *The orbits of a two-dimensional cooperative system converge either to a rest point or to infinity.*

Proof Let us denote the orthants of $\mathbb{R}^2$ by $C_1 = \mathbb{R}_+^2$, $C_2 = \{(x_1, x_2) : x_1 \leq 0, x_2 \geq 0\}$, $C_3 = -\mathbb{R}_+^2$ and $C_4 = -C_2$. If $\dot{x}(t_0) \in C_1$ for some $t_0 \in \mathbb{R}$, then

$\mathbf{x}(t) \in C_1$ for all $t \geq t_0$: if for example $\dot{x}_1(t) = 0$ but $\dot{x}_2(t) \geq 0$ then

$$\ddot{x}_1 = \dot{f}_1(\mathbf{x}(t)) = \frac{\partial f_1}{\partial x_1}\dot{x}_1 + \frac{\partial f_1}{\partial x_2}\dot{x}_2 \geq 0$$

and hence $\dot{x}_1$ cannot become negative. The same argument shows that $\dot{\mathbf{x}}(t_0) \in C_3$ implies $\dot{\mathbf{x}}(t) \in C_3$ for all $t \geq t_0$. If $\dot{\mathbf{x}}(t_0) \in C_2$, then either $\dot{\mathbf{x}}(t)$ enters C_1 or C_3 for some subsequent time, or it remains in C_2 for all $t \geq t_0$; similarly for $\dot{\mathbf{x}}(t_0) \in C_4$. In every case, the components $x_1(t)$ and $x_2(t)$ are ultimately monotone and hence converge either to a limit or to infinity. $\quad\square$

The same result holds for *competitive systems*, defined by $\frac{\partial f_i}{\partial x_j}(\mathbf{x}) \leq 0$ for all $i \neq j$. The orbits of a two-dimensional competitive system converge either to a rest point or to infinity. One has only to replace C_1 and C_3 by C_2 and C_4.

Let us assume that two species (with densities x and y) are competing with each other. Rather than assume linear interactions as in the Lotka–Volterra equation (3.8), we shall describe the interaction by the more general differential equation

$$\begin{aligned} \dot{x} &= xS(x, y) \\ \dot{y} &= yW(x, y) \end{aligned} \tag{3.13}$$

where we only assume that the growth of each species affects the other one adversely, i.e. that

$$\frac{\partial S}{\partial y} \leq 0 \quad \text{and} \quad \frac{\partial W}{\partial x} \leq 0 \tag{3.14}$$

and that growth to infinity is impossible. Equation (3.13), then, is competitive and has no orbits converging to infinity: it follows that every orbit must converge to a rest point.

The same is true for mutualism, where (3.14), of course, is replaced by

$$\frac{\partial S}{\partial y} \geq 0 \quad \text{and} \quad \frac{\partial W}{\partial x} \geq 0. \tag{3.15}$$

Exercise 3.4.2 Prove directly that (3.14) implies that all orbits of (3.13) converge to rest points, using the fact that the isoclines are given by graphs of functions of y (resp. x). These graphs divide $\mathbb{R}_+^2$ into subsets which are either forward or backward invariant.

Exercise 3.4.3 Show that in $\mathbb{R}^2$, the change of coordinates $(y_1, y_2) = (x_1, -x_2)$ transforms a cooperative system into a competitive system.

Exercise 3.4.4 Show that in $\mathbb{R}^n$, the time reversal of a cooperative system yields a competitive system.

3.5 Notes

Good introductions to ecological population dynamics are Freedman (1980) and Hallam and Levin (1986). For more on cooperative and competitive systems, we refer to Hirsch (1982, 1988), Smale (1976), and Smith (1995). These methods play an important role in the study of spatiotemporal ecological models, see Czárán (1997), and the competition in the chemostat, see Smith and Waltman (1995).

4

Ecological equations for two species

With the help of Dulac's criterion and the Poincaré–Bendixson theorem, two-dimensional Lotka–Volterra equations can be completely analysed. In particular, they admit no limit cycles. But other models of predator–prey interactions readily lead to Hopf bifurcations and periodic attractors.

4.1 The Poincaré–Bendixson theorem

Two-dimensional differential equations cannot behave very wildly. The reason for this lies in the *Jordan curve theorem*, which implies that any periodic orbit γ in the plane divides the plane into two disjoint connected sets, the interior and the exterior. Two points in the interior (or two points in the exterior) can be connected by a path which never intersects γ, but every path between the interior and the exterior has to cross γ. One of the consequences is the *Poincaré–Bendixson theorem*:

Theorem 4.1.1 *Let $\dot{\mathbf{x}} = \mathbf{f}(\mathbf{x})$ be an ODE defined on an open set $G \subseteq \mathbb{R}^2$. Let $\omega(\mathbf{x})$ be a nonempty compact ω-limit set. Then, if $\omega(\mathbf{x})$ contains no rest point, it must be a periodic orbit.*

It is, incidentally, quite possible that $\omega(\mathbf{x})$ is empty, or unbounded. It may also happen that $\omega(\mathbf{x})$ consists neither of one rest point nor of one periodic orbit.

As an immediate consequence of the Poincaré–Bendixson theorem, one obtains that *if $K \subseteq G$ is nonempty, compact and forward invariant, then K must contain a rest point or a periodic orbit. If γ is a periodic orbit which, together with its interior Γ, is contained in G, then Γ contains a rest point.*

An important technique for proving that periodic orbits do not exist is the *Bendixson–Dulac method*: a differential equation $\dot{\mathbf{x}} = \mathbf{f}(\mathbf{x})$ defined on a simply connected subset G of $\mathbb{R}^2$ (i.e. a set G without holes) and satisfying $\text{div}\,\mathbf{f}(\mathbf{x}) > 0$ for all $\mathbf{x} \in G$ admits no periodic orbit. Here,

$$\text{div}\,\mathbf{f}(\mathbf{x}) = \frac{\partial f_1}{\partial x_1}(\mathbf{x}) + \frac{\partial f_2}{\partial x_2}(\mathbf{x}) \tag{4.1}$$

is the *divergence* of $\mathbf{f}$. It is the trace of the Jacobian.

Indeed, if γ is such a periodic orbit and Γ its interior, then Γ is an invariant set and has its area preserved under the flow: but this cannot be, since $\text{div}\,\mathbf{f}(\mathbf{x}) > 0$ means that the flow is area-expanding. Indeed, *Green's theorem* implies

$$\int_{\Gamma} \text{div}\,\mathbf{f}(\mathbf{x})d(x_1, x_2) = \pm \int_0^T [f_2(\mathbf{x}(t))\dot{x}_1(t) - f_1(\mathbf{x}(t))\dot{x}_2(t)]\,dt$$

where T is the period of γ. The left hand side is positive, while the right hand side is 0 since $f_1 = \dot{x}_1$ and $f_2 = \dot{x}_2$. This is a contradiction.

As a corollary, one obtains: *if there exists a positive function B on G such that the vector field Bf has positive divergence at every point, then $\dot{\mathbf{x}} = \mathbf{f}(\mathbf{x})$ admits no periodic orbit.* (Such a function B is said to be a *Dulac function*.) Indeed, $\mathbf{f}$ differs from $B\mathbf{f}$ only by a change in velocity, which does not affect the orbits, as we can see in the following exercises.

Exercise 4.1.2 If the function $B(\mathbf{x}, t)$ is strictly positive, then show that the solutions of the two differential equations $\dot{\mathbf{x}} = \mathbf{f}(\mathbf{x}, t)$ and $\dot{\mathbf{x}} = B(\mathbf{x}, t)\mathbf{f}(\mathbf{x}, t)$ can be transformed into each other by a strictly monotonic change in the time scale $\tau = \phi(t)$.

Exercise 4.1.3 In the time-independent case, show that the result of exercise 4.1.2 is equivalent to the fact that these two differential equations have the same orbits. (Hint: use the implicit function theorem to express the n coordinates of a solution as functions of one of them.)

Exercise 4.1.4 If there is a positive function $B(\mathbf{x})$ such that $\text{div}\,B(\mathbf{x})\mathbf{f}(\mathbf{x}) = 0$ for all $\mathbf{x} \in G \subseteq \mathbb{R}^2$, then show that there exists a constant of motion for $\dot{\mathbf{x}} = \mathbf{f}(\mathbf{x})$.

4.2 Periodic orbits for two-dimensional Lotka–Volterra equations

Theorem 4.2.1 *The two-dimensional Lotka–Volterra equation*

$$\begin{aligned} \dot{x} &= x(a + bx + cy) \\ \dot{y} &= y(d + ex + fy) \end{aligned} \qquad (4.2)$$

admits no isolated periodic orbits.

Proof Let γ be a periodic solution of (4.2). Then there must be a rest point in the interior of γ (and hence of $\mathbb{R}^2_+$). The two lines

$$\begin{aligned} a + bx + cy &= 0 \\ d + ex + fy &= 0 \end{aligned}$$

must therefore intersect in a unique point of $\mathrm{int}\,\mathbb{R}^2_+$. In particular

$$\Delta = bf - ce \neq 0 \quad . \qquad (4.3)$$

We shall now apply the Bendixson–Dulac theorem, using the Dulac function

$$B(x, y) = x^{\alpha - 1} y^{\beta - 1} \qquad (4.4)$$

with coefficients α and β which will be specified later on. Denoting the right hand sides of (4.2) by $P(x, y)$ and $Q(x, y)$, we compute the divergence of the vector field (BP, BQ) as

$$\frac{\partial}{\partial x}(BP) + \frac{\partial}{\partial y}(BQ) =$$

$$= \frac{\partial}{\partial x}\left[x^{\alpha}y^{\beta - 1}(a + bx + cy)\right] + \frac{\partial}{\partial y}\left[x^{\alpha - 1}y^{\beta}(d + ex + fy)\right]$$

$$= \alpha x^{\alpha - 1}y^{\beta - 1}(a + bx + cy) + x^{\alpha}y^{\beta - 1}b + \beta x^{\alpha - 1}y^{\beta - 1}(d + ex + fy) + x^{\alpha - 1}y^{\beta}f$$

$$= B\left[\alpha(a + bx + cy) + bx + \beta(d + ex + fy) + fy\right] \quad .$$

We choose α and β such that

$$\begin{aligned} \alpha b + \beta e &= -b \\ \alpha c + \beta f &= -f \quad . \end{aligned} \qquad (4.5)$$

This is possible by (4.3) and leads to

$$\frac{\partial}{\partial x}(BP) + \frac{\partial}{\partial y}(BQ) = \delta B \qquad (4.6)$$

with

$$\delta = a\alpha + d\beta \quad . \qquad (4.7)$$

Since we have assumed that a periodic orbit γ exists, the Bendixson–Dulac test implies $\delta = 0$. But then (4.6) becomes

$$-\frac{\partial}{\partial x}(BP) = \frac{\partial}{\partial y}(BQ) \quad . \tag{4.8}$$

This is just the *integrability condition* for the two-dimensional vector field $(BQ, -BP)$. Hence there exists a function $V = V(x, y)$ on int $\mathbb{R}^2_+$ such that

$$\frac{\partial V}{\partial x} = BQ \qquad \frac{\partial V}{\partial y} = -BP \quad . \tag{4.9}$$

The derivative of $t \to V(x(t), y(t))$ then satisfies

$$\dot{V} = \frac{\partial V}{\partial x}\dot{x} + \frac{\partial V}{\partial y}\dot{y} = PQ(B - B) \equiv 0 \tag{4.10}$$

and hence V is an *invariant of motion*, just as in section 2.3. If a periodic orbit exists, it is surrounded by other periodic orbits: no periodic orbit is isolated. $\qquad\qquad\square$

Exercise 4.2.2 Show that the constant δ from (4.7) is the trace of the Jacobian at the interior fixed point.

Exercise 4.2.3 Show that (4.2) has a continuum of periodic orbits if and only if the eigenvalues at the interior rest point are purely imaginary (i.e. $\delta = 0$ and $\Delta > 0$). (Hint: compute the Hessian of V.) Show that three phase portraits are possible.

Exercise 4.2.4 Compute the constant of motion V from (4.9) explicitly. Show that it is in general of the form

$$V(x, y) = x^p y^q (A + Bx + Cy) \quad .$$

Exercise 4.2.5 Give an alternative proof of theorem 4.2.1 by showing that every two-dimensional Lotka–Volterra equation (4.2) admits a global Lyapunov function of the above form.

4.3 Limit cycles and the predator–prey model of Gause

There is a fair amount of empirical evidence for periodic attractors in predator–prey systems. A periodic orbit γ is an *attractor* if $\omega(\mathbf{x}) = \gamma$ for all initial conditions $\mathbf{x}$ in some neighbourhood of γ; it is a *limit cycle* if $\omega(\mathbf{x}) = \gamma$ for at least one $\mathbf{x} \notin \gamma$.

Exercise 4.3.1 Show that the unit circle is a periodic attractor for

$$\begin{aligned}\dot{x} &= x - y - x(x^2 + y^2) \\ \dot{y} &= x + y - y(x^2 + y^2) \ .\end{aligned} \tag{4.11}$$

(Hint: use $V(x, y) = (1 - x^2 - y^2)^2$ as a Lyapunov function.)

Such limit cycles do not occur for linear systems: as we have seen in the previous section, they do not exist for two-dimensional Lotka–Volterra systems either. The periodic orbits in the classical predator–prey equation (2.1) are not structurally stable. To get a more robust cycling, we must look for nonlinear interactions. The following model of Gause provides a good example.

Let x and y denote the densities of prey and predator, respectively. In the absence of predators, the prey population should converge towards a limit $K > 0$. Thus $\dot{x} = xg(x)$ with

(a) $g(x) > 0$ for $x < K$, $g(x) < 0$ for $x > K$ and $g(K) = 0$.

The predators will reduce the rate of increase $\dot{x}$ of the prey by $yp(x)$, where $p(x)$ is the amount of prey killed by one predator. Thus

(b) $p(0) = 0$ and $p(x) > 0$ for $x > 0$.

The rate of increase $\dot{y}/y$ of the predators, finally, shall be given by $-d + q(x)$. The constant $d > 0$ corresponds to the mortality of predators in the absence of prey, and $q(x)$ is a positive, monotonically increasing function. Thus

(c) $q(0) = 0$ and $\frac{d}{dx}q(x) > 0$ for $x > 0$.

This yields the equation

$$\begin{aligned}\dot{x} &= xg(x) - yp(x) \\ \dot{y} &= y(-d + q(x)) \ .\end{aligned} \tag{4.12}$$

Once more, $x \equiv 0$ and $y \equiv 0$ are solutions of (4.12). Hence $\mathbb{R}_+^2$ is invariant. Since q increases monotonically, there is at most one $\bar{x} > 0$ with $q(\bar{x}) = d$. If $\bar{x}$ does not exist, the predators vanish. We shall exclude this case and assume that $\bar{x}$ exists. The y-isocline, then, is the vertical axis $x = \bar{x}$. The x-isocline is given by the equation

$$y = \frac{xg(x)}{p(x)} , \tag{4.13}$$

i.e. by the graph of a function of x defined on the interval $]0, K[$. Obviously, the isoclines intersect at most once. If there is no intersection (i.e. if $\bar{x} \geq K$),

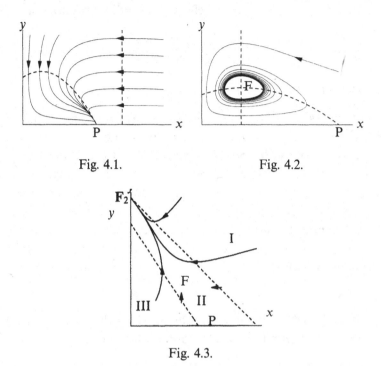

Fig. 4.1.   Fig. 4.2.

Fig. 4.3.

the predators will vanish, as may be seen from the signs of $\dot{x}$ and $\dot{y}$ (see fig. 4.1). But if the intersection $\mathbf{F} = (\bar{x}, \bar{y})$ exists (i.e. if $\bar{x} < K$), then $\mathbf{F}$ is the unique rest point in $\operatorname{int} \mathbb{R}^2_+$. This is the case which we shall consider from now on.

Exercise 4.3.2 Compute the Jacobian at $\mathbf{F}$. Show that $\mathbf{F}$ is a sink if and only if

$$\frac{d}{dx}\left(\frac{xg(x)}{p(x)}\right) \tag{4.14}$$

is negative at $x = \bar{x}$, i.e. if the slope of the x-isocline at $\mathbf{F}$ is negative. If the slope is positive, $\mathbf{F}$ is unstable.

On $\operatorname{bd} \mathbb{R}^2_+$, there are two fixed points, namely $(0,0)$ and $\mathbf{P} = (K,0)$ (see figs. 4.2 and 4.3). We check immediately that both are saddles. The stable manifold of $\mathbf{P}$ is given by the positive x-axis. The unstable manifold consists of two orbits of which only one lies in $\mathbb{R}^2_+$.

Let $\mathbf{x}$ be a point on this orbit. Clearly $\omega(\mathbf{x})$ is nonempty and compact. Using the Poincaré–Bendixson theorem, we obtain the following two alternatives:

(a) if $\omega(\mathbf{x})$ contains no rest point, then $\omega(\mathbf{x})$ is a periodic orbit γ (see fig. 4.2). This orbit must surround a rest point, i.e. the point **F**. Obviously γ is a limit cycle: in fact every orbit in the exterior spirals towards γ;

(b) if $\omega(\mathbf{x})$ contains a rest point, this must be the point **F**, since $(0,0)$ and $(K,0)$ are out of question. With the help of the signs of $\dot{x}$ and $\dot{y}$, one checks that every orbit in int $\mathbb{R}^2_+$ converges to **F**: such an orbit, indeed, cannot cross the unstable manifold of **P** (see fig. 4.3). In this case **F** is globally stable.

4.4 Saturated response

The response functions $p(x)$ and $q(x)$ in (4.12) are often assumed to be proportional to $x/(a+x)$ (with $a > 0$): this displays a saturation effect for large x which is very common in ecology and chemical kinetics. If $g(x)$, furthermore, is linear, then (4.12) becomes

$$
\begin{aligned}
\dot{x} &= rx\left(1 - \tfrac{x}{K}\right) - y\tfrac{cx}{a+x} \\
\dot{y} &= y\left(-d + \tfrac{bx}{a+x}\right)
\end{aligned}
\qquad (4.15)
$$

where all parameters are positive.

Exercise 4.4.1 If either $b \le d$ or $K \le ad/(b-d)$, then show that all orbits of (4.15) in int $\mathbb{R}^2_+$ converge to the fixed point $\mathbf{P} = (K,0)$.

Let us assume for the rest of the section that $b > d$ and $K > ad/(b-d)$. Then (4.15) admits a unique interior fixed point $\mathbf{F} = (\bar{x}, \bar{y})$ with

$$
\bar{x} = \frac{ad}{b-d} \quad . \qquad (4.16)
$$

Exercise 4.4.2 Using exercise 4.3.2, show that **F** is a sink if and only if

$$
K < a + 2\bar{x} \quad . \qquad (4.17)
$$

Theorem 4.4.3 *The rest point* **F** *is globally stable for* (4.15) *if and only if* $K \le a + 2\bar{x}$.

Proof We consider the Dulac function

$$
B(x,y) = \frac{a+x}{x} y^{\alpha-1} \qquad (4.18)
$$

where α will be chosen later. Denoting the right hand sides of (4.15) by P and Q, we obtain

$$\frac{\partial}{\partial x}(BP) + \frac{\partial}{\partial y}(BQ) = y^{\alpha-1}x^{-1}\left(rx\left(1 - \frac{a}{K} - \frac{2x}{K}\right) + \alpha(-ad + (b-d)x)\right).$$
(4.19)

At $x = \bar{x}$, (4.19) is negative in view of our assumption $K < a + 2\bar{x}$. Hence we can find a constant α such that the parabola $rx\left(1 - (a+2x)/K\right)$ lies below the line $\alpha((b-d)x - ad)$. If we choose, for example, $\alpha = r(K-a)/K(b-d)$, then the bracket on the right hand side of (4.19) can be rewritten as

$$-\frac{2r}{K}\left(x - \frac{K-a}{2}\right)^2 + r\left(1 - \frac{a}{K}\right)\left(-\bar{x} + \frac{K-a}{2}\right) \leq 0 \quad . \qquad (4.20)$$

Hence periodic orbits are excluded.

If $K > a + 2\bar{x}$ then by the criterion in exercise 4.3.2 the point $\mathbf{F}$ is a source and the previous section shows that there must be a limit cycle around $\mathbf{F}$.

□

Let us take a look at what happens if we let the parameter K increase. For $K \leq a + 2\bar{x}$ all orbits in int $\mathbb{R}_+^2$ spiral towards $\mathbf{F}$. At $K = a + 2\bar{x}$, the point $\mathbf{F}$ stops being a sink but is still asymptotically stable. For $K > a + 2\bar{x}$, the point $\mathbf{F}$ is a source, and in a neighbourhood of $\mathbf{F}$ everything flows away. However, it is as if the differential equation did not know this at points which are further off: there, the orbits are still approaching $\mathbf{F}$. It is intuitively obvious that this must lead to at least one periodic orbit in the intermediate zone.

Behind this example there lurks a general principle.

4.5 Hopf bifurcations

Let G be an open subset of $\mathbb{R}^n$, and

$$\dot{\mathbf{x}} = \mathbf{f}_\mu(\mathbf{x}) \qquad (4.21)$$

a family of differential equations depending on some parameter $\mu \in \,]{-\varepsilon}, \varepsilon[$. Let $\mathbf{P}_\mu$ be a rest point of (4.21). Let us assume that all eigenvalues of the Jacobian J_μ have negative real parts, with the exception of one pair of complex conjugate eigenvalues

$$\alpha(\mu) \pm i\beta(\mu)$$

(with $\alpha(\mu), \beta(\mu) \in \mathbb{R}$) for which the sign of the real part $\alpha(\mu)$ is that of μ, and which satisfies $\beta(0) \neq 0$. In particular, then, $\mathbf{P}_\mu$ is a sink (and hence asymptotically stable) for $\mu < 0$, but unstable for $\mu > 0$.

Let us add to this three 'technical conditions':

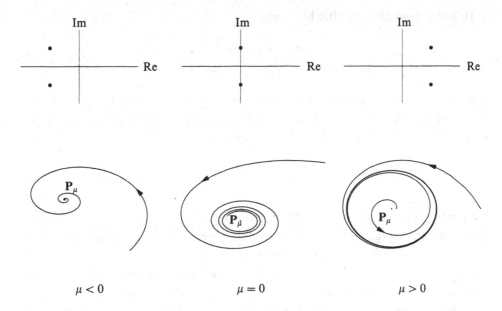

Fig. 4.4. A supercritical Hopf bifurcation

(a) the components of $\mathbf{f}_\mu(\mathbf{x})$ are analytic (i.e. given by power series);

(b) $\frac{d}{d\mu}\alpha(0) > 0$;

(c) $\mathbf{P}_0$ is asymptotically stable.

Hopf's theorem asserts that under these conditions, and for sufficiently small positive values of the parameter μ, the unstable rest point $\mathbf{P}_\mu$ is surrounded by a periodic attractor.

This theorem states that under the given conditions (which are quite often satisfied, but sometimes rather difficult to check), a periodic solution splits off from the rest point if the parameter value μ crosses a critical threshold. The rest point, formerly asymptotically stable, becomes unstable: in its place, the periodic orbit is now the attractor. A stable rest point turns into a stable oscillation (see fig. 4.4). The example given in the previous section was characteristic.

In order to better understand the 'technical conditions', let us return to the planar case. By a smooth change in coordinates, the Taylor series of degree 3 of $\mathbf{f}_\mu(\mathbf{x})$ can be brought into the *normal form*

$$
\begin{aligned}
\dot{x}_1 &= \left(d\mu + a(x_1^2 + x_2^2)\right) x_1 - \left(\omega + c\mu + b(x_1^2 + x_2^2)\right) x_2 \\
\dot{x}_2 &= \left(\omega + c\mu + b(x_1^2 + x_2^2)\right) x_1 + \left(d\mu + a(x_1^2 + x_2^2)\right) x_2 \ .
\end{aligned}
\tag{4.22}
$$

In polar coordinates, this becomes

$$\begin{aligned}
\dot{r} &= \left(d\mu + ar^2\right) r \\
\dot{\theta} &= \omega + c\mu + br^2 \quad .
\end{aligned} \tag{4.23}$$

The first of these equations does not depend on θ. If its right hand side is 0 for some $r > 0$, this corresponds to a periodic orbit of this radius. If $d \neq 0$ and $a \neq 0$, there are periodic orbits of radius

$$r = \sqrt{\frac{-d\mu}{a}} \quad .$$

Depending on the signs of d and a, they occur for $\mu > 0$ or $\mu < 0$. If $a < 0$ they are attracting, and if $a > 0$ repelling; one speaks of *supercritical* and *subcritical* Hopf bifurcations, respectively. If $d \neq 0$ and $a = 0$, there are no periodic orbits except for $\mu = 0$ (for which parameter value the plane is filled with them). The Hopf bifurcation, in this case, is *degenerate*.

Note that from (4.23) $d = \frac{d}{d\mu}\alpha(0)$. The second technical condition means $d > 0$, therefore, and the third technical condition $a < 0$.

In higher dimensions, the same phenomena occur (roughly speaking) on the two-dimensional *centre manifold* (an invariant manifold tangential to the eigenspace of the eigenvalues crossing the imaginary axis). Since the other eigenvalues have negative real part, all nearby orbits converge to this centre manifold. With a supercritical bifurcation, a stable fixed point becomes unstable, while a stable periodic orbit splits off. With a subcritical bifurcation, an unstable fixed point becomes stable, but with a very small basin of attraction bounded by an unstable periodic orbit.

Exercise 4.5.1 Analyse the behaviour of (a) $\dot{x} = \mu x - x^3$; (b) $\dot{x} = \mu x - x^2$; and (c) $\dot{x} = -\mu + x^2$ as μ changes from negative to positive values. These are typical examples of (a) a pitchfork bifurcation (a stable fixed point becomes unstable and a pair of stable fixed points splits off; (b) a transcritical bifurcation (two fixed points collide and exchange their stability properties); and (c) a saddle-node bifurcation (a stable and an unstable fixed point collide and vanish).

4.6 Notes

A comprehensive treatment of periodic orbits is given in Farkas (1994). Theorem 4.2.1 was first proved by Moiseev in 1939, see Andronov *et al.* (1973). A classification of all two-dimensional phase portraits of Lotka–Volterra equations can be found in Bomze (1983, 1995) and Reyn (1987). The constant of motion in exercise 4.2.3 and corresponding Hamiltonian structure of (4.9) has been generalized to higher-dimensional Lotka–Volterra equations by Plank (1995). The model (4.12) was proposed by Gause, and another general predator–prey model was considered by Kolmogorov, see Freedman (1980). The stability criterion from exercise 4.3.2 is from Rosenzweig and MacArthur (1963). Theorem 4.4.3 is due to Hsu *et al.* (1978). More general global stability results were shown in Cheng *et al.* (1981), and Cheng (1981) proved the uniqueness of the limit cycle of (4.15). Hofbauer and So (1990) showed that (4.12) can admit several limit cycles, even if (4.13) is a concave function. Excellent books on bifurcation theory are Kuznetsov (1995) and Troger and Steindl (1991).

5

Lotka–Volterra equations for more than two populations

A general introduction to Lotka–Volterra equations in higher dimensions emphasizes the importance of the behaviour close to the boundary of the state space and stresses global features such as heteroclinic cycles.

5.1 The general Lotka–Volterra equation

The general Lotka–Volterra equation for n populations is of the form

$$\dot{x}_i = x_i \left(r_i + \sum_{j=1}^{n} a_{ij} x_j \right) \qquad i = 1, \ldots, n . \tag{5.1}$$

The x_i denote the densities; the r_i are intrinsic growth (or decay) rates, and the a_{ij} describe the effect of the j-th upon the i-th population, which is positive if it enhances and negative if it inhibits the growth. All sorts of interactions can be modelled in this way, as long as one is prepared to assume that the influence of every species upon the growth rates is linear. The matrix $A = (a_{ij})$ is called the *interaction matrix*.

The state space is, of course, the nonnegative orthant

$$\mathbb{R}_+^n = \{ \mathbf{x} = (x_1, \ldots, x_n) \in \mathbb{R}^n : x_i \geq 0 \text{ for } i = 1, \ldots, n \} .$$

The boundary points of $\mathbb{R}_+^n$ lie on the coordinate planes $x_i = 0$, which correspond to the states where species i is absent. These *faces* are invariant, since $x_i(t) = 0$ is the unique solution of the i-th equation of (5.1) satisfying $x_i(0) = 0$. In such a model, a missing species cannot 'immigrate'. Thus the boundary bd $\mathbb{R}_+^n$, and consequently $\mathbb{R}_+^n$ itself, are invariant under (5.1). So is the interior int $\mathbb{R}_+^n$, which means that if $x_i(0) > 0$ then $x_i(t) > 0$ for all t. The density $x_i(t)$ may approach 0, however, which means extinction.

42

Whereas all possible two-dimensional Lotka–Volterra equations can be classified, many questions remain wide open for higher dimensions. In particular, numerical simulation shows that even the case of three populations may lead to some kind of chaotic motion (see Fig. 16.1): the asymptotic behaviour of the solutions consists of highly irregular oscillations and depends in a very sensitive way upon the initial conditions. The long-term outcome, in such a case, is unpredictable.

We are far from an understanding of this type of chaos, which in its erratic nature resembles that of the discrete system $x \to 4x(1 - x)$ described in section 1.7. In this chapter, we shall describe a few general results about (5.1) and then turn to some special cases of biological interest.

5.2 Interior rest points

The rest points of (5.1) in int $\mathbb{R}^n_+$ are the solutions of the linear equations

$$r_i + \sum_{j=1}^{n} a_{ij}x_j = 0 \qquad i = 1, \ldots, n \tag{5.2}$$

whose components are positive. (The rest points on the boundary faces of $\mathbb{R}^n_+$ can be found in a similar way: one has only to note that the restriction of (5.1) to any such face is again of Lotka–Volterra type.)

Theorem 5.2.1 *The interior of $\mathbb{R}^n_+$ contains α- or ω-limit points if and only if* (5.1) *admits an interior rest point.*

Proof One direction of this proposition is trivial, because a rest point coincides with its own α- and ω-limit. It is the converse which is of interest, since it is (in principle) not hard to check whether (5.2) admits positive solutions. If it does not, then every orbit has to converge to the boundary, or to infinity. In particular, if int $\mathbb{R}^n_+$ contains a periodic orbit, it must also contain a rest point.

In order to prove this converse, let $L : \mathbf{x} \to \mathbf{y}$ be defined by

$$y_i = r_i + \sum_{j=1}^{n} a_{ij}x_j \qquad i = 1, \ldots, n \;.$$

If (5.1) admits no interior rest point, the set $K = L(\text{int } \mathbb{R}^n_+)$ is disjoint from **0**. A well known theorem from convex analysis implies that there exists a hyperplane H through **0** which is disjoint from the convex set K. Thus there exists a vector $\mathbf{c} = (c_1, \ldots, c_n) \neq \mathbf{0}$ orthogonal to H (i.e. $\mathbf{c} \cdot \mathbf{x} = 0$ for all $\mathbf{x} \in H$)

such that $\mathbf{c} \cdot \mathbf{y}$ is positive for all $\mathbf{y} \in K$. (Here, $\mathbf{x} \cdot \mathbf{y} = \sum x_i y_i$ is the usual inner product in $\mathbb{R}^n$.) Setting

$$V(\mathbf{x}) = \sum c_i \log x_i , \tag{5.3}$$

we see that V is defined on $\operatorname{int} \mathbb{R}^n_+$. If $\mathbf{x}(t)$ is a solution of (5.1) in $\operatorname{int} \mathbb{R}^n_+$, then the time derivative of $t \to V(\mathbf{x}(t))$ satisfies

$$\dot{V} = \sum c_i \frac{\dot{x}_i}{x_i} = \sum c_i y_i = \mathbf{c} \cdot \mathbf{y} > 0 . \tag{5.4}$$

Thus V is increasing along each orbit. But then no point $\mathbf{z} \in \operatorname{int} \mathbb{R}^n_+$ may belong to its ω-limit: indeed, by Lyapunov's theorem 2.6.1, the derivative $\dot{V}$ would have to vanish there. This contradiction completes the proof. □

As a consequence, we see that *if* (5.1) *admits no interior rest point, then it is gradient-like in* $\operatorname{int} \mathbb{R}^n_+$.

In general, (5.2) will admit at most one solution in $\operatorname{int} \mathbb{R}^n_+$. It is only in the degenerate case $\det A = 0$ that (5.2) can have more than one solution: these will form a continuum of rest points.

Exercise 5.2.2 Construct an invariant of motion in the case of a continuum of fixed points. (Hint: try (5.3) for suitable $\mathbf{c}$.)

If there exists a unique interior rest point $\mathbf{p}$, and if the solution $\mathbf{x}(t)$ converges neither to the boundary nor to infinity, then its time average converges to $\mathbf{p}$.

Theorem 5.2.3 *If there exist positive constants a and A such that $a < x_i(t) < A$ for all i and all $t > 0$, and $\mathbf{p}$ is the unique rest point in $\operatorname{int} \mathbb{R}^n_+$, then*

$$\lim_{T \to \infty} \frac{1}{T} \int_0^T x_i(t) dt = p_i \qquad i = 1, \dots, n . \tag{5.5}$$

Proof Let us write (5.1) in the form

$$(\log x_i)^{\cdot} = r_i + \sum_j a_{ij} x_j \tag{5.6}$$

and integrate it from 0 to T. After division by T, we obtain

$$\frac{\log x_i(T) - \log x_i(0)}{T} = r_i + \sum a_{ij} z_j(T) \tag{5.7}$$

where

$$z_j(T) = \frac{1}{T} \int_0^T x_j(t)dt . \tag{5.8}$$

Clearly $a < z_j(T) < A$ for all j and all $T > 0$. Now consider any sequence T_k converging to $+\infty$. The bounded sequence $z_j(T_k)$ admits a convergent subsequence. By diagonalization we obtain a subsequence — which we are going to denote by T_k again — such that $z_j(T_k)$ converges for every j towards some limit which we shall denote by $\bar{z}_j$. The sequences $\log x_i(T_k) - \log x_i(0)$ are also bounded. Passage to the limit in (5.7) thus leads to

$$0 = r_i + \sum a_{ij}\bar{z}_j .$$

The point $\bar{z} = (\bar{z}_1, \ldots, \bar{z}_n)$ is therefore a rest point. Since $\bar{z}_j \geq a > 0$, it belongs to int $\mathbb{R}_+^n$. Hence it coincides with **p**. This implies (5.5). $\qquad\square$

Exercise 5.2.4 Give another proof of theorem 5.2.1, using a similar time-average argument. (This will work at least in the generic case.)

Exercise 5.2.5 Show that a similar averaging principle holds for the difference equation $\mathbf{x} \rightarrow \mathbf{x}'$ with

$$x_i' = x_i \exp(r_i + \sum_j a_{ij}x_j).$$

Exercise 5.2.6 What happens with theorem 5.2.3 if the assumption concerning the uniqueness of the rest point is dropped?

5.3 The Lotka–Volterra equations for food chains

Let us investigate food chains with n members (chains with up to six members are found in nature). The first population is the prey for the second, which is the prey for the third, and so on up to the n-th, which is at the top of the food pyramid. Taking competition within each species into account, and assuming constant interaction terms, we obtain

$$\begin{aligned}
\dot{x}_1 &= x_1(r_1 - a_{11}x_1 - a_{12}x_2) \\
\dot{x}_j &= x_j(-r_j + a_{j,j-1}x_{j-1} - a_{jj}x_j - a_{j,j+1}x_{j+1}) \quad j = 2, \ldots, n-1 \\
\dot{x}_n &= x_n(-r_n + a_{n,n-1}x_{n-1} - a_{nn}x_n)
\end{aligned} \tag{5.9}$$

with all $r_j, a_{ij} > 0$. The case $n = 2$ is just (2.15). We shall presently see that the general case leads to nothing new:

Theorem 5.3.1 *If (5.9) admits an interior rest point* **p**, *then* **p** *is globally stable in the sense that all orbits in* int $\mathbb{R}_+^n$ *converge to* **p**.

Proof In order to prove this we write (5.9) as $\dot{x}_i = x_i w_i$ and try

$$V(x) = \sum c_i(x_i - p_i \log x_i), \tag{5.10}$$

for suitably chosen c_i, as a Lyapunov function in int $\mathbb{R}_+^n$. Clearly

$$\dot{V} = \sum c_i\left(\dot{x}_i - p_i \frac{\dot{x}_i}{x_i}\right) = \sum c_i(x_i w_i - p_i w_i) = \sum c_i(x_i - p_i)w_i . \tag{5.11}$$

Since **p** is a rest point, we have

$$r_j = a_{j,j-1}p_{j-1} - a_{jj}p_j - a_{j,j+1}p_{j+1}$$

for $j = 2,\ldots,n-1$, and similar equations for $j = 1$ and $j = n$. This implies

$$w_j = a_{j,j-1}(x_{j-1} - p_{j-1}) - a_{jj}(x_j - p_j) - a_{j,j+1}(x_{j+1} - p_{j+1}).$$

Writing $y_j = x_j - p_j$, we obtain from (5.11)

$$\dot{V} = -\sum_{j=1}^{n} c_j a_{jj} y_j^2 + \sum_{j=1}^{n-1} y_j y_{j+1}(-c_j a_{j,j+1} + c_{j+1} a_{j+1,j}). \tag{5.12}$$

We are still free to choose the constants $c_j > 0$. Let us do it in such a way that

$$\frac{c_{j+1}}{c_j} = \frac{a_{j,j+1}}{a_{j+1,j}} \tag{5.13}$$

for $j = 1,\ldots,n$. (5.12) then implies

$$\dot{V} = -\sum c_j a_{jj}(x_j - p_j)^2 \le 0 . \tag{5.14}$$

By Lyapunov's theorem the ω-limit of every orbit in int $\mathbb{R}_+^n$ consists of the rest point **p**.   □

Exercise 5.3.2 Show that in the absence of competition within the predators (i.e. $a_{jj} = 0$ for $j = 2,\ldots,n$), **p** is still a global attractor in int $\mathbb{R}_+^n$. (Hint: an ω-limit is an invariant set.)

Exercise 5.3.3 Discuss the different phase portraits for (5.9). Show that with increasing r_1 more and more predators can subsist. What are the bifurcation values?

Exercise 5.3.4 Study food chains with recycling:

$$\dot{x}_1 = Q - a_{12}x_1x_2 + \sum_{k=2}^{n} b_k x_k$$

$$\dot{x}_j = x_j(-r_j + a_{j,j-1}x_{j-1} - a_{j,j+1}x_{j+1}) \qquad j = 2,\ldots,n \; .$$

5.4 The exclusion principle

The *exclusion principle* states that if n populations depend linearly on m resources, with $m < n$, then at least one of the populations will vanish. More populations than there are resources (or *ecological niches*, in another interpretation) cannot subsist in the long run.

The assumption on the linear dependence of the resources is crucial. It means that the growth rate of the i-th population is of the form

$$\frac{\dot{x}_i}{x_i} = b_{i1}R_1 + \ldots + b_{im}R_m - \alpha_i \tag{5.15}$$

($i = 1,\ldots,n$). The constant $\alpha_i > 0$ indicates the rate of decline in the absence of any resource. R_k is the abundance of the k-th resource, and the coefficients b_{ik} describe the efficiency of the i-th species in making use of the k-th resource. The abundance of the resources depends of course on the population densities. If the dependence is linear, i.e. if

$$R_k = \bar{R}_k - \sum x_i a_{ki} \tag{5.16}$$

with positive constants $\bar{R}_k$ and a_{ki}, then (5.15) is a special case of the Lotka–Volterra equation (5.1). Assumption (5.16) is not needed, however. It is enough to postulate that the resources can be exhausted, i.e. that the densities x_i cannot grow to infinity.

Since $n > m$, the system of equations

$$\sum_{i=1}^{n} c_i b_{ij} = 0 \qquad j = 1,\ldots,m$$

admits a nontrivial solution $(c_1,\ldots,c_n)$. Let $\alpha = \sum_{i=1}^{n} c_i \alpha_i$.

We shall only consider the general case where $\alpha \neq 0$. By an appropriate choice of the c_i, we can even assume $\alpha > 0$. (5.15) implies

$$\sum c_i(\log x_i)^{\cdot} = \sum c_i \frac{\dot{x}_i}{x_i} = -\alpha \; .$$

Integrating from 0 to T we obtain

$$\prod_{i=1}^{n} x_i(T)^{c_i} = C e^{-\alpha T} \tag{5.17}$$

for some constant C. For $T \to +\infty$ the right hand side converges to 0. Since all $x_i(T)$ are bounded, there must be at least one index i such that $\liminf x_i(T) = 0$ for $T \to \infty$, which spells extinction for the corresponding species.

5.5 A model for cyclic competition

Within the class of Lotka–Volterra equations, the existence of an interior rest point implies its stability for models of food chains. This is not so, however, for models of competitive interactions: as we have seen in section 3.3, competition between two species can lead to the extinction of one of them even if an interior rest point exists (in the case of bistability).

If three or more species compete, another and rather curious thing can occur. It may look for some time as if species 1 were bound to be the unique survivor; then, suddenly, its density drops, species 2 takes its place and seems to dominate the 'ecosystem'; after some time, it in turn collapses, and leaves the field to species 3, which appears to be the ultimate winner; but then, species 1 suddenly rallies and outcompetes its rivals, and so another round starts. The species supersede each other in cyclic fashion: the time spans during which one species predominates grow larger and larger. An observer may get the impression that one species is better adapted and the other two doomed to extinction, until suddenly, without exterior cue, another revolution occurs.

We shall examine such a behaviour for the equation

$$
\begin{aligned}
\dot{x}_1 &= x_1(1 - x_1 - \alpha x_2 - \beta x_3) \\
\dot{x}_2 &= x_2(1 - \beta x_1 - x_2 - \alpha x_3) \\
\dot{x}_3 &= x_3(1 - \alpha x_1 - \beta x_2 - x_3)
\end{aligned}
\tag{5.18}
$$

with $0 < \beta < 1 < \alpha$ and $\alpha + \beta > 2$. Of course, the assumptions behind this equation are so artificial that we are never going to find them 'in the field'. But they do help with computations and display features which we must be prepared to meet in more general situations.

The special symmetry assumption behind the model is that of a cyclic interaction between the species: if we replace 1 by 2, 2 by 3 and 3 by 1, equation (5.18) will remain unchanged. This cyclic symmetry leads to a drastic simplification of some computations, which we shall use a few more times later on.

In most cases it is a difficult task to find the eigenvalues of a Jacobian. For cyclic symmetry, however, it is child's play. An $n \times n$ matrix is said to be *circulant* if it is of the form

$$\begin{bmatrix} c_0 & c_1 & c_2 & \cdots & c_{n-1} \\ c_{n-1} & c_0 & c_1 & \cdots & c_{n-2} \\ \vdots & \vdots & \vdots & & \vdots \\ c_1 & c_2 & c_3 & \cdots & c_0 \end{bmatrix} \tag{5.19}$$

where a cyclic permutation sends the elements of each row into those of the next.

Exercise 5.5.1 Check that the eigenvalues of (5.19) are

$$\gamma_k = \sum_{j=0}^{n-1} c_j \lambda^{jk} \qquad k = 0,\ldots,n-1 \tag{5.20}$$

and the eigenvectors are

$$\mathbf{y}_k = (1, \lambda^k, \lambda^{2k}, \ldots, \lambda^{(n-1)k}), \tag{5.21}$$

where λ is the n-th root of unity

$$\lambda = \exp(2\pi i/n) \quad . \tag{5.22}$$

Now let us return to (5.18) and note that it admits a unique interior rest point $\mathbf{m}$ given by

$$m_1 = m_2 = m_3 = \frac{1}{1+\alpha+\beta} . \tag{5.23}$$

The Jacobian at the point $\mathbf{m}$ is

$$\frac{1}{1+\alpha+\beta} \begin{bmatrix} -1 & -\alpha & -\beta \\ -\beta & -1 & -\alpha \\ -\alpha & -\beta & -1 \end{bmatrix} . \tag{5.24}$$

This matrix is circulant. By (5.20) its eigenvalues are $\gamma_0 = -1$ (with eigenvector $(1,1,1)$) and

$$\gamma_1 = \bar\gamma_2 = \frac{1}{1+\alpha+\beta}(-1 - \alpha e^{2\pi i/3} - \beta e^{4\pi i/3}) \quad .$$

The real part of γ_1 and γ_2 is thus

$$\frac{1}{1+\alpha+\beta}\left(-1 + \frac{\alpha+\beta}{2}\right) \tag{5.25}$$

which is positive by assumption. Hence $\mathbf{m}$ is a saddle.

There are four further rest points, all on $\mathrm{bd}\,\mathbb{R}_+^3$, namely $\mathbf{0}$ (which is a source) and the saddles $\mathbf{e}_1, \mathbf{e}_2, \mathbf{e}_3$ (the standard unit vectors).

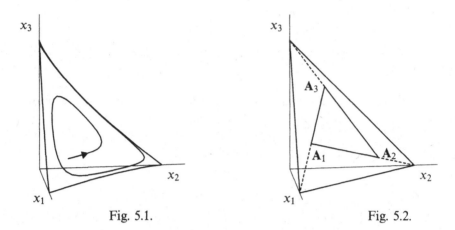

Fig. 5.1.   Fig. 5.2.

The restriction of (5.18) to the face $x_3 = 0$ yields a competition equation for x_1 and x_2, which we have studied in section 3.3. In the absence of species 3, species 2 will outcompete species 1 (see fig. 5.1). This implies that the stable manifold of $\mathbf{e}_2$ is the two-dimensional set $\{(x_1, x_2, x_3) : x_1 \geq 0, x_2 > 0, x_3 = 0\}$, while the unstable manifold of $\mathbf{e}_1$ consists of a single orbit $\mathbf{o}_2$ converging to $\mathbf{e}_2$.

On the other boundary faces, the situation is similar: on the plane $x_1 = 0$, there is an orbit $\mathbf{o}_3$ with α-limit $\mathbf{e}_2$ and ω-limit $\mathbf{e}_3$, and on $x_2 = 0$ an orbit $\mathbf{o}_1$ from $\mathbf{e}_3$ to $\mathbf{e}_1$. We denote by F the set consisting of the three saddles $\mathbf{e}_1$, $\mathbf{e}_2$ and $\mathbf{e}_3$ and the three connecting orbits $\mathbf{o}_1, \mathbf{o}_2$ and $\mathbf{o}_3$ (see fig. 5.1). Such an invariant set F is called a *heteroclinic cycle*. We shall show that all orbits in int $\mathbb{R}_+^3$ (with the exception of those on the diagonal) have F as ω-limit. The state thus remains for a long time close to the rest point $\mathbf{e}_1$, then travels along $\mathbf{o}_2$ to the vicinity of the rest point $\mathbf{e}_2$, lingers there for a still longer time, then jumps over to the rest point $\mathbf{e}_3$ and so on, in cyclic fits and starts.

To verify this, we shall use the functions

$$S = x_1 + x_2 + x_3 \tag{5.26}$$

and

$$P = x_1 x_2 x_3 . \tag{5.27}$$

One has

$$\dot{S} = x_1 + x_2 + x_3 - \left[x_1^2 + x_2^2 + x_3^2 + (\alpha + \beta)(x_1 x_2 + x_2 x_3 + x_3 x_1) \right] \tag{5.28}$$

and

$$\dot{P} = \dot{x}_1 x_2 x_3 + x_1 \dot{x}_2 x_3 + x_1 x_2 \dot{x}_3 = P(3 - (1 + \alpha + \beta)S) \quad . \qquad (5.29)$$

Equation (5.28) implies $\dot{S} \leq S(1 - S)$, so that no population can explode. A straightforward computation yields

$$\left(\frac{P}{S^3}\right)^{\cdot} = S^{-4}P\left(1 - \frac{\alpha+\beta}{2}\right)\left[(x_1 - x_2)^2 + (x_2 - x_3)^2 + (x_3 - x_1)^2\right] \leq 0 .$$

The theorem of Lyapunov then implies that any orbit which is not on the (invariant) diagonal $x_1 = x_2 = x_3$ converges to the boundary (the set where P vanishes). But we have already investigated the behaviour there: the only candidate for an ω-limit is the set F.

Of course such an ω-limit can never occur 'in reality'. Since $\liminf x_i(t) = 0$, one of the species will sooner or later vanish, and then one of the remaining species will outcompete the other one. Still, the model is of biological interest, as it suggests a surprising mechanism for sudden upheavals in ecological communities.

It is instructive to study the behaviour of the time averages for this system. We have seen in section 5.2 that time averages converge to an interior rest point whenever the orbit stays away from the coordinate planes. In (5.18), however, orbits converge to $\operatorname{bd} \mathbb{R}_+^3$: as we shall see in a moment, their time averages no longer converge.

Since the orbits of (5.18) spend most of their time near the fixed points $\mathbf{e}_1, \mathbf{e}_2$ and $\mathbf{e}_3$, their time averages

$$\mathbf{z}(T) = \frac{1}{T}\int_0^T \mathbf{x}(t)dt$$

will converge to the plane spanned by these three points, which is given by

$$z_1 + z_2 + z_3 = 1 . \qquad (5.30)$$

Now consider again equation (5.7) for the time averages. Since $x_i(T)$ is bounded from above, every accumulation point of the left hand side is nonpositive. Thus every limit point $\mathbf{z}$ of the time averages $\mathbf{z}(T)$ satisfies

$$r_i + \sum_{j=1}^n a_{ij}z_j \leq 0 . \qquad (5.31)$$

To be more precise, consider a sequence $T_k \to +\infty$ such that $\mathbf{z}(T_k) \to \mathbf{z}$, and let $\bar{\mathbf{x}}$ be a limit point of $\mathbf{x}(T_k)$. There are two possibilities:

(a) $\bar{\mathbf{x}}$ lies on one of the three connecting orbits $\mathbf{o}_1, \mathbf{o}_2$ or $\mathbf{o}_3$. Then only one of the coordinates of $\bar{\mathbf{x}}$ is zero; say $\bar{x}_1 = 0, \bar{x}_2 > 0, \bar{x}_3 > 0$. (5.31) yields one inequality and two equations to be satisfied by $\mathbf{z}$:

$$
\begin{aligned}
1 - z_1 - \alpha z_2 - \beta z_3 &\leq 0 \\
1 - \beta z_1 - z_2 - \alpha z_3 &= 0 \\
1 - \alpha z_1 - \beta z_2 - z_3 &= 0 \ .
\end{aligned}
\tag{5.32}
$$

(5.32) determines a line segment between the interior fixed point $\mathbf{m}$ and the intersection of the x_2- and x_3-isoclines in the plane $x_1 = 0$. Together with (5.30) this determines the position of $\mathbf{z}$. Let us call this point A_2, and define A_3 and A_1 similarly.

(b) $\bar{\mathbf{x}}$ is one of the three boundary rest points, say $\mathbf{e}_1$. Then (5.31) yields one equality and two inequalities: this is just like (5.32) with equality and inequality signs interchanged. This shows that any such limit point $\mathbf{z}$ lies on the line segment between A_3 and A_1. Since the set of all limit points of $\mathbf{z}(T)$ is obviously connected, every point on this segment will be a limit point.

This shows that the set of all limit points of the time averages $\mathbf{z}(T)$ is just the boundary of the triangle $A_1 A_2 A_3$, which is given by the intersection of the plane (5.30) with the region (5.31), see fig. 5.2.

Exercise 5.5.2 Compute A_1, A_2 and A_3 explicitly.

Exercise 5.5.3 Show that the time intervals which a solution spends near the rest points $\mathbf{e}_1, \mathbf{e}_2, \mathbf{e}_3, \mathbf{e}_1, \dots$ increase geometrically with factor $\frac{\alpha-1}{1-\beta} > 1$. Using this fact, find another (independent and even more general) proof of the previous result.

Exercise 5.5.4 Analyse (5.18) for other values of α and β.

Exercise 5.5.5 Let $\hat{\mathbf{x}}$ be an interior fixed point of (5.1). Let $\mathbf{c}$ be a left eigenvector of the Jacobian at $\hat{\mathbf{x}}$, i.e.

$$
\sum_{i=1}^{n} c_i \hat{x}_i a_{ij} = \lambda c_j, \quad \lambda \neq 0
$$

and let $\mathbf{y} = \mathbf{0}$ be the only solution of the system

$$
\sum_{i=1}^{n} c_i y_i = 0 \text{ and } \sum_{i,j} c_i a_{ij} y_i y_j = 0 \ .
$$

Show that under these assumptions, all ω-limits of solutions of (5.1) are either $\hat{\mathbf{x}}$ or contained in bd $\mathbb{R}^n_+$.

5.6 Notes

The model (5.18) for cyclic competition is due to May and Leonard (1975). It was discussed in subsequent papers by Chenciner (1977), which contains the result in exercise 5.5.5, Coste *et al.* (1979) and Schuster *et al.* (1979b). For more general results on time averages, see Gaunersdorfer (1992). The results on food chains are due to Harrison (1978), So (1979) and Gard and Hallam (1979). Examples of 'chaotic' systems were discovered by Gilpin (1979) and by Arneodo *et al.* (1980), and further studied by Gardini *et al.* (1989). The exclusion principle is due to Volterra; it need not be valid for nonlinear dependence on the resources. For theorem 5.2.3 we refer to Coste *et al.* (1979) and Schuster *et al.* (1980). The difference equation of exercise 5.2.5 was analysed in Hofbauer *et al.* (1987) and Rand *et al.* (1994). The result on separating hyperplanes used in the proof of theorem 5.2.1 can be found in Nikaido (1968).

Part two

Game Dynamics and Replicator Equations

6

Evolutionarily stable strategies

Motivated by some biological examples, we introduce the game-theoretical notions of Nash equilibrium and evolutionarily stable strategy. An evolutionarily stable population is proof against invading minorities.

6.1 Hawks and doves

A long-standing theme of ethology is the prevalence of *conventional fights*. Conflicts among animals (especially within heavily armed species) are often settled by displays rather than all-out fighting. A whole gamut of threatening signals and harmless assessments of strength serves to settle contests for food, territory and mates, so that escalated fights leading to injury or death are relatively rare. Stags, for example, engage in roaring matches, starting slowly at first and increasing the rate. If the intruder does not quit at that stage, this is followed by a parallel walk. This in turn may end with a retreat, or with a direct contest in strength: the stags clash head on, as if on a cue, interlock antlers and push against each other. If one stag turns earlier, and faces the flank of his opponent with the lethal points of his antlers, he halts his attack and resumes the parallel walk. Only a few contests lead to serious injuries.

Such conventional fights have been compared with boxing tournaments, and the restrained nature of animal aggression has often been stressed. It is obviously all to the good of the species, but needs an explanation from an evolutionary point of view. A stag ruthlessly killing his rivals should inherit their harems and increase his offspring. His sons will inherit his fighting behaviour, and multiply faster than stags sticking to conventional displays. Why, then, do escalated contests remain rare?

The biologist John Maynard Smith used *game theory* to explain the high frequency of conventional contests. It takes the form of a thought experiment: suppose there are only two possible behavioural types: one escalates the conflict until injury or the flight of the opponent settles the issue; the other sticks to displays and retreats if the opponent escalates. These two types of behaviour are usually described as 'hawks' and 'doves' , although this is somewhat misleading. The conflicts, after all, are supposed to take place within one species and not between two; furthermore, real doves do escalate — we shall return to this in a moment.

The contests may take place over a morsel of food, the boundary line between territories or a potential mate. The prize corresponds to a gain in fitness G, while an injury reduces fitness by C. Fitness here means simply reproductive success.

If two doves meet, they posture, glare at each other, swell up, change colour or what not: but eventually, one of them retreats. The winner obtains G, the loser gets nothing, so that the average increase in fitness, for a dove meeting another dove, is $G/2$. A dove meeting a hawk flees and its fitness remains unchanged, while that of the hawk increases by G. Finally, if a hawk meets a hawk, they escalate until one of the two gets knocked out. The fitness of the winner is increased by G, that of the loser reduced by C, so that the average increase in fitness is $(G - C)/2$, which is negative if the cost of the injury exceeds the prize of the fight (as we shall always assume). This is encapsulated in a *payoff matrix*:

	if it meets a hawk	if it meets a dove	
a hawk receives	$\frac{G-C}{2}$	G	(6.1)
a dove receives	0	$\frac{G}{2}$	

In a population consisting mostly of doves, hawks will spread, for they are likely to meet only doves and get a gain G out of each contest, while a dove will only get $G/2$. But in a population of mostly hawks, it is the dove which is better off: it avoids every fight and keeps its fitness unchanged, while the hawks are at each other's throat with an average loss in fitness of $(G - C)/2$. Neither type of behaviour is better than the other: their success depends on the composition of the population. If the frequency of hawks is x, and that of doves correspondingly $1 - x$, then the average increase in fitness will be $x(G - C)/2 + (1 - x)G$ for the hawks and $(1 - x)G/2$ for the doves . Equality holds if and only if $x = G/C$. If the frequency of hawks is less than G/C, they fare better than the doves and should therefore spread; if their frequency is larger, they do less well and should diminish. Evolution should lead to a stable rest point with G/C as the frequency of the hawks .

In particular, if the cost of injury C is very large, the hawk frequency will be small. It is well supported by observations that the 'gloved fist' type of contest prevails especially among the heavily armed species. Apparently harmless animals, on the other hand, have not developed traits for avoiding escalation. Real doves, for example, which under natural conditions cannot inflict serious injuries upon each other, will fight to death when confined in a cage.

Exercise 6.1.1 If the display implies a decrease of fitness E (through loss of time and energy), compute the payoff matrix, the stable equilibrium and the frequency of hawks which maximizes the group benefit.

Exercise 6.1.2 Extend the previous analysis by allowing two other types of behaviour: (a) 'retaliators' who stick to displays unless the opponent starts escalating, in which case they escalate too; (b) 'bullies' who fake an escalation, but run off if their opponent escalates too.

6.2 Evolutionary stability

A type of behaviour is said to be *evolutionarily stable* if, whenever all members of the population adopt it, no dissident behaviour could invade the population under the influence of natural selection.

In order to formalize this, let us denote by $W(I, Q)$ the fitness of an individual of type I in a population of composition Q, and by $xJ + (1 - x)I$ the mixed population where x is the frequency of J-types and $1 - x$ that of I-types. A population consisting of I-types will be *evolutionarily stable* if, whenever a small amount of deviant J-types is introduced, the old type I fares better than the newcomers J. This means that for all $J \neq I$,

$$W(J, \varepsilon J + (1 - \varepsilon)I) < W(I, \varepsilon J + (1 - \varepsilon)I) \tag{6.2}$$

for all sufficiently small $\varepsilon > 0$.

We may obviously assume that $W(I, Q)$ is continuous in the second variable: a small change in the composition of the population will have only a slight effect on the fitness of I. By letting $\varepsilon \to 0$, we obtain from (6.2) that

$$W(J, I) \leq W(I, I) \tag{6.3}$$

for all J, which means that no type fares better against a population of I-types than the I-type itself. (The converse is not true in general: (6.3) does not imply (6.2).) In the hawk–dove game, neither of the two behavioural types is evolutionarily stable: each can be invaded by the other. But we shall presently see that the type of behaviour which consists in escalating with probability $p = G/C$ cannot be invaded by any other type, and hence is evolutionarily stable.

As a further example, let us consider the 'sex ratio' game: here, different types of behaviour correspond to different sex ratios (i.e. frequencies of males among offspring). Already Darwin was puzzled by the prevalence of the sex ratio $1/2$ in animal populations. It is not, as one may first think, an immediate result of the sex determination through X- and Y-chromosomes. In fact, at conception the ratio may be quite different, and subsequently change to yield a value near $1/2$ at birth. What is the evolutionary reason of this? After all, as animal breeders know, a female biased sex ratio leads to a higher overall growth rate. Why are there so many males?

We shall presently see that although the number of children is not affected by the sex ratio, the number of grandchildren is. Roughly speaking, if there were more females, then males would have good prospects. Since the same holds vice versa, this leads to a sex ratio of $1/2$.

To check this, let us denote by p the sex ratio of a given individual, and by m the average sex ratio in the population. Let N_1 be the population number in the daughter generation F_1 (of which mN_1 will be male and $(1 - m)N_1$ female) and N_2 the number in the following generation F_2. Each member of F_2 has one mother and one father: the probability that a given male in the F_1 generation is its father is $\frac{1}{mN_1}$, and the expected number of children produced by a male in the F_1 generation is therefore $\frac{N_2}{mN_1}$ (assuming random mating). Similarly, a female in the F_1 generation contributes an average of $\frac{N_2}{(1-m)N_1}$ children. The expected number of grandchildren of an individual with sex ratio p will be proportional to

$$p\frac{N_2}{mN_1} + (1 - p)\frac{N_2}{(1 - m)N_1}$$

i.e. its fitness is proportional to

$$w(p, m) = \frac{p}{m} + \frac{1 - p}{1 - m}. \tag{6.4}$$

(We may clearly exclude the cases $m = 0$ and $m = 1$ which lead to immediate extinction.) For given $m \in]0, 1[$ the function $p \to w(p, m)$ is affine linear, increasing for $m < 1/2$, decreasing for $m > 1/2$ and constant for $m = 1/2$.

Let us now consider a sex ratio q, and ask whether it is evolutionarily stable in the sense that no other sex ratio p can invade. If such a deviant sex ratio is introduced in a small proportion ε, the average sex ratio is $r = \varepsilon p + (1 - \varepsilon)q$. The q-ratio fares better than the p-ratio if and only if

$$w(p,r) < w(q,r) . \tag{6.5}$$

This is obviously the case for every p when $q = 1/2$ (it is enough to note that p and r are either both smaller, or both larger than $1/2$). For $q < 1/2$, a sex ratio $p > q$ will do better, however, and consequently spread; similarly, any $q > 1/2$ can be invaded by a smaller p. Thus $1/2$ is the unique sex ratio which is evolutionarily stable in the sense of (6.2).

6.3 Normal form games

Let us now introduce some game-theoretical terminology and assume that we model a situation where the behaviour used by each individual 'player' can be described in terms of a finite set of *pure strategies*. This could be the hawk and the dove strategy, or — in the context of the sex ratio game — 'produce only sons' and 'produce only daughters'. In general, we shall assume that there are N such pure strategies R_1 to R_N, and allow the players to use *mixed strategies* too: these will consist in playing the pure strategies R_1 to R_N with some pre-assigned probabilities p_1 to p_N. (In the hawk–dove game, this would mean escalating with a certain probability; in the sex ratio game, mixing male and female offspring with some given ratio). Since $p_i \geq 0$ and $\sum p_i = 1$, a strategy corresponds to a point $\mathbf{p}$ in the simplex

$$S_N = \left\{ \mathbf{p} = (p_1,\ldots,p_N) \in \mathbb{R}^N : p_i \geq 0 \text{ and } \sum_{i=1}^{N} p_i = 1 \right\}. \tag{6.6}$$

The corners of the simplex are the standard unit vectors $\mathbf{e}_i$ (where the i-th component is 1 and all others are 0) and correspond to the N pure strategies R_i. The interior int S_N consists of the *completely mixed strategies* $\mathbf{p}$, i.e. those for which $p_i > 0$ for all i, and the boundary bd S_N consists of all $\mathbf{p} \in S_N$ for which the *support* supp($\mathbf{p}$) $= \{i : 1 \leq i \leq N \text{ and } p_i > 0\}$ is a proper subset J of $\{1,\ldots,N\}$. Conversely, each such subset J defines a *boundary face* spanned by the corresponding $\mathbf{e}_i$.

Let us assume first that the game involves two players only, and let us denote by u_{ij} the payoff for a player using the pure strategy R_i against a player using the pure strategy R_j. The $N \times N$ matrix $U = (u_{ij})$ is said to be the *payoff matrix*. An R_i-strategist then obtains the (expected) payoff $(U\mathbf{q})_i = \sum u_{ij}q_j$ against a $\mathbf{q}$-strategist (since q_j is the probability that he is met with strategy R_j). The payoff for a **p**-strategist against a **q**-strategist is given by

$$\mathbf{p} \cdot U\mathbf{q} = \sum_{ij} u_{ij} p_i q_j. \tag{6.7}$$

By $\beta(\mathbf{q})$, we denote the set of *best replies* to **q**, i.e. of strategies for which the map $\mathbf{p} \to \mathbf{p} \cdot U\mathbf{q}$ attains its maximal value. Obviously, this set must contain some pure strategies: more precisely, it is a nonempty boundary face of S_N.

The most important notion in game theory is that of a *Nash equilibrium*: this is a strategy which is a best reply to itself. Every *normal form game* (defined by an $N \times N$ payoff matrix U) admits at least one Nash equilibrium, as we shall see in section 13.4. A strategy **q** is said to be a *strict Nash equilibrium* if it is the *unique* best reply to itself, i.e. if

$$\mathbf{p} \cdot U\mathbf{q} < \mathbf{q} \cdot U\mathbf{q} \tag{6.8}$$

holds for all strategies $\mathbf{p} \neq \mathbf{q}$. It is easy to see that it has to be pure, in that case.

Exercise 6.3.1 Compute the Nash equilibria for all 2×2 matrices. When are they strict?

6.4 Evolutionarily stable strategies

Let us now consider a large population of players who encounter randomly chosen opponents. If all players use the same *strict* Nash equilibrium **q**, every individual deviating from it will be penalized. Hence, such dissident behaviour will not spread. In the hawk–dove example, however, there is no strict Nash equilibrium. We cannot assume that every *nonstrict* Nash equilibrium is proof against invasion by a dissident minority: for the invaders may use a strategy which does just as well, and may spread — unless, as we shall see, they encounter an evolutionarily stable population.

Indeed, let us recall the definition of evolutionary stability given in (6.2). The two types I and J now correspond to two strategies $\hat{\mathbf{p}}$ and $\mathbf{p}$ and the population $\varepsilon J + (1 - \varepsilon)I$ to the strategy mix $\varepsilon \mathbf{p} + (1 - \varepsilon)\hat{\mathbf{p}}$. Thus the strategy $\hat{\mathbf{p}} \in S_N$ will be said to be *evolutionarily stable* if for all $\mathbf{p} \in S_N$ with $\mathbf{p} \neq \hat{\mathbf{p}}$, the inequality

$$\mathbf{p} \cdot U(\varepsilon \mathbf{p} + (1 - \varepsilon)\hat{\mathbf{p}}) < \hat{\mathbf{p}} \cdot U(\varepsilon \mathbf{p} + (1 - \varepsilon)\hat{\mathbf{p}}) \tag{6.9}$$

holds for all $\varepsilon > 0$ that are sufficiently small, i.e. smaller than some appropriate *invasion barrier* $\bar{\varepsilon}(\mathbf{p}) > 0$.

Equation (6.9) may be written as

$$(1 - \varepsilon)(\hat{\mathbf{p}} \cdot U\hat{\mathbf{p}} - \mathbf{p} \cdot U\hat{\mathbf{p}}) + \varepsilon(\hat{\mathbf{p}} \cdot U\mathbf{p} - \mathbf{p} \cdot U\mathbf{p}) > 0 . \tag{6.10}$$

This implies that $\hat{\mathbf{p}}$ is an evolutionarily stable strategy (or ESS) if and only if the following two conditions are satisfied:

(a) *equilibrium condition*

$$\mathbf{p} \cdot U\hat{\mathbf{p}} \leq \hat{\mathbf{p}} \cdot U\hat{\mathbf{p}} \text{ for all } \mathbf{p} \in S_N \tag{6.11}$$

(b) *stability condition*

$$\text{if } \mathbf{p} \neq \hat{\mathbf{p}} \text{ and } \mathbf{p} \cdot U\hat{\mathbf{p}} = \hat{\mathbf{p}} \cdot U\hat{\mathbf{p}}, \text{ then } \mathbf{p} \cdot U\mathbf{p} < \hat{\mathbf{p}} \cdot U\mathbf{p} . \tag{6.12}$$

(6.11) is just the definition of a *Nash equilibrium*. Strategy $\hat{\mathbf{p}}$ is a best reply against itself. This property alone, however, does not guarantee non-invadability, since it allows the existence of another strategy $\mathbf{p}$ which is an *alternative best reply*. The stability condition states that, in such a case, $\hat{\mathbf{p}}$ fares better against $\mathbf{p}$ than $\mathbf{p}$ against itself. Clearly, strict Nash implies ESS and ESS implies Nash.

(6.11) implies, by setting $\mathbf{p} = \mathbf{e}_i$, that

$$(U\hat{\mathbf{p}})_i \leq \hat{\mathbf{p}} \cdot U\hat{\mathbf{p}} \tag{6.13}$$

for $i = 1, \ldots, N$. By summation, one sees that for all i with $p_i > 0$ (corresponding to the pure strategies which are effectively played) equality must hold. The Nash equilibrium condition (6.11) means that there exists a constant c such that $(U\hat{\mathbf{p}})_i \leq c$ for all i, with equality if $i \in \text{supp}(\mathbf{p})$. In particular, $\hat{\mathbf{p}} \in \text{int} S_N$ is a Nash equilibrium if and only if its coordinates satisfy the linear system of equations

$$(U\mathbf{p})_1 = \ldots = (U\mathbf{p})_N \tag{6.14}$$
$$p_1 + \cdots + p_N = 1 . \tag{6.15}$$

Theorem 6.4.1 *The strategy $\hat{\mathbf{p}} \in S_N$ is an ESS if and only if*

$$\hat{\mathbf{p}} \cdot U\mathbf{q} > \mathbf{q} \cdot U\mathbf{q} \tag{6.16}$$

for all $\mathbf{q} \neq \hat{\mathbf{p}}$ in some neighbourhood of $\hat{\mathbf{p}}$ in S_N.

Proof Let us start by assuming that $\hat{\mathbf{p}}$ is evolutionarily stable. We show first that every $\mathbf{q}$ close to $\hat{\mathbf{p}}$ can be written as $\varepsilon\mathbf{p} + (1 - \varepsilon)\hat{\mathbf{p}}$, for small ε. Actually we can choose $\mathbf{p}$ in the compact set $C = \{\mathbf{x} \in S_N : x_i = 0 \text{ for some } i \in \text{supp}(\hat{\mathbf{p}})\}$, which consists of the faces that do not contain $\hat{\mathbf{p}}$. For every $\mathbf{p} \in C$, (6.9) holds for all $\varepsilon < \bar{\varepsilon}(\mathbf{p})$. It is easy to see that $\bar{\varepsilon}(\mathbf{p})$ can be chosen to be continuous. Then $\bar{\varepsilon} := \min\{\bar{\varepsilon}(\mathbf{p}) : \mathbf{p} \in C\}$ is strictly positive, and (6.9) holds for all $\varepsilon \in]0, \bar{\varepsilon}[$. It is enough, then, to multiply (6.9) by ε, and to add

$$(1 - \varepsilon)\hat{\mathbf{p}} \cdot U((1 - \varepsilon)\hat{\mathbf{p}} + \varepsilon\mathbf{p}) \tag{6.17}$$

to both sides. With $\mathbf{q} := (1 - \varepsilon)\hat{\mathbf{p}} + \varepsilon\mathbf{p}$, this yields (6.16) for all $\mathbf{q} \neq \hat{\mathbf{p}}$ in an appropriate neighbourhood of $\mathbf{p}$. The converse is similar. □

It follows that if $\hat{\mathbf{p}} \in \text{int } S_N$ is evolutionarily stable, then there is no other ESS, and in fact no other Nash equilibrium: indeed, (6.16) must then hold for all $\mathbf{q} \in S_N$. This also shows that the support of one ESS cannot be contained in the support of another. There are games without ESS, however, and games with several ESS (which have then to lie on bd S_N).

In the hawk–dove game, one easily computes that the strategy $\hat{\mathbf{p}} = (\frac{G}{C}, \frac{C-G}{C})$ is an ESS. Indeed

$$\hat{\mathbf{p}} \cdot U\mathbf{p} - \mathbf{p} \cdot U\mathbf{p} = \frac{1}{2C}(G - Cp_1)^2 \tag{6.18}$$

is strictly positive for all $p_1 \neq G/C$, so that (6.16) is satisfied. Since $\hat{\mathbf{p}} \in \text{int } S_2$, $\hat{\mathbf{p}}$ is the unique ESS.

Exercise 6.4.2 Show that the games given by

$$(a) \begin{bmatrix} 1 & 0 & 0 \\ 0 & 1 & 0 \\ 0 & 0 & 1 \end{bmatrix}; \qquad (b) \begin{bmatrix} 0 & 1 & -1 \\ -1 & 0 & 1 \\ 1 & -1 & 0 \end{bmatrix}$$

have three, resp. no, ESS.

Exercise 6.4.3 Let $\hat{\mathbf{p}} \in \text{int } S_N$ be a Nash equilibrium. Show that $\hat{\mathbf{p}}$ is an ESS if and only if

$$\boldsymbol{\xi} \cdot U\boldsymbol{\xi} < 0 \text{ for all } \boldsymbol{\xi} \neq \mathbf{0} \text{ with } \sum_{i=1}^{N} \xi_i = 0 . \tag{6.19}$$

Exercise 6.4.4 Show that in the proof of Theorem 6.4.1, one can use as the invasion barrier

$$\begin{aligned} \bar{\varepsilon}(\mathbf{p}) &= \frac{(\hat{\mathbf{p}} - \mathbf{p}) \cdot U\hat{\mathbf{p}}}{(\hat{\mathbf{p}} - \mathbf{p}) \cdot U(\hat{\mathbf{p}} - \mathbf{p})} \quad \text{if} \quad \mathbf{p} \cdot U\mathbf{p} > \hat{\mathbf{p}} \cdot U\mathbf{p} \quad (\text{so that} \quad 0 < \bar{\varepsilon}(\mathbf{p}) < 1) \\ &= 1 \quad \text{otherwise} . \end{aligned}$$

Exercise 6.4.5 A closed nonempty subset E of S_N is said to be an *evolutionarily* *stable set* (or ES set) if for each $\mathbf{x} \in E$ there exists a neighborhood W such that

$$\mathbf{x} \cdot U\mathbf{y} \geq \mathbf{y} \cdot U\mathbf{y}, \tag{6.20}$$

for all $\mathbf{y} \in W$, with strict inequality if $\mathbf{y} \notin E$. Show that E must consist of Nash equilibria, and that it is enough to restrict attention, in (6.20), to those $\mathbf{y}$ which are best replies to $\mathbf{x}$. Show that the singleton $\{\mathbf{x}\}$ is an ES set if and only if $\mathbf{x}$ is an ESS. Describe the set of Nash equilibria for the game with payoff matrix

$$\begin{bmatrix} 0 & 2 & 0 \\ 2 & 0 & 0 \\ 1 & 1 & 0 \end{bmatrix}$$

and show that it is an ES set. Show that this game has no ESS.

6.5 Population games

So far, we have assumed that the success of a strategy depends on the outcome of pairwise encounters with randomly chosen opponents. This need not always be the case: the success of a sex ratio, for instance, depends not on the sex ratio of a particular opponent, but on the average sex ratio in the population. The ESS theory can be extended to such situations. Let us assume that the payoff f_i for the pure strategy R_i depends on the state of the population, or more precisely that f_i is a function of the frequencies m_j of the strategies R_j *in the population*. The strategy mix of the population is given by a point $\mathbf{m}$ which also belongs to S_N. Since a $\mathbf{p}$-strategist plays R_i with probability p_i, his payoff will be given by

$$\sum p_i f_i(\mathbf{m}) = \mathbf{p} \cdot \mathbf{f}(\mathbf{m}) . \tag{6.21}$$

We define $\hat{\mathbf{p}}$ to be a *local* ESS if $\hat{\mathbf{p}} \cdot \mathbf{f}(\mathbf{q}) > \mathbf{q} \cdot \mathbf{f}(\mathbf{q})$ holds for all $\mathbf{q} \neq \hat{\mathbf{p}}$ in some neighbourhood of $\hat{\mathbf{p}}$.

Exercise 6.5.1 Show that this is the case if and only if
(a) $\mathbf{q} \cdot \mathbf{f}(\hat{\mathbf{p}}) \leq \hat{\mathbf{p}} \cdot \mathbf{f}(\hat{\mathbf{p}})$ for all $\mathbf{q} \in S_N$, and
(b) $\hat{\mathbf{p}} \cdot \mathbf{f}(\mathbf{q}) > \mathbf{q} \cdot \mathbf{f}(\mathbf{q})$ whenever equality holds in (a) and $\mathbf{q} \neq \hat{\mathbf{p}}$ is sufficiently close to $\hat{\mathbf{p}}$.
(Hint: show first the analogue of (6.9).) Show by examples that several ESS can coexist in int S_2. Show that the characterization given by exercise 6.4.3 is still valid.

If the payoff functions f_i are linear in $\mathbf{m}$, i.e. of the form $(U\mathbf{m})_i$ for some matrix U, then we are back to the case studied in section 6.4.

6.6 Notes

The basic equilibrium notion in game theory is due to Nash, see e.g. Nash (1996). The hawk–dove game and the notion of evolutionarily stable strategy were introduced in Maynard Smith and Price (1973) and Maynard Smith (1974). A similar notion, the unbeatable strategy, was used by Hamilton (1967), see also the comments in Hamilton (1996). A good survey of the development during the first decade of Evolutionary Game Theory is offered by Maynard Smith (1982). Other surveys are Hines (1987) and Akin (1990). For biological applications see Riechert and Hammerstein (1983), and for applications to human societies see Sugden (1986). The connection between Nash equilibria and ESS is discussed in Bomze (1986) and van Damme (1991). We have only considered the case of finitely many pure strategies. For the general case see Bomze and Pötscher (1989). The notion of local ESS (section 6.5) was first formulated by Pohley and Thomas (1983). For the sex ratio game in section 6.2 we refer to Maynard Smith (1982), for the escalation of fights to Enquist and Leimar (1983) and Houston and McNamara (1988). An analysis of evolutionary game theory in terms of informations and actions can be found in Selten (1983). There are many modifications of the notion of ESS, see e.g. Binmore and Samuelson (1992), van Damme (1994). We refer to Thomas (1984), Mesterton-Gibbons (1992) and Binmore (1992) for very readable introductions to game theory. Advanced texts are Fudenberg and Tirole (1991) and van Damme (1991). A good account of evolutionary game theory and its connection with 'rational' game theory is Weibull (1995), see also Vega–Redondo (1996), Samuelson (1997) and the survey of Kandori (1996). Game-theoretical ideas are increasingly seen as being more appropriate than certain optimization arguments used in evolutionary biology, see Metz *et al.* (1996b).

7

Replicator dynamics

The replicator dynamics describes the evolution of the frequencies of strategies in a population. We discuss Nash equilibria and evolutionarily stable strategies in terms of the replicator dynamics, and show that the replicator equation is equivalent to the Lotka–Volterra equation. Furthermore, we analyse rock–scissors–paper games and partnership games.

7.1 The replicator equation

The notion of evolutionary stability relies upon implicit dynamical considerations. In certain situations, the underlying dynamics can be modelled by a differential equation on the simplex S_n.

Thus let us assume that the population is divided into n types E_1 to E_n with frequencies x_1 to x_n. The fitness f_i of E_i will be a function of the composition of the population, i.e. of the state $\mathbf{x}$. If the population is very large, and if the generations blend continuously into each other, we may assume that the state $\mathbf{x}(t)$ evolves in S_n as a differentiable function of t. The rate of increase $\dot{x}_i/x_i$ of type E_i is a measure of its evolutionary success. Following the basic tenet of Darwinism, we may express this success as the difference between the fitness $f_i(\mathbf{x})$ of E_i and the average fitness $\bar{f}(\mathbf{x}) = \sum x_i f_i(\mathbf{x})$ of the population. Thus we obtain

$$\frac{\dot{x}_i}{x_i} = \text{fitness of } E_i - \text{average fitness},$$

which yields the *replicator equation*

$$\dot{x}_i = x_i(f_i(\mathbf{x}) - \bar{f}(\mathbf{x})) \qquad i = 1,\ldots,n. \tag{7.1}$$

The simplex S_n is invariant under (7.1): if $\mathbf{x} \in S_n$ then $\mathbf{x}(t) \in S_n$ for all $t \in \mathbb{R}$.

Indeed, the sum $S = x_1 + \cdots + x_n$ satisfies

$$\dot{S} = (1 - S)\bar{f}$$

which has $S(t) = 1$ as a solution (thus if the solution of (7.1) starts on the plane $\sum x_i = 1$, it remains there). Furthermore, if $x_i(0) = 0$ then $x_i(t) = 0$ for all t, so that the faces of the simplex S_n, and therefore S_n itself, are invariant.

From now on, we shall only consider the restriction of (7.1) to S_n.

Let us stress that we derived (7.1) by assuming that the success of E_i corresponds to its rate of increase, and thus tacitly that 'like begets like'. We shall discuss in chapter 22 how (7.1) relates to the Mendelian machinery of inheritance.

Exercise 7.1.1 Check the 'quotient rule': for $x_j > 0$

$$\left(\frac{x_i}{x_j}\right)^{\cdot} = \left(\frac{x_i}{x_j}\right)(f_i(\mathbf{x}) - f_j(\mathbf{x})) . \tag{7.2}$$

The addition of a function $\Psi : S_n \to \mathbb{R}$ to all the f_i does not affect equation (7.1) on S_n. Indeed, with $g_i(\mathbf{x}) = f_i(\mathbf{x}) + \Psi(\mathbf{x})$, one has $\bar{g}(\mathbf{x}) = \sum x_i g_i(\mathbf{x}) = \bar{f}(\mathbf{x}) + \Psi(\mathbf{x})$ (on S_n) and therefore $g_i(\mathbf{x}) - \bar{g}(\mathbf{x}) = f_i(\mathbf{x}) - \bar{f}(\mathbf{x})$ for all $\mathbf{x} \in S_n$ and all i.

Of particular interest is the case of linear f_i. There exists, then, an $n \times n$ matrix $A = (a_{ij})$ such that $f_i(\mathbf{x}) = (A\mathbf{x})_i$. Equation (7.1) takes the form

$$\dot{x}_i = x_i((A\mathbf{x})_i - \mathbf{x} \cdot A\mathbf{x}) . \tag{7.3}$$

The rest points of (7.3) in $\mathrm{int}S_n$ are the solutions of

$$(A\mathbf{x})_1 = \cdots = (A\mathbf{x})_n \tag{7.4}$$
$$x_1 + \cdots + x_n = 1 \tag{7.5}$$

satisfying $x_i > 0$ for $i = 1, \ldots, n$. In general, there exists one or no such solution. In degenerate cases, the solutions form a linear manifold in $\mathrm{int}S_n$. The rest points on the subfaces of S_n are obtained in a similar way.

Exercise 7.1.2 The addition of a constant c_j to the j-th column of A does not change (7.3) on S_n. By adding appropriate constants, show that one may transform A into a simpler form, having 0 in the diagonal, for example, or 0 in the last row.

Exercise 7.1.3 Apply the projective transformation $\mathbf{x} \to \mathbf{y}$ with

$$y_i = \frac{x_i c_i}{\sum_j x_j c_j}$$

(with $c_j > 0$). Show that this transforms (7.3) into the replicator equation with matrix $(a_{ij}c_j^{-1})$. A rest point $\mathbf{p} \in \text{int } S_n$ can thereby be moved into the barycentre $\frac{1}{n}\mathbf{1} = (\frac{1}{n}, \ldots, \frac{1}{n})$ of S_n.

7.2 Nash equilibria and evolutionarily stable states

Let us return to the game-theoretical interpretation of the replicator dynamics. There is an underlying normal form game with N pure strategies R_1 to R_N and a payoff function given by an $N \times N$ matrix U. A strategy is defined by a point in S_N: the types E_1 to E_n correspond therefore to n points $\mathbf{p}^1, \ldots, \mathbf{p}^n \in S_N$. The *state* of the population is defined by the frequencies x_i of the types E_i, i.e. by a point $\mathbf{x} \in S_n$. With $a_{ij} = \mathbf{p}^i \cdot U \mathbf{p}^j$ (the payoff obtained by a $\mathbf{p}^i$-strategist against $\mathbf{p}^j$-opponents), we obtain for the fitness $f_i(\mathbf{x})$ of the type E_i the expression

$$f_i(\mathbf{x}) = \sum_j a_{ij}x_j = (A\mathbf{x})_i \qquad (7.6)$$

so that the replicator equation (7.1) reduces to (7.3).

Although N and n are, a priori, quite unrelated, it will be convenient to stress the parallels between the pure strategies R_1 to R_N, points in S_N and the $N \times N$ payoff matrix U on the one hand and the types of players E_1 to E_n, points in S_n and the $n \times n$ fitness matrix A on the other. In particular, we shall say that a point $\hat{\mathbf{x}} \in S_n$ is a (symmetric) *Nash equilibrium* if

$$\mathbf{x} \cdot A\hat{\mathbf{x}} \leq \hat{\mathbf{x}} \cdot A\hat{\mathbf{x}} \qquad (7.7)$$

for all $\mathbf{x} \in S_n$, and an *evolutionarily stable state* (state, not strategy!) if

$$\hat{\mathbf{x}} \cdot A\mathbf{x} > \mathbf{x} \cdot A\mathbf{x} \qquad (7.8)$$

for all $\mathbf{x} \neq \hat{\mathbf{x}}$ in a neighbourhood of $\hat{\mathbf{x}}$ (cf. (6.11) and (6.16)).

Theorem 7.2.1

(a) If $\hat{\mathbf{x}} \in S_n$ is a Nash equilibrium of the game described by the payoff matrix A, then $\hat{\mathbf{x}}$ is a rest point of (7.3).

(b) If $\hat{\mathbf{x}}$ is the ω-limit of an orbit $\mathbf{x}(t)$ in int S_n, then $\hat{\mathbf{x}}$ is a Nash equilibrium.

(c) If $\hat{\mathbf{x}}$ is Lyapunov stable, then it is a Nash equilibrium.

Proof

(a) If $\hat{\mathbf{x}}$ is a Nash equilibrium, there exists a constant c such that $(A\hat{\mathbf{x}})_i = c$ for all i with $\hat{x}_i > 0$. Hence $\hat{\mathbf{x}}$ satisfies the equations corresponding to (7.4–5) for a rest point in the face spanned by the $\mathbf{e}_i$ with $i \in \text{supp}(\hat{\mathbf{x}})$.

(b) Let us assume now that $x(t) \in S_n$ converges to $\hat{x}$, but that $\hat{x}$ is *not* a Nash equilibrium, i.e. not a best reply to itself. Then there exists an i and an $\varepsilon > 0$ such that $e_i \cdot A\hat{x} - \hat{x} \cdot A\hat{x} > \varepsilon$, and hence such that $\dot{x}_i / x_i > \varepsilon$, for t sufficiently large, which is impossible.

(c) Suppose that $\hat{x}$ is not a Nash equilibrium. Then, by continuity, there exists an index i and an $\varepsilon > 0$ such that $(Ax)_i - x \cdot Ax > \varepsilon$ for all x in a neighbourhood of $\hat{x}$. For such x, the component x_i increases exponentially, which contradicts the Lyapunov stability of $\hat{x}$.

□

Exercise 7.2.2 Show that not every rest point $\hat{x}$ of (7.3) is a Nash equilibrium (there may be indices i with $\hat{x}_i = 0$ but $(A\hat{x})_i > c$, where c is as in the proof above). Show that not every Nash equilibrium is Lyapunov stable, and that not every Nash equilibrium is the limit of an interior orbit. Show that if the time average of an orbit $x(t)$ in int S_n converges to $\hat{x}$, then $\hat{x}$ is a Nash equilibrium.

Exercise 7.2.3 Show that a state $\hat{x}$ can be the ω-limit point of all orbits $x(t)$ in the interior of S_n without being Lyapunov stable. (Hint: try the matrix

$$A = \begin{bmatrix} 0 & 1 & 0 \\ 0 & 0 & 2 \\ 0 & 0 & 1 \end{bmatrix} \quad .) \tag{7.9}$$

Theorem 7.2.4 *If $\hat{x} \in S_n$ is an evolutionarily stable state for the game with payoff matrix A, then it is an asymptotically stable rest point of (7.3).*

Proof Let us first check that the function

$$P(x) = \prod_i x_i^{\hat{x}_i} \tag{7.10}$$

has its unique maximum (on S_n) at the point $\hat{x}$. This follows from *Jensen's inequality*: if f is a strictly convex function defined on some interval I, then

$$f\left(\sum p_i x_i\right) \leq \sum p_i f(x_i) \tag{7.11}$$

for all $x_1, \ldots, x_n \in I$ and every $p = (p_1, \ldots, p_n) \in$ int S_n, with equality if and only if all x_i coincide. We apply this to the strictly convex function $f = -\log$ on $I = [0, +\infty]$, setting as usual $0 \log 0 = 0 \log \infty = 0$. For $\hat{x}, x \in S_n$, we obtain

$$\sum_{i=1}^{n} \hat{x}_i \log \frac{x_i}{\hat{x}_i} = \sum_{\hat{x}_i > 0} \hat{x}_i \log \frac{x_i}{\hat{x}_i} \leq \log \sum_{\hat{x}_i > 0} x_i \leq \log \sum_{i=1}^{n} x_i = \log 1 = 0 ,$$

hence

$$\sum_{i=1}^{n} \hat{x}_i \log x_i \leq \sum_{i=1}^{n} \hat{x}_i \log \hat{x}_i,$$

and thus $P(\mathbf{x}) \leq P(\hat{\mathbf{x}})$ with equality if and only if $\mathbf{x} = \hat{\mathbf{x}}$.

If $P > 0$ (i.e. for all $\mathbf{x} \in S_n$ with $x_i > 0$ whenever $\hat{x}_i > 0$) one has

$$\frac{\dot{P}}{P} = (\log P)' = \left(\sum \hat{x}_i \log x_i\right)' = \sum_{\hat{x}_i > 0} \hat{x}_i \frac{\dot{x}_i}{x_i} = \sum \hat{x}_i ((A\mathbf{x})_i - \mathbf{x} \cdot A\mathbf{x})$$

$$= \hat{\mathbf{x}} \cdot A\mathbf{x} - \mathbf{x} \cdot A\mathbf{x} .$$

Since $\hat{\mathbf{x}}$ is evolutionarily stable, (7.8) implies $\dot{P} > 0$ for all $\mathbf{x} \neq \hat{\mathbf{x}}$ in some neighbourhood of $\hat{\mathbf{x}}$. The function P, then, is a *strict local Lyapunov function* for (7.3), and all orbits near $\hat{\mathbf{x}}$ converge to $\hat{\mathbf{x}}$. □

In fact, we have just shown that $\hat{\mathbf{x}}$ *is evolutionarily stable iff P is a strict local Lyapunov function for* (7.3).

If $\hat{\mathbf{x}} \in$ int S_n is evolutionarily stable, then it is a globally stable rest point for (7.3) (in the sense that every orbit in int S_n converges to $\hat{\mathbf{x}}$) since $\dot{P}(\mathbf{x}) > 0$ for all $\mathbf{x} \in$ int S_n (with $\mathbf{x} \neq \hat{\mathbf{x}}$).

An example for an equilibrium in int S_3 which is asymptotically stable without being evolutionarily stable is obtained with the matrix

$$A = \begin{bmatrix} 0 & 6 & -4 \\ -3 & 0 & 5 \\ -1 & 3 & 0 \end{bmatrix} . \tag{7.12}$$

The point $\mathbf{m} = (\frac{1}{3}, \frac{1}{3}, \frac{1}{3})$ is a rest point and asymptotically stable, since its eigenvalues $\frac{1}{3}(-1 \pm i\sqrt{2})$ have negative parts. But $\mathbf{e}_1 = (1,0,0)$ is an evolutionarily stable state on bd S_3 (see fig. 7.1). It follows that $\mathbf{m}$ is not evolutionarily stable (being in the interior, it would have to be the unique ESS).

Exercise 7.2.5 Check all this and find a global Lyapunov function.

In section 7.8, we shall see that if the matrix A is symmetric, the ESS are precisely the asymptotically stable states.

Exercise 7.2.6 What does the game-dynamical equation look like for nonlinear games (see section 6.5)? Show that theorem 7.2.4 is still valid.

Exercise 7.2.7 If the payoff matrix A satisfies (6.19), then the game has a *unique* ESS. (Hint: you may use the fact that every game has a Nash equilibrium.) This ESS is globally stable. Show that (6.19) is immune against adding arbitrary constants to the *rows* of A.

Exercise 7.2.8 Show that every ES set (see exercise 6.4.5) is asymptotically stable (with the obvious definition for the asymptotic stability of a closed set).

7.3 Strong stability

How does the replicator equation (7.3) with matrix A agree with the underlying normal form game with payoff matrix U? (Recall that U is an $N \times N$ matrix, where N is the number of pure strategies, whereas A is an $n \times n$ matrix, where n is the number of different types in the population, each corresponding to a strategy $\mathbf{p}^i$, and $a_{ij} = \mathbf{p}^i \cdot U\mathbf{p}^j$). The dynamics given by (7.3) describes the evolution of the frequencies of the n types $E_1, \ldots, E_n$ given by the strategies $\mathbf{p}^1, \ldots, \mathbf{p}^n$. The state $\mathbf{x} \in S_n$ corresponds to a *mean population strategy* $\mathbf{p} = \sum x_i \mathbf{p}^i$. Depending on the evolution of the frequencies x_i, this strategy $\mathbf{p}$ describes a path in the strategy space S_N. We note that the set of all $\mathbf{x} \in S_n$ for which $\mathbf{p}$ equals a given value (for instance $\hat{\mathbf{p}}$) is, in general, the intersection of a linear manifold with S_n.

To begin with, take $n = 2$ and let E_1 and E_2 be two types corresponding to the strategies $\mathbf{p}$ and $\hat{\mathbf{p}}$ in S_N. If x_1 and x_2 are their respective frequencies, then their payoffs are given by $f_1(\mathbf{x}) = (A\mathbf{x})_1 = \mathbf{p} \cdot U(x_1\mathbf{p} + x_2\hat{\mathbf{p}})$ and $f_2(\mathbf{x}) = (A\mathbf{x})_2 = \hat{\mathbf{p}} \cdot U(x_1\mathbf{p} + x_2\hat{\mathbf{p}})$.

Since $x_2 = 1 - x_1$, it is enough to describe the evolution of x_1, which we denote by x. By (7.1), the dynamics on the interval $[0, 1]$ is given by

$$\dot{x} = x\big(f_1(\mathbf{x}) - \bar{f}(\mathbf{x})\big) \tag{7.13}$$

$$= x(1 - x)\big(f_1(\mathbf{x}) - f_2(\mathbf{x})\big) \tag{7.14}$$

$$= x(1 - x)\Big[x(\mathbf{p} \cdot U\mathbf{p} - \hat{\mathbf{p}} \cdot U\mathbf{p}) - (1 - x)(\hat{\mathbf{p}} \cdot U\hat{\mathbf{p}} - \mathbf{p} \cdot U\hat{\mathbf{p}})\Big]. \tag{7.15}$$

The endpoints $x = 0$ and $x = 1$ of the interval are fixed points which correspond to one or other of the two types. According to Maynard Smith, the $\hat{\mathbf{p}}$-type is evolutionarily stable if it cannot be invaded by any of the $\mathbf{p}$-types. This means that $x = 0$ is asymptotically stable: the frequency x of $\mathbf{p}$ is decreasing whenever it is sufficiently small. But this is obviously the case if and only if either (a) $\mathbf{p} \cdot U\hat{\mathbf{p}} < \hat{\mathbf{p}} \cdot U\hat{\mathbf{p}}$ or (b) $\mathbf{p} \cdot U\hat{\mathbf{p}} = \hat{\mathbf{p}} \cdot U\hat{\mathbf{p}}$ and $\mathbf{p} \cdot U\mathbf{p} < \hat{\mathbf{p}} \cdot U\mathbf{p}$. This corresponds precisely to the notion of ESS as given in (6.11) and (6.12).

Exercise 7.3.1 Give a similar interpretation of the notion of (local) ESS for nonlinear games described in exercise 6.5.1.

We shall say that a strategy $\hat{\mathbf{p}}$ is *strongly stable* if, whenever $\hat{\mathbf{p}}$ is feasible in the sense that it is a convex combination of the initial phenotypes $\mathbf{p}^1, \ldots, \mathbf{p}^n$, the mean population strategy $\sum x_i \mathbf{p}^i$ converges under (7.3) to $\hat{\mathbf{p}}$ for every initial value sufficiently close to $\hat{\mathbf{p}}$. The following theorem is probably the best validation for the concept of evolutionary stability.

Theorem 7.3.2 *The strategy $\hat{\mathbf{p}}$ is evolutionarily stable if and only if it is strongly stable.*

Proof In a way, we have proved this result already. Let us assume first that $\hat{\mathbf{p}} \in S_N$ is strongly stable. Let $\mathbf{p} \in S_N$ be any other strategy. Let us consider a population consisting of only two phenotypes, namely $\hat{\mathbf{p}}$ and $\mathbf{p}$. Using the result from the beginning of this section, we see that asymptotic convergence implies that $\hat{\mathbf{p}}$ is an evolutionarily stable strategy.

Conversely, let us assume that $\hat{\mathbf{p}}$ is an evolutionarily stable strategy, and let $\mathbf{p}^1, \ldots, \mathbf{p}^n$ be strategies in S_N whose convex hull contains $\hat{\mathbf{p}}$. Then $\hat{\mathbf{p}} = \sum \hat{x}_i \mathbf{p}^i$ for some $\hat{\mathbf{x}} \in S_n$. The function P given by (7.10) has its unique maximum at $\hat{\mathbf{x}}$, as we have seen in the proof of theorem 7.2.4. Let $\mathbf{x} \in S_n$ be such that the corresponding mean strategy $\mathbf{p} = \sum x_i \mathbf{p}^i$ is close to $\hat{\mathbf{p}}$. The function $t \to P(\mathbf{x}(t))$ satisfies

$$(\log P)^{\cdot} = \hat{\mathbf{x}} \cdot A\mathbf{x} - \mathbf{x} \cdot A\mathbf{x}, \tag{7.16}$$

as we have seen already in the proof of theorem 7.2.4. Since $a_{ij} = \mathbf{p}^i \cdot U\mathbf{p}^j$, this last expression is equal to $\hat{\mathbf{p}} \cdot U\mathbf{p} - \mathbf{p} \cdot U\mathbf{p}$, and by (6.16) we obtain a Lyapunov function. The ω-limit of $\mathbf{x}(t)$ is in the set where $\dot{P} = 0$, and hence $\mathbf{p}(\mathbf{x}(t))$ converges to $\hat{\mathbf{p}}$. $\qquad\square$

Exercise 7.3.3 Compare this theorem with the result in exercise 7.2.5.

Exercise 7.3.4 Show that if $\mathbf{p}^1$ is evolutionarily stable, but *not* a convex combination of $\mathbf{p}^2, \ldots, \mathbf{p}^n$, then the corresponding state $\mathbf{e}_1 \in S_n$ is asymptotically stable.

7.4 Examples of replicator dynamics

In the following, if we do not explicitly specify the different types making up the population, we shall assume that they correspond to the pure strategies of the underlying game. Thus we associate with every game a 'pure strategist dynamics'. The next examples are mainly intended to illustrate mathematical aspects of (7.3).

In the case $n = 2$, by setting $x = x_1$ and $1 - x = x_2$, we obtain from (7.3) a one-dimensional differential equation on $[0,1]$:

$$\dot{x} = x(1 - x)((A\mathbf{x})_1 - (A\mathbf{x})_2) . \tag{7.17}$$

In the case of the hawk–dove game (see section 6.1) this yields

$$\dot{x} = \frac{1}{2}x(1 - x)(G - Cx) .$$

The point G/C is a global attractor in $]0,1[$.

Exercise 7.4.1 Describe all phase portraits of (7.17), assuming that

$$A = \begin{bmatrix} a & b \\ c & d \end{bmatrix} .$$

Show that A admits an ESS unless $a = c$ and $b = d$ (in which case no strategy has an advantage). If $a > c$ (or $a = c$ and $b > d$), the strategy $\mathbf{e}_1$ is an ESS. If $d > b$ (or $d = b$ and $a < c$), the strategy $\mathbf{e}_2$ is an ESS. If $a < c$ and $d < b$, there is a unique ESS (which one?), which lies in the interior of S_2. Show that the asymptotically stable rest points are just the evolutionarily stable states. Compare with the competition between two species discussed in section 3.3: generically, we have either dominance or coexistence or bistability.

If the gain of one player is always the loss of the other, i.e. if $a_{ij} = -a_{ji}$ holds for all i and j, then the game is called a *zero-sum game*. In that case

$$\mathbf{x} \cdot A\mathbf{x} = -\mathbf{x} \cdot A\mathbf{x} = 0$$

and (7.3) becomes

$$\dot{x}_i = x_i(A\mathbf{x})_i . \tag{7.18}$$

Exercise 7.4.2 The rock–scissors–paper game with payoff matrix

$$A = \begin{bmatrix} 0 & 1 & -1 \\ -1 & 0 & 1 \\ 1 & -1 & 0 \end{bmatrix} \tag{7.19}$$

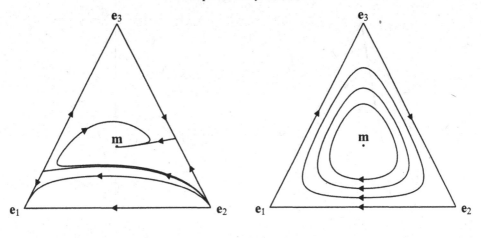

Fig. 7.1.   Fig. 7.2.

is a zero-sum game. Show that $x_1 x_2 x_3$ is a constant of motion and that the phase portrait is given by fig. 7.2.

Exercise 7.4.3 Analyse the dynamics of zero-sum games (7.18) (with $a_{ij} = -a_{ji}$). (If there is an interior fixed point $\mathbf{p}$, then (7.18) has a constant of motion.)

Let us now return to the hawk–dove game and assume that there occurs, in addition to the types E_1 (hawk) and E_2 (dove), a third type E_3 playing hawk and dove with probabilities $p_1 = G/C$ and $p_2 = (C - G)/C$ given by the equilibrium frequencies (see section 6.1). The payoff matrix then is

$$A = \begin{bmatrix} \frac{G-C}{2} & G & \frac{G(C-G)}{2C} \\[2mm] 0 & \frac{G}{2} & \frac{G(C-G)}{2C} \\[2mm] \frac{G(G-C)}{2C} & \frac{G(G+C)}{2C} & \frac{G(C-G)}{2C} \end{bmatrix}. \tag{7.20}$$

The type E_3 cannot be invaded by E_1 or E_2, since the corresponding *strategy* is evolutionarily stable. The corresponding *state* $\mathbf{e_3}$ is not evolutionarily stable, however, and neither is the mixture $\mathbf{p} = p_1 \mathbf{e}_1 + p_2 \mathbf{e}_2$ in this extended game. Indeed, one has the following result:

Exercise 7.4.4 There exists a line F of rest points through $\mathbf{e}_3$ and $\mathbf{p}$ (see fig. 7.3). All orbits converge to F, remaining on the constant level curves of

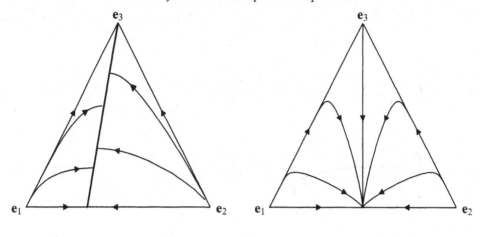

Fig. 7.3.   Fig. 7.4.

the function

$$Q(\mathbf{x}) = x_1^{p_1} x_2^{p_2} x_3^{-1} .$$

(7.21)

Show that F is an ES set (see exercise 6.4.5).

Exercise 7.4.5 Show that for the game with matrix

$$\begin{bmatrix} 0 & 2 & 0 \\ 2 & 0 & 2 \\ 1 & 1 & 1 \end{bmatrix}$$

the set of Nash equilibria is asymptotically stable, but not an ES set.

Exercise 7.4.6 Analyse (7.3) with the matrix

$$A = \begin{bmatrix} 0 & 10 & 1 \\ 10 & 0 & 1 \\ 1 & 1 & 1 \end{bmatrix} .$$

(7.22)

Show that the type E_3 is stable against invasion by E_1 alone, or by E_2 alone, but not if both types invade simultaneously. See fig. 7.4.

Exercise 7.4.7 For discrete generations, the *difference equation* $\mathbf{x} \to T\mathbf{x}$ with

$$(T\mathbf{x})_i = x_i \frac{(A\mathbf{x})_i + C}{\mathbf{x} \cdot A\mathbf{x} + C}$$

(7.23)

is a candidate for a discrete time replicator dynamics. Here, C is a positive constant corresponding to the fitness in the absence of interaction, such that $(A\mathbf{x})_i + C$ is always positive (recall that the payoff a_{ij} is the *increment* of fitness). Prove the following:

(a) Nash equilibria are fixed points of (7.23).
(b) An evolutionarily stable state need not be asymptotically stable for (7.23) (hint: use a rock–scissors–paper game as in exercise 7.7.3.); but it is for C sufficiently large.
(c) For zero-sum games, all orbits of (7.23) (up to a possible interior rest point) converge to the boundary. (Hint: if there is an interior rest point $\mathbf{p}$, then $\prod x_i^{p_i}$ is strictly decreasing. If there is no interior rest point then proceed as in theorem 7.6.1.)

7.5 Replicator dynamics and the Lotka–Volterra equation

The replicator equation is a cubic equation on the compact set S_n, while the Lotka–Volterra equation is quadratic on $\mathbb{R}^n_+$. We shall presently see, however, that the replicator equation in n variables $x_1, \ldots, x_n$ is equivalent to the Lotka–Volterra equation in $n-1$ variables $y_1, \ldots, y_{n-1}$.

Theorem 7.5.1 *There exists a differentiable, invertible map from $\hat{S}_n = \{\mathbf{x} \in S_n : x_n > 0\}$ onto $\mathbb{R}^{n-1}_+$ mapping the orbits of the replicator equation*

$$\dot{x}_i = x_i \left((A\mathbf{x})_i - \mathbf{x} \cdot A\mathbf{x} \right) \tag{7.24}$$

onto the orbits of the Lotka–Volterra equation

$$\dot{y}_i = y_i \left(r_i + \sum_{j=1}^{n-1} a'_{ij} y_j \right) \qquad i = 1, \ldots, n-1 \tag{7.25}$$

where $r_i = a_{in} - a_{nn}$ and $a'_{ij} = a_{ij} - a_{nj}$.

Proof Let us define $y_n \equiv 1$ and consider the transformation $\mathbf{y} \mapsto \mathbf{x}$ given by

$$x_i = \frac{y_i}{\sum_{j=1}^n y_j} \qquad i = 1, \ldots, n \tag{7.26}$$

which maps $\{\mathbf{y} \in \mathbb{R}^n_+ : y_n = 1\}$ onto $\hat{S}_n$. The inverse $\mathbf{x} \mapsto \mathbf{y}$ is given by

$$y_i = \frac{y_i}{y_n} = \frac{x_i}{x_n} \qquad i = 1, \ldots, n . \tag{7.27}$$

Now let us consider the replicator equation in n variables given by (7.24). We shall assume that the last row of the $n \times n$ matrix $A = (a_{ij})$ consists of zeros: by exercise 7.1.2, this is no restriction of generality. By the quotient rule (7.2)

$$\dot{y}_i = \left(\frac{x_i}{x_n}\right)^{\cdot} = \left(\frac{x_i}{x_n}\right)\left[(A\mathbf{x})_i - (A\mathbf{x})_n\right]. \qquad (7.28)$$

Since $(A\mathbf{x})_n = 0$, this implies

$$\dot{y}_i = y_i\left(\sum_{j=1}^{n} a_{ij}x_j\right) = y_i\left(\sum_{j=1}^{n} a_{ij}y_j\right)x_n . \qquad (7.29)$$

By a change in velocity (see exercises 4.1.2, 4.1.3), we can remove the term x_n. Since $y_n = 1$, this yields

$$\dot{y}_i = y_i\left(a_{in} + \sum_{j=1}^{n-1} a_{ij}y_j\right) \qquad (7.30)$$

or (with $r_i = a_{in}$) equation (7.25). The converse direction from (7.25) to (7.24) is analogous.   □

Results about Lotka–Volterra equations can therefore be carried over to the replicator equation and vice versa. For instance, we immediately get from theorem 4.2.1 that the replicator equation (7.3) admits no limit cycle for $n = 3$. Frequently, some properties are simpler to prove (or more natural to formulate) for one equation and some for the other.

Exercise 7.5.2 Show that if all r_i are equal in the n-dimensional Lotka–Volterra equation (5.1) then the $x_i = y_i(y_1 + \cdots + y_n)^{-1}$ satisfy the replicator equation (7.24) on S_n with the same interaction matrix (a_{ij}). (Hence the dynamics of (5.1) is reduced by one dimension.) Conversely, every replicator equation on S_n can be imbedded in a (competitive) Lotka–Volterra equation in $\mathbb{R}^n$ with all $r_i = r > 0$ (and all $a_{ij} < 0$).

7.6 Time averages and an exclusion principle

We have seen in theorem 5.2.1 that a Lotka–Volterra equation admits an ω-limit point in $\operatorname{int}\mathbb{R}_+^n$ if and only if it has a rest point in $\operatorname{int}\mathbb{R}_+^n$. From this and the equivalence theorem of last section, we can infer the following *exclusion principle*:

Theorem 7.6.1 *If the replicator equation (7.3) has no rest point in* $\operatorname{int} S_n$, *then every orbit converges to* $\operatorname{bd} S_n$.

Exercise 7.6.2 Prove this exclusion principle directly: (a) show that there exists a $\mathbf{c} \in \mathbb{R}^n$ such that $\mathbf{c} \cdot \mathbf{z} > \mathbf{c} \cdot \mathbf{y}$ for all $\mathbf{z} \in W$ and all $\mathbf{y} \in D$, where $W = A(\text{int } S_n)$ and $D = \{\mathbf{y} \in \mathbb{R}^n : y_1 = \cdots = y_n\}$ (W and D are convex); (b) show that $\sum c_i = 0$; (c) show that $V(\mathbf{x}) = \sum c_i \log x_i$ is strictly increasing along the orbits in int S_n.

Exercise 7.6.3 Show that the game with payoff matrix A admits a Nash equilibrium in int S_n if and only if there is no strategy $\mathbf{u}$ dominating a strategy $\mathbf{v}$ in the sense that $\mathbf{u} \cdot A\mathbf{x} > \mathbf{v} \cdot A\mathbf{x}$ for all $\mathbf{x} \in \text{int } S_n$. (Hint: the vector $\mathbf{c}$ from the previous exercise can be written as the difference $\mathbf{u} - \mathbf{v}$ of two strategies.)

Theorem 7.6.4 *If (7.3) admits a unique rest point* $\mathbf{p}$ *in* int S_n, *and if the ω-limit of the orbit of* $\mathbf{x}(t)$ *is in* int S_n, *then*

$$\lim_{t \to \infty} \frac{1}{T} \int_0^T x_i(t)dt = p_i \qquad i = 1, \ldots, n . \tag{7.31}$$

This does not follow immediately from the corresponding theorem 5.2.3 for Lotka–Volterra equations, since a coordinate transformation or a change in velocity could affect the time average. Hence, we have the following:

Exercise 7.6.5 Prove the previous theorem. (Hint: proceed along the same lines as in the Lotka–Volterra case.)

Exercise 7.6.6 Show that the solutions of the discrete time version

$$x_i \to x_i \frac{e^{(A\mathbf{x})_i}}{\sum x_k e^{(A\mathbf{x})_k}} \tag{7.32}$$

also have this averaging property.

7.7 The rock–scissors–paper game

We shall now analyse the general rock–scissors–paper game, which is characterized by having three pure strategies such that R_1 is beaten by R_2, which is beaten by R_3, which is beaten by R_1. Since we can normalize the payoff matrix such that the diagonal terms are 0, we obtain

$$A = \begin{bmatrix} 0 & -a_2 & b_3 \\ b_1 & 0 & -a_3 \\ -a_1 & b_2 & 0 \end{bmatrix} \tag{7.33}$$

where $a_i, b_i > 0$. It is easy to see that apart from the vertices, there are no rest points on bd S_3. On the other hand, there always exists an interior rest point

$$\mathbf{p} = \frac{1}{\Sigma}(b_2b_3 + b_2a_3 + a_3a_2, b_3b_1 + b_3a_1 + a_1a_3, b_1b_2 + b_1a_2 + a_1a_2) \quad (7.34)$$

with $\Sigma > 0$ such that $\mathbf{p} \in S_3$. Clearly $\mathbf{p}$ satisfies

$$\mathbf{p} \cdot A\mathbf{p} = \frac{\det A}{\Sigma}, \quad (7.35)$$

so that the average payoff at the Nash equilibrium has the same sign as $\det A$. The vertices $\mathbf{e}_i$ are saddle points with eigenvalues b_i and $-a_i$ for $i = 1, 2, 3$. Together with the edges, they form a heteroclinic cycle (as in section 5.5). We denote by l_i the lines $\{\mathbf{x} \in S_3 : (A\mathbf{x})_{i-1} = (A\mathbf{x})_{i+1}\}$ which intersect in $\mathbf{p}$ (we count indices mod 3, of course).

Exercise 7.7.1 Show that $\mathbf{p}$ is an ESS if and only if $b_i > a_{i+1}$ and the three numbers $\sqrt{b_i - a_{i+1}}$ correspond to the lengths of the sides of a triangle.

Theorem 7.7.2 *The following conditions are equivalent for* (7.33):

(a) $\mathbf{p}$ *is asymptotically stable;*
(b) $\mathbf{p}$ *is globally stable;*
(c) $\det A > 0$, *i.e.* $a_1a_2a_3 < b_1b_2b_3$;
(d) $\mathbf{p} \cdot A\mathbf{p} > 0$.

Proof There exists an $\mathbf{m} \in$ int S_3 such that

$$b_1m_1 - a_2m_2 = b_2m_2 - a_3m_3 = b_3m_3 - a_1m_1. \quad (7.36)$$

Indeed, $\mathbf{m}$ is just the interior rest point of the replicator equation with matrix

$$B = \begin{bmatrix} 0 & b_2 & -a_3 \\ -a_1 & 0 & b_3 \\ b_1 & -a_2 & 0 \end{bmatrix}. \quad (7.37)$$

The common value c of the expressions in (7.36), i.e. $\mathbf{m} \cdot B\mathbf{m}$, is strictly positive if and only if $\det B > 0$, i.e. $b_1b_2b_3 > a_1a_2a_3$ by (7.35). Let us now consider the map sending $\mathbf{x}$ to $\mathbf{y}$ where $y_i = x_i m_i^{-1}(x_1 m_1^{-1} + x_2 m_2^{-1} + x_3 m_3^{-1})^{-1}$. This is a projective bijection from S_3 onto itself which transforms the replicator equation with matrix A into the replicator equation with matrix

$$\begin{bmatrix} 0 & -m_2a_2 & m_3b_3 \\ m_1b_1 & 0 & -m_3a_3 \\ -m_1a_1 & m_2b_2 & 0 \end{bmatrix}$$

(see exercise 7.1.3.). Thus we have obtained a rock–scissors–paper game like (7.33), but now satisfying $b_i - a_{i+1} = c$ for $i = 1, 2, 3$. (For $c = 0$, this means that we have a zero-sum game.) In this case the function $P(\mathbf{x}) := \prod x_i^{p_i}$ satisfies

$$\dot{P} = P(\mathbf{p}\cdot A\mathbf{x} - \mathbf{x}\cdot A\mathbf{x}) = -P(\boldsymbol{\xi}\cdot A\boldsymbol{\xi}) = -cP(\xi_1\xi_2 + \xi_2\xi_3 + \xi_3\xi_1) \qquad (7.38)$$

if we set $\boldsymbol{\xi} = \mathbf{x} - \mathbf{p}$ (we note that $(A\mathbf{p})_i$ does not depend on i, so that $\boldsymbol{\xi}\cdot A\mathbf{p} = 0$). Now $\xi_1\xi_2 + \xi_2\xi_3 + \xi_3\xi_1 = \frac{1}{2}(\xi_1 + \xi_2 + \xi_3)^2 - (\xi_1^2 + \xi_2^2 + \xi_3^2) \leq 0$, with equality if and only if $\mathbf{x} = \mathbf{p}$. This implies that for $c = 0$, P is an invariant of motion, and $\mathbf{p}$ therefore a centre. For $c > 0$, $\mathbf{p}$ is evolutionarily (and hence globally) stable and for $c < 0$, all orbits in int S_3 converge to the boundary, with the exception of $\mathbf{p}$. $\qquad\square$

Exercise 7.7.3 Analyse the rock–scissors–paper game

$$A = \begin{bmatrix} a & b & c \\ c & a & b \\ b & c & a \end{bmatrix}$$

(with $b > a > c \geq 0$) for the discrete dynamics (7.23) (set $C = 0$). Show that $(x_1 x_2 x_3)^{-1}\mathbf{x}\cdot A\mathbf{x}$ is a global Lyapunov function, and the centre point is asymptotically stable if and only if $bc > a^2$.

As in the case of the Lotka–Volterra equation (section 5.5), the time averages of a rock–scissors–paper game will converge, not to the rest point, but to a triangle $A_1A_2A_3$, if the heteroclinic cycle on the boundary is an attractor. This can be shown in much the same way. Let

$$\mathbf{z}(T) := \frac{1}{T}\int_0^T \mathbf{x}(t)dt \qquad (7.39)$$

be the time average of an interior orbit converging to the boundary, and $\mathbf{z} = \lim \mathbf{z}(T_k)$ an accumulation point (for some sequence $T_k \to \infty$). By refining this sequence, we can assume that $\bar{\mathbf{x}} := \lim_{k\to\infty} \mathbf{x}(T_k)$ also exists. Dividing (7.3) by x_i and integrating from 0 to T_k, we obtain $\log xxxxx$

$$\lim_{k\to\infty} \frac{\log x_i(T_k) - \log x_i(0)}{T_k}$$
$$= \lim \frac{1}{T_k}\sum a_{ij}\int_0^{T_k} x_j(t)dt - \lim \frac{1}{T_k}\int_0^{T_k} \mathbf{x}\cdot A\mathbf{x}dt . \qquad (7.40)$$

The second limit on the right hand side is zero. Indeed, the orbit spends most of the time near the vertices of S_3 (where the mean payoff $\mathbf{x}\cdot A\mathbf{x}$ is 0 because we have assumed $a_{ii} = 0$). Thus the right hand side is $\sum a_{ij}z_j$. If $\bar{\mathbf{x}}$

is on an edge, for instance if $\bar{x}_1 = 0$ but $\bar{x}_2, \bar{x}_3 \neq 0$, then (7.40) yields

$$\sum_{j=1}^{3} a_{1j}z_j \leq 0 \qquad \sum_{j=1}^{3} a_{2j}z_j = 0 \qquad \sum_{j=1}^{3} a_{3j}z_j = 0 \qquad (7.41)$$

which, together with $z_1 + z_2 + z_3 = 1$, yields the solution

$$A_2 = \frac{1}{b_1 b_2 + b_2 a_3 + a_1 a_2}(b_2 a_3, a_1 a_2, b_1 b_2) . \qquad (7.42)$$

Similarly, one obtains A_1 or A_3 if $\bar{x}$ is another edge. If $\bar{x}$ is on one of the vertices, for instance e_1, then the equality and inequality signs of (7.41) are interchanged and one obtains the line segment $A_3 A_1$.

Exercise 7.7.4 Show that A_i, A_{i+1} and e_{i+1} are collinear and that A_{i+1} lies on the segment $l_i = \{x \in S_3 : (Ax)_{i-1} = (Ax)_{i+1}\}$. In the limiting case $\prod a_i = \prod b_i$, the A_i coincide with the Nash equilibrium p. On the other hand, if $b_i \to 0$, then A_i converges to e_i. Show that the interior of the triangle $A_1 A_2 A_3$ is the region where all payoffs $(Ax)_i$ are negative.

7.8 Partnership games and gradients

If the payoff matrix A is symmetric, i.e. if $a_{ij} = a_{ji}$ for all i and j, then the interests of the two players coincide. We speak in this case of *partnership games*. Not surprisingly, the average payoff increases over time, but the replicator dynamics need not lead to a *global* maximum.

Theorem 7.8.1 *For partnership games, the average payoff $x \cdot Ax$ is a strict Lyapunov function. The asymptotically stable rest points of (7.3) are precisely the evolutionarily stable states, and are given by the (local strict) maxima of the average payoff. An asymptotically stable rest point in $\text{int} \, S_n$ is globally stable.*

Proof Clearly $(x \cdot Ax)^{\cdot} = \dot{x} \cdot Ax + x \cdot A\dot{x}$. Since A is symmetric, the terms on the right hand side are equal. For $V(x) = x \cdot Ax$ this yields

$$\dot{V}(x) = 2 \sum \dot{x}_i (Ax)_i = 2 \sum x_i [(Ax)_i - x \cdot Ax] (Ax)_i$$

and therefore, since $\sum x_i = 1$,

$$\dot{V}(x) = 2 \sum x_i [(Ax)_i - x \cdot Ax]^2 \geq 0 . \qquad (7.43)$$

Equality holds if and only if all terms $(Ax)_i$ for which $x_i > 0$ take the same value. This condition, as we have seen in section 7.1, characterizes the rest points of (7.3). Thus V is a strict Lyapunov function.

Now let **p** be asymptotically stable. Then

$$\mathbf{x} \cdot A\mathbf{x} < \mathbf{p} \cdot A\mathbf{p} \qquad (7.44)$$

for all $\mathbf{x} \neq \mathbf{p}$ in some neighbourhood of **p**. Replacing **x** by $2\mathbf{x} - \mathbf{p}$ (which is also near **p**) we get

$$\mathbf{x} \cdot A\mathbf{x} < \mathbf{p} \cdot A\mathbf{x} \qquad (7.45)$$

which is just condition (6.16) for an ESS. The converse follows from theorem 7.2.4. □

Exercise 7.8.2 For partnership games, the average payoff $\mathbf{x} \cdot A\mathbf{x}$ is constant along every connected component of equilibria. Show that for general games, this is not true.

We shall presently see that for partnership games, the average payoff $\mathbf{x} \cdot A\mathbf{x}$ is not only a Lyapunov function, but even a *potential* for (7.3). For this, we have to introduce the notion of a *gradient system*.

Let M be a manifold (we shall only have to consider open subsets of $\mathbb{R}^n$ or $\operatorname{int} S_n$). Let G be a general Riemannian metric which associates (in a smooth way) with each **x** a symmetric positive definite matrix $G(\mathbf{x}) = (g_{ij}(\mathbf{x}))$, such that the inner product of two vectors $\boldsymbol{\xi}, \boldsymbol{\eta}$ in the tangent space at **x** is given by

$$\langle \boldsymbol{\eta}, \boldsymbol{\xi} \rangle_{\mathbf{x}} = \sum g_{ij}(\mathbf{x}) \eta_i \xi_j = \boldsymbol{\eta} \cdot G(\mathbf{x}) \boldsymbol{\xi}. \qquad (7.46)$$

If M is an open subset of $\mathbb{R}^n$, the tangent space at **x** is $\mathbb{R}^n$ itself; if $M = \operatorname{int} S_n$ then the tangent space at $\mathbf{x} \in \operatorname{int} S_n$ is $\mathbb{R}^n_0 = \{ \boldsymbol{\xi} \in \mathbb{R}^n : \sum \xi_i = 0 \}$.

Let $V(\mathbf{x})$ be a differentiable function on M with values in $\mathbb{R}$. Then the derivative $DV(\mathbf{x})$ is a linear map from the tangent space at **x** into $\mathbb{R}$. Hence there exists a vector **v** in the tangent space such that

$$\langle \boldsymbol{\eta}, \mathbf{v} \rangle_{\mathbf{x}} = DV(\mathbf{x})(\boldsymbol{\eta}) \qquad (7.47)$$

for all $\boldsymbol{\eta}$ in the tangent space at **x**. This vector **v** is the *G-gradient* of V at **x**. The vector field $\dot{\mathbf{x}} = \mathbf{v}(\mathbf{x})$ defines a dynamical system, and the function V is said to be a *potential* of it. If G is the unit matrix, we obtain the usual (Euclidean) gradient $\operatorname{grad} V(\mathbf{x})$ with components $\frac{\partial V}{\partial x_i}(\mathbf{x})$. If M is an open subset of $\mathbb{R}^n$, then the G-gradient of V is given by $G(\mathbf{x})^{-1} \operatorname{grad} V(\mathbf{x})$, where $\operatorname{grad} V(\mathbf{x})$ is the Euclidean gradient.

For $\mathbf{x} \in \operatorname{int} S_n$, we introduce a particularly interesting inner product by

$$\langle \boldsymbol{\xi}, \boldsymbol{\eta} \rangle_{\mathbf{x}} = \sum_{i=1}^n \frac{1}{x_i} \xi_i \eta_i \qquad (7.48)$$

which means that we set $g_{ij}(\mathbf{x}) = \delta_{ij}\frac{1}{x_i}$. This is said to be the *Shahshahani inner product*.

Theorem 7.8.3 *If A is symmetric, the replicator equation (7.3) is a Shahshahani gradient having as potential one half of the mean payoff $V(\mathbf{x}) = \frac{1}{2}\mathbf{x}\cdot A\mathbf{x}$.*

Proof For $\mathbf{x} \in \text{int } S_n$ and $\xi \in \mathbb{R}_0^n$ we compute

$$\langle \dot{\mathbf{x}}, \xi \rangle_{\mathbf{x}} = \sum \frac{1}{x_i} x_i\big((A\mathbf{x})_i - 2V\big)\xi_i = \sum (A\mathbf{x})_i \xi_i - 2V \sum x_{i_i}$$
$$= \sum \frac{\partial V}{\partial x_i}\xi_i = D_{\mathbf{x}}V(\xi) . \tag{7.49}$$

By (7.47), $\dot{\mathbf{x}}$ is the Shahshahani gradient of V. $\square$

Exercise 7.8.4 Show that the change of coordinates $x_i = y_i^2/4$ transforms S_n with the Shahshahani metric into the part of the $(n-1)$-dimensional sphere of radius 2 lying in the positive orthant, equipped with the usual Euclidean metric.

Exercise 7.8.5 Show that the geodesic distance, i.e. the length of the shortest path, between two states $\mathbf{p}, \mathbf{q} \in \text{int } S_n$ with respect to the Shahshahani metric is given by

$$d(\mathbf{p}, \mathbf{q}) = 2 \arccos\left(\sum_{i=1}^{n} \sqrt{p_i q_i} \right) .$$

(Hint: the shortest path between two points on a sphere is on a great circle.)

Exercise 7.8.6 Show that the Jacobian matrix of a gradient system at any fixed point has only real eigenvalues.

Exercise 7.8.7 The replicator equation (7.3) is a Shahshahani gradient if and only if for all i, j, k

$$a_{ij} + a_{jk} + a_{ki} = a_{ik} + a_{kj} + a_{ji}. \tag{7.50}$$

Show that (7.50) implies that analogous conditions hold for all p-cycles, $p > 3$. It holds if and only if there exists a symmetric matrix B and c_j such that $a_{ij} = b_{ij} + c_j$, or equivalently if and only if there exist $\mathbf{u}, \mathbf{v} \in \mathbb{R}^n$ such that $a_{ij} - a_{ji} = u_i + v_j$ for all i, j.

Exercise 7.8.8 If (7.3) is a Shahshahani gradient then so is the replicator equation describing the frequencies of the mixed strategies $\mathbf{p}^1, \ldots, \mathbf{p}^M \in S_n$. What is the potential?

Exercise 7.8.9 It was shown in exercise 7.6.2 that a replicator equation (7.3) with no interior rest point has a global Lyapunov function of the form $V(\mathbf{x}) = \sum_{i=1}^{n} c_i \log x_i$, with $\mathbf{c} \in \mathbb{R}_0^n$. Find a Riemannian metric on int S_n which makes (7.3) the gradient of $V(\mathbf{x})$.

Exercise 7.8.10 We know that if $\mathbf{p}$ is an interior ESS, then $V(\mathbf{x}) = \sum_{i=1}^{n} p_i \log x_i$ is a Lyapunov function for the replicator equation (7.3). Show that one cannot, in general, find a Riemannian metric on int S_n which makes the replicator equation the gradient of $V(\mathbf{x})$.

7.9 Notes

The replicator equation (7.3) was introduced by Taylor and Jonker (1978). The term was coined by Schuster and Sigmund (1983). The equation was generalized to n-person games by Palm (1984) and Ritzberger (1995). Part (b) of theorem 7.2.1 is from Nachbar (1990), and part (c) from Bomze (1986), see also Weibull (1995). Theorem 7.2.4 is from Zeeman (1980) and Hofbauer *et al.* (1979), for extensions to infinitely many strategies see Bomze (1991). Examples (7.12) and (7.33) are due to Zeeman (1980), where a classification of the stable phase portraits for $n = 3$ can be found. Difference equations are investigated in Bishop and Cannings (1978) (for the 'war of attrition'), Akin and Losert (1984) (for zero-sum games), Hofbauer (1984) and Weissing (1991) (for rock–scissors–paper games), Dekel and Scotchmer (1992) and Cabrales and Sobel (1992). The equivalence between game dynamics and the Lotka–Volterra equation (theorem 7.5.1) was shown by Hofbauer (1981a). Examples (7.20) and (7.22) are from Maynard Smith (1982). The result described in exercise 7.6.3 is due to Akin (1980). Theorem 7.6.1 is from Hofbauer (1981b) and theorem 7.6.4 is from Schuster *et al.* (1981a). Theorem 7.3.2 on strong stability is due to Cressman (1990), whose book *The Stability Concept of Evolutionary Game Theory* (1992) is basic to the field. For exercise 7.3.4, we refer to Bomze and van Damme (1992). For the time averages in the rock–scissors–paper game, see Gaunersdorfer and Hofbauer (1995). The Riemannian metric on S_n was introduced for a genetic system in Shahshahani (1979), see also section 19.5.

8

Other game dynamics

Many of the properties of the replicator equation are valid for other game dynamics which may serve, for instance, to model imitation processes. We describe a large class of game dynamics that eliminate pure strategies which are iteratively strictly dominated, and discuss some instances of the best-reply dynamics motivated by fictitious play. We show that no reasonable dynamics can converge to equilibrium for all games.

8.1 Imitation dynamics

The replicator dynamics mimics the effect of natural selection (although it blissfully disregards the complexities of sexual reproduction). In the context of games played in human societies, however, the spreading of successful strategies is more likely to occur through imitation than through inheritance. How should we model this imitation processes?

Let us start with symmetric games defined by an $n \times n$ payoff matrix A, and assume that the pure strategies R_1 to R_n are adopted by a (large) population of players with frequencies $x_i(t)$ at time t, so that the state is given, at any instant, by a point $\mathbf{x} \in S_n$. Strategy R_i then earns $(A\mathbf{x})_i = \sum a_{ij}x_j$ as expected payoff, and the average payoff in the population is given by $\mathbf{x} \cdot A\mathbf{x}$. We shall suppose that occasionally a player is picked out of the population and afforded the opportunity to change his strategy. He samples another player at random and adopts his strategy with a certain probability. A general ansatz for such an imitation dynamics is given by the input–output model

$$\dot{x}_i = x_i \sum_j \left[f_{ij}(\mathbf{x}) - f_{ji}(\mathbf{x}) \right] x_j. \tag{8.1}$$

Since $\sum \dot{x}_i = 0$, the simplex S_n is invariant under (8.1). In a small time interval Δt the flow from R_j to R_i is given by $x_i x_j f_{ij} \Delta t$, where the product $x_i x_j$ gives the probability for randomly and independently sampling the revising player (who uses strategy R_j) and the potentially imitated player (who uses R_i), and where f_{ij} is the rate at which the R_j-strategist switches to R_i. We shall assume that this rate depends on the current payoffs $(A\mathbf{x})_i$ and $(A\mathbf{x})_j$ obtained by the two strategies in the population:

$$f_{ij}(\mathbf{x}) = f\left((A\mathbf{x})_i, (A\mathbf{x})_j\right) \tag{8.2}$$

where the function $f = f(u,v)$ defines the *imitation rule* (the same for all players, and in particular independent of i and j). The simplest rule would be *'imitate the better'*, i.e. $f(u,v) = 0$ if $u < v$ and $f(u,v) = 1$ if $u > v$. This, however, has the technical disadvantage of being discontinuous.

If we assume that the switching rate is a function of the payoff difference, i.e. if $f(u,v) = \phi(u-v)$ with monotonically increasing ϕ, (8.1) yields (with $\psi(u) = \phi(u) - \phi(-u)$) the dynamics

$$\dot{x}_i = x_i \sum_j x_j \psi\left((A\mathbf{x})_i - (A\mathbf{x})_j\right) \tag{8.3}$$

for some odd and strictly increasing function ψ. If we consider, in particular, $\phi(u) = u^\alpha_+$, for $\alpha \geq 0$, we obtain $\psi(u) = |u|^\alpha_+ \operatorname{sgn} u$. For $\alpha = 1$, this yields $\psi(u) = u$. With this *proportional imitation rule*, which effectively says *'imitate actions that perform better, with a probability proportional to the expected gain'*, (8.3) reduces to the usual replicator dynamics

$$\dot{x}_i = x_i\left((A\mathbf{x})_i - \mathbf{x}\cdot A\mathbf{x}\right) . \tag{8.4}$$

In the limiting case $\alpha \to 0$, on the other hand, we are back to the rule 'imitate the better'.

Other versions of (8.1) take the form

$$\dot{x}_i = x_i\left[f((A\mathbf{x})_i) - \bar{f}\right] \quad \text{with} \quad \bar{f} = \sum x_i f((A\mathbf{x})_i) \tag{8.5}$$

where $f(u)$ is increasing in u. (If f is a linear function, this is again the replicator equation.) (8.5) occurs, for instance, if players switch only to better strategies, with a rate proportional to the difference $[f((A\mathbf{x})_i) - f((A\mathbf{x})_j)]_+$ in a 'fitness function' $f(u)$ which is monotonically increasing in the payoff u. (8.5) also occurs if $f_{ij} = f((A\mathbf{x})_i)$, i.e. if the switching rate depends only on the success of the imitated player. Furthermore, (8.5) also occurs if $f_{ij} = -f((A\mathbf{x})_j)$, i.e. the rate of switching depends only on the payoff of the imitating player (this can be interpreted as *imitation driven by dissatisfaction*, if we assume that less successful individuals review more often, but imitate blindly).

Exercise 8.1.1 Equations (8.3–5) have the same interior rest points, and linearization leads to Jacobian matrices which are proportional. Show that the local behaviour is the same as for the replicator equation, if the fixed point is hyperbolic. In particular, an interior ESS is locally stable for these dynamics.

Exercise 8.1.2 Is there a connection between 'imitate the better' dynamics and the replicator equation? (Hint: smooth pieces of orbits of this imitation dynamics are parts of orbits of the replicator dynamics, but for a different game.)

Exercise 8.1.3 Draw the phase portrait of the 'imitate the better' dynamics for the rock–scissors–paper game (7.19).

Exercise 8.1.4 Show that in the 'imitate the better' dynamics, a strategy i increases its population share if and only if its payoff $(A\mathbf{x})_i$ is larger than the *median* of $(A\mathbf{x})_1, \ldots, (A\mathbf{x})_n$ (instead of the mean, as for the replicator dynamics).

8.2 Monotone selection dynamics

Frequently there is no compelling reason for a specific imitation dynamics. Fortunately, it turns out that many dynamics share essential features. Let us stick to *game dynamics* of the form

$$\dot{x}_i = x_i g_i(\mathbf{x}) \tag{8.6}$$

where the C^1 functions g_i have the property that $\sum x_i g_i(\mathbf{x}) = 0$ for all $\mathbf{x} \in S_n$, so that S_n and its boundary faces are invariant.

One large class of game dynamics is defined by the property that pure strategies with higher payoff grow at a higher rate. More precisely, a game dynamics is said to be *payoff monotonic* if the growth rates of the different strategies are ranked according to their payoff:

$$g_i(\mathbf{x}) > g_j(\mathbf{x}) \Leftrightarrow (A\mathbf{x})_i > (A\mathbf{x})_j . \tag{8.7}$$

Obviously, the replicator dynamics is payoff monotonic, since we even have $g_i(\mathbf{x}) - g_j(\mathbf{x}) = (A\mathbf{x})_i - (A\mathbf{x})_j$ (see exercise 7.1.1). It is easy to see that every payoff monotonic dynamics has the same rest points as the replicator equation, namely all Nash equilibria of subgames. By essentially using the same arguments as for the replicator dynamics, one shows that for a payoff monotonic dynamics, the Lyapunov stable rest points are Nash equilibria, the strict Nash equilibria are asymptotically stable and the rest points which are ω-limits of interior orbits are Nash equilibria (cf. theorem 7.2.1). In fact, these properties even hold for the considerably larger class of game dynamics which are *weakly payoff positive*: whenever there exist pure strategies which earn above the population average, then *some* such strategy has strictly positive growth rate. Formally:

$$B(\mathbf{x}) := \{i : (A\mathbf{x})_i > \mathbf{x} \cdot A\mathbf{x}\} \neq \emptyset \Rightarrow g_i(\mathbf{x}) > 0 \quad \text{for some} \quad i \in B(\mathbf{x}) . \quad (8.8)$$

Exercise 8.2.1 Check that the rest points of weakly payoff positive dynamics satisfy the properties mentioned above (this is sometimes called the 'folk theorem of evolutionary game theory').

Exercise 8.2.2 Show that all the imitation dynamics in section 8.1 are payoff monotonic.

Exercise 8.2.3 Show that if the matrix A is symmetric, the average payoff $\mathbf{x} \cdot A\mathbf{x}$ is monotonically increasing for every payoff monotonic dynamics.

Exercise 8.2.4 For every rock–scissors–paper game given by (7.33), find two payoff monotonic dynamics, one for which the interior rest point $\mathbf{p}$ is globally stable and one for which all interior orbits (with the exception of $\mathbf{p}$ itself) converge to the boundary.

Exercise 8.2.5 Consider the game with payoff matrix

$$A = \begin{bmatrix} 0 & -1 & \varepsilon \\ \varepsilon & 0 & -1 \\ -1 & \varepsilon & 0 \end{bmatrix} . \quad (8.9)$$

Show that for $\varepsilon = 0$, all interior orbits (except the interior rest point $\mathbf{p} = (\frac{1}{3}, \frac{1}{3}, \frac{1}{3})$) converge to the boundary, for every payoff monotonic dynamics. It follows that if the dynamics depends continuously on the payoff matrix, then for small $\varepsilon > 0$ all orbits (outside a neighbourhood of $\mathbf{p}$) end up in a small neighbourhood of the boundary.

Exercise 8.2.6 Show that theorem 7.6.1 is valid for general monotone selection dynamics if $n = 3$, but not if $n \geq 4$.

8.3 Selection against iteratively dominated strategies

The pure strategy R_i is said to be *strictly dominated* if there is some (pure or mixed) strategy $y \in S_n$ such that

$$(Ax)_i < y \cdot Ax \tag{8.10}$$

for all $x \in S_n$. If the game is played by 'rational' players, one can assume that such players will never use R_i (this is sometimes used as the definition of 'rational players'). Thus one may as well remove all strictly dominated strategies from the game. After this is done, there may remain some pure strategies which are strictly dominated in the new game. For instance, if

$$A = \begin{bmatrix} 5 & 2 & 7 \\ 0 & 0 & 4 \\ 1 & 5 & 5 \end{bmatrix} \tag{8.11}$$

then strategy R_2 is strictly dominated. Once it is removed, strategy R_3 (which was not strictly dominated in the original game) becomes strictly dominated, and hence can be removed. This procedure can be repeated only finitely often, and leads to a nonempty subset of pure strategies, which does not depend on the details of the elimination procedure and which is the complement of all *iteratively strictly dominated* strategies.

We note that this reduction of the game is based on the assumption that *all* players engaged in it are rational, and know that the others are rational, and that the others know that everybody knows etc. Seen from this angle, it is gratifying that the replicator dynamics, which does not presuppose any rationality, eliminates all iteratively strictly dominated strategies, in the sense that their frequencies converge to 0 (compare exercise 7.6.3). We shall show that this extends to a large class of game dynamics. If the dominating strategy is also pure, then it actually holds for all payoff monotonic dynamics.

Exercise 8.3.1 If a strategy i is strictly dominated by another *pure* strategy j, i.e. $(Ax)_i < (Ax)_j$ for all $x \in S_n$, then show that $x_i \to 0$ along all interior solutions of (8.6) satisfying (8.7). (Hint: quotient rule, see exercise 7.1.1.)

A game dynamics (8.6) is said to be *convex monotone* if

$$y \cdot Ax > (Ax)_i \Rightarrow \sum y_j g_j(x) > g_i(x) \tag{8.12}$$

for all i and $y, x \in S_n$.

Theorem 8.3.2 *If the game dynamics (8.6) is convex monotone and the pure strategy R_i is iteratively strictly dominated, then its frequency $x_i(t)$ converges to 0.*

Proof Assume first that R_i is strictly dominated by some $\mathbf{y} \in S_n$. By continuity, there exists a $\delta > 0$ such that

$$g_i(\mathbf{x}) - \sum y_j g_j(\mathbf{x}) < -\delta \tag{8.13}$$

for all $\mathbf{x} \in S_n$. With $P(\mathbf{x}) := x_i \prod_j x_j^{-y_j}$ we obtain for any interior solution $t \to \mathbf{x}(t)$ that

$$\dot{P}(\mathbf{x}) = \sum \frac{\partial P(\mathbf{x})}{\partial x_j} \dot{x}_j = P(\mathbf{x}) \left(g_i(\mathbf{x}) - \sum y_j g_j(\mathbf{x}) \right). \tag{8.14}$$

Hence $\dot{P}(\mathbf{x}) < -\delta P(\mathbf{x})$. This implies that $x_i(t)$, which is smaller than $P(\mathbf{x}(t))$, decreases exponentially. It is enough, now, to repeat this argument: once the x_i are small enough, an inequality analoguous to (8.12) holds for all the strategies which are eliminated in the next round, etc. Thus the frequencies of all iteratively strictly dominated pure strategies converge to 0. $\square$

Let us consider now the special class of game dynamics (8.5), with f increasing, so that it satisfies (8.7).

Exercise 8.3.3 The name 'convex monotone' for (8.12) is motivated by the fact that (8.5) is convex monotone if and only if f is a convex function. Check this fact.

Exercise 8.3.4 The game dynamics (8.6) is said to be *aggregate monotonic* if

$$\mathbf{y} \cdot A\mathbf{x} > \mathbf{z} \cdot A\mathbf{x} \Leftrightarrow \sum y_j g_j(\mathbf{x}) > \sum z_j g_j(\mathbf{x}). \tag{8.15}$$

Show that aggregate monotonic implies convex monotonic, which implies payoff monotonic in turn. Show that an aggregate monotonic equation (8.6) is, up to a change in velocity, nothing but the replicator equation.

Theorem 8.3.2 is sharp in the sense that if the function f in (8.5) is not convex, then there is a game with strictly dominated strategies surviving along interior solutions. Indeed, let us consider a game with payoff matrix

$$A = \begin{bmatrix} a & c & b & \gamma \\ b & a & c & \gamma \\ c & b & a & \gamma \\ a+\beta & a+\beta & a+\beta & 0 \end{bmatrix} \tag{8.16}$$

where $c < a < b$, $0 < \beta < b - a$ and $\gamma > 0$. The three pure strategies R_1, R_2, R_3 form a rock–scissors–paper cycle of best replies, and the face $x_4 = 0$ is bounded by a heteroclinic orbit Γ_1 from $\mathbf{e}_1$ to $\mathbf{e}_2$ to $\mathbf{e}_3$ and back to $\mathbf{e}_1$. This face also contains the rest point $\mathbf{p} = (\frac{1}{3}, \frac{1}{3}, \frac{1}{3}, 0)$. Since R_4 can

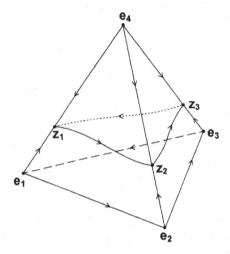

Fig. 8.1.

invade each of the other pure strategies, e_i is unstable in the e_4-direction, for $i = 1, 2, 3$. Conversely, each of the other R_i can invade R_4. Hence we have three more fixed points on bd S_4, namely $z_1 = (\frac{\gamma}{\beta+\gamma}, 0, 0, \frac{\beta}{\beta+\gamma})$, and similarly z_2 and z_3, which attract all orbits on the boundary faces $x_2 = 0$, $x_3 = 0$ and $x_1 = 0$, respectively. and hence are connected by another heteroclinic cycle Γ_2 on the union of these faces (see fig. 8.1).

It is easy to see that for

$$a + b + c > 3(a + \beta) \tag{8.17}$$

the strategy p is a Nash equilibrium in the full game, and moreover dominates R_4, so that $x_4(t) \to 0$ in the replicator dynamics (8.4), and for (8.5) for convex f. Inequality (8.17) implies $2a < b + c$, so that for the replicator dynamics p is globally stable: in particular, it attracts all orbits in the interior of the boundary face $x_4 = 0$.

For (8.5), p is still locally stable, see exercise 8.1.1. But if f is not convex, global stability does not always hold for (8.5). Indeed, we may choose a, b and c in (8.16) such that $2a < b + c$ and $2f(a) > f(b) + f(c)$. The eigenvalues of the vertex e_i ($i = 1, 2, 3$) in the face $x_4 = 0$ are given by $\rho = f(b) - f(a) > 0$ and $-\tau = f(c) - f(a) < 0$. We have studied the rock–scissors–paper game for the replicator dynamics in section 7.7, where we have seen that if the product of the three 'incoming' eigenvalues at the vertices e_1, e_2, e_3 is larger than the product of the three 'outgoing' eigenvalues, then the heteroclinic cycle on the boundary of S_3 is attracting. Since orbits close to the boundary spend most of their time in the vicinity of the vertices, where their behaviour is determined by their linearization, i.e. their eigenvalues, the result remains valid for the nonreplicator dynamics (for details see section 16.3 or 17.5): in particular, if the 'incoming' eigenvalue at each of the three vertices $\mathbf{e}_i$ is larger than the 'outgoing' eigenvalue, the heteroclinic cycle Γ_1 is locally attracting in the face $x_4 = 0$, with respect to the game dynamics specified by (8.5) – and this is actually the case since $2f(a) > f(b) + f(c)$ implies $\tau > \rho$. However, Γ_1 is unstable in the x_4 direction.

If $\beta > 0$ is chosen small enough, (8.17) will be valid, so that R_4 is strictly dominated by $\mathbf{p}$. Furthermore, the heteroclinic cycle Γ_2 will be asymptotically stable in S_4, since there again the incoming eigenvalues are larger than the outgoing eigenvalues. Hence along an open set of interior solutions, x_4 is close to $\frac{\beta}{\beta+\gamma}$ most of the time.

Exercise 8.3.5 Compute the eigenvalues at $\mathbf{z}_i$, and check that the incoming eigenvalue along Γ_2 is larger than the outgoing eigenvalue, if β is small enough.

Exercise 8.3.6 Show that the discrete time replicator dynamics given by (7.23) need not eliminate strictly dominated pure strategies. (Hint: use a payoff matrix of type (8.16), and make sure that the region where $(A\mathbf{x})_4 > \mathbf{x} \cdot A\mathbf{x}$ contains Γ_1, but not $\mathbf{p}$.)

Exercise 8.3.7 Show that the discrete time replicator dynamics given by (7.32) eliminates strictly dominated pure strategies.

8.4 Best-response dynamics

The very first game dynamics ever studied was given by the method of *fictitious play*. Let us consider discrete generations and assume that in each generation a new player enters the population and adopts — as a rational agent — a best reply to the strategy mix in the current population. Thus in generation $k + 1$, the new entrant chooses a strategy $\mathbf{r}_{k+1} \in \{R_1, \ldots, R_n\}$ which maximizes his expected payoff against the mean strategy, which is $\mathbf{s}_k = \frac{1}{k} \sum_{i=1}^{k} \mathbf{r}_i$. For the rest of the game, this player has to stick to $\mathbf{r}_{k+1}$. The change in the mean strategy is given by the difference equation

$$\mathbf{s}_{k+1} - \mathbf{s}_k = \frac{\mathbf{r}_{k+1} - \mathbf{s}_k}{k + 1} \quad \text{with} \quad \mathbf{r}_{k+1} \in \beta(\mathbf{s}_k) \tag{8.18}$$

where $\beta(\mathbf{x})$ denotes the set of *best replies* against strategy $\mathbf{x} \in S_n$. Going from discrete generations to a continuous dynamics, we obtain the *continuous fictitious play process*

$$\dot{\mathbf{s}}(t) = \frac{1}{t}(\mathbf{r}(t) - \mathbf{s}(t)) \quad \text{with} \quad \mathbf{r}(t) \in \beta(\mathbf{s}(t)) \tag{8.19}$$

or equivalently $\mathbf{s}(t) = t^{-1} \int_0^t \mathbf{r}(\tau)d\tau$.

A problem with (8.18) and (8.19) is that $\beta(\mathbf{x})$ is not always a singleton, so that we enter the realm of *differential inclusions*. There is no need to engage in it, since one can construct (at least one) piecewise linear solutions through each initial condition explicitly (see below). We shall also omit the factor $1/t$, which corresponds to a change in velocity (see exercise 4.1.2) and thus we end up with the *best-response dynamics*

$$\dot{\mathbf{x}} = \beta(\mathbf{x}) - \mathbf{x}. \tag{8.20}$$

An alternative interpretation of (8.20) is that in a very large population, a small fraction of players revise their strategy, choosing best replies to the current mean population strategy $\mathbf{x}$. Such players are rational, but myopic: they do not anticipate the results of their actions.

Let us consider, for instance, a game with payoff matrix

$$A = \begin{bmatrix} 0 & 1 & 1 \\ 1 & 0 & 1 \\ 1 & 1 & 0 \end{bmatrix}. \tag{8.21}$$

Every solution of (8.20) reaches in finite time one of the three lines of symmetry, where two strategies obtain the same (maximal) payoff. The further evolution is constrained to these lines and leads (again in finite time) to the Nash equilibrium $(\frac{1}{3}, \frac{1}{3}, \frac{1}{3})$, where it stops (see fig. 8.2). The same happens whenever all off-diagonal terms of A are positive. This shows that we cannot expect to trace orbits back in time.

Exercise 8.4.1 Compare this with the behaviour of the replicator equation. Do the same for the payoff matrix $A = \begin{bmatrix} 1 & 0 \\ 0 & 2 \end{bmatrix}$. Show that the constant solution $\mathbf{x}(t) = \mathbf{p}$ is not the only solution starting at the interior equilibrium $\mathbf{p}$.

As another example, let us consider the general 'rock–scissors–paper' game described in section 7.7, i.e. with

$$A = \begin{bmatrix} 0 & -a_2 & b_3 \\ b_1 & 0 & -a_3 \\ -a_1 & b_2 & 0 \end{bmatrix} \tag{8.22}$$

where $a_i, b_i > 0$. The lines $l_i = \{\mathbf{x} \in S_3 : (A\mathbf{x})_{i-1} = (A\mathbf{x})_{i+1}\}$ divide S_3 into three regions (see fig. 8.3). In the region containing $\mathbf{e}_i$, the best-response path

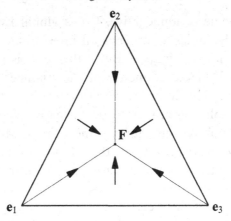

Fig. 8.2. The best-response dynamics for game (8.21)

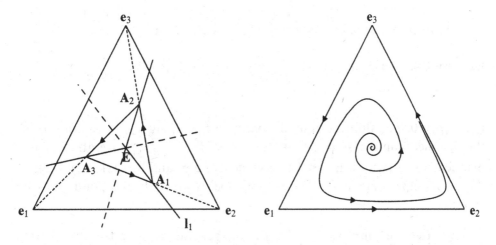

Fig. 8.3. The best response dynamics and replicator dynamics of the rock–scissors–paper game

points towards $\mathbf{e}_{i+1}$: as soon as the path reaches l_i, it turns abruptly towards $\mathbf{e}_{i+2}$, etc. Let us denote this turning point by $\mathbf{x}^i$. Clearly $\mathbf{x}^i, \mathbf{x}^{i+1}$ and $\mathbf{e}_{i+2}$ are collinear.

The piecewise linear function $V(\mathbf{x}) := \max_i (A\mathbf{x})_i$ satisfies for the best-response dynamics (8.20) the relation $\dot{V}(\mathbf{x}) = -V(\mathbf{x})$. (In the region where the maximum is attained by strategy R_i we have $\dot{V} = (A\dot{\mathbf{x}})_i = a_{ii} - (A\mathbf{x})_i$.) Hence $V(\mathbf{x}(t))$ decreases along the piecewise linear solutions as long as it is positive, and increases when it is negative. The minimum of $V(\mathbf{x})$ is attained at the unique Nash equilibrium $\mathbf{p}$ given by (7.34) (its value there is given by (7.35)). If this value is nonnegative, V is a global Lyapunov function for the best-response dynamics, and hence all interior solutions converge to **p**, just as with the replicator equation. But if the value is strictly negative, then V takes the value 0 on the boundary of the triangle $A_1A_2A_3$ described in section 7.7, and is negative in its interior. (We just have to remember that A_2, for instance, is the solution of (7.41).) In this case, $|V|$ is a global Lyapunov function which attains its minimum at the triangle $A_1A_2A_3$. This triangle, the so-called *Shapley triangle*, attracts the interior orbits of the best-response dynamics, just as it attracts the *time averages* of the replicator dynamics!

Exercise 8.4.2 Show that there exists at most one triangle $\mathbf{y}^1, \mathbf{y}^2, \mathbf{y}^3$ such that the $\mathbf{y}^i$ lie on the lines l_i and the points $\mathbf{y}^i, \mathbf{y}^{i+1}$ and $\mathbf{e}_{i+2}$ are collinear. (Hint: if the $\mathbf{z}^i$ form another such triangle, the two triangles are perspective from **p**. Since their sides $\mathbf{y}^1\mathbf{y}^2$ and $\mathbf{z}^1\mathbf{z}^2$ intersect in $\mathbf{e}_3$ etc., the three points $\mathbf{e}_1, \mathbf{e}_2$ and $\mathbf{e}_3$ are collinear by Desargues' theorem, which is absurd.) Thus the triangle $A_1A_2A_3$ (if it exists) is the only triangle with this property.

Exercise 8.4.3 Let s_i denote the distance from $\mathbf{x}^i$ to **p**. Show that

$$s_{i+1} = (\alpha_i s_i)(1 + \beta_i s_i)^{-1} \tag{8.23}$$

for $\alpha_i, \beta_i > 0$. The full return map from l_1 to l_1 sending $\mathbf{x}^i$ to $\mathbf{x}^{i+3}$ transforms the distance s into $\tau(s) = \rho s(1 + \beta s)^{-1}$ with $\rho := \prod_{i=1}^{3} \frac{a_i}{b_i}$. Iteration of this map leads to $\tau^t(s) \to 0$ if $\rho \leq 1$ (convergence to **p**) and to $\tau^t(s) \to (\rho - 1)\beta^{-1}$ if $\rho > 1$. Show that for $i = 3k + 1$ this yields precisely the point A_2 from (7.42).

After these examples we turn to the general case and show how to construct all piecewise linear solutions of (8.20) through a given initial point **x**. A strategy **b** serves as a best response for some time, if and only if $\mathbf{b} \in \beta((1 - \varepsilon)\mathbf{x} + \varepsilon\mathbf{b})$ holds for small $\varepsilon \geq 0$. This is equivalent to the fact that **b** is a best reply against **x**, and among all these best replies, it is a best reply against itself, i.e. **b** is a Nash equilibrium of the restricted game $A_\mathbf{x} = (a_{ij})_{i,j\in\beta(\mathbf{x})}$. Since at least one such Nash equilibrium exists there is at least one direction **b** in which the population can evolve, obeying (8.20). In example (8.21) and the rock–scissors–paper game, this direction is unique, but in general, there may be more than one way to proceed (as in exercise 8.4.1). By iterating this process one obtains at least one piecewise linear solution through **x** defined for all times $t > 0$.

Theorem 8.4.4 *Let* **p** *be an interior ESS for the game with payoff matrix A.*

Then **p** *is globally asymptotically stable for the best-response dynamics: All piecewise linear paths reach* **p** *in finite time.*

Proof Consider the function $V(\mathbf{x}) = \max_i (A\mathbf{x})_i - \mathbf{x} \cdot A\mathbf{x}$ which satisfies $V(\mathbf{x}) \geq 0$, and $V(\mathbf{x}) = 0$ if and only if $\mathbf{x} = \mathbf{p}$. Along a linear piece $\dot{\mathbf{x}} = \mathbf{b} - \mathbf{x}$, one has $V = (\mathbf{b} - \mathbf{x}) \cdot A\mathbf{x}$ and $\dot{V} = -\dot{\mathbf{x}} \cdot A\mathbf{x} + (\mathbf{b} - \mathbf{x}) \cdot A\dot{\mathbf{x}} = -(\mathbf{b} - \mathbf{x}) \cdot A\mathbf{x} + (\mathbf{b} - \mathbf{x}) \cdot A(\mathbf{b} - \mathbf{x})$. For $\mathbf{x} \neq \mathbf{p}$, the first term is negative by definition, and the second term is negative because of (6.19), and even bounded away from 0. Hence $V(\mathbf{x}(t))$ decreases strictly, and reaches the value 0 in finite time. $\qquad\square$

A limit case is the following:

Exercise 8.4.5 For a zero-sum game, show that all piecewise linear solutions of (8.20) converge to the set of equilibria.

Exercise 8.4.6 Consider a zero-sum game with $n = 4$ and a line of equilibria. Show that all nonconstant solutions converge to one and the same equilibrium on this line.

Exercise 8.4.7 Investigate the best-response dynamics (8.20) for the matrix (7.12).

Exercise 8.4.8 Show that if strategy R_i is strictly dominated, its frequency $x_i(t)$ converges to 0 for every solution of (8.20).

Exercise 8.4.9 Is the exclusion principle (theorem 7.6.1) valid for the best-response dynamics?

8.5 Adjustment dynamics

We have seen in section 8.2 that the payoff monotonic dynamics given by (8.6) and (8.7) are a natural class of dynamics large enough to contain the replicator equation and the imitation dynamics. The best-response dynamics does not belong to this class, however, since it does not leave the boundary of S_N invariant (it is an 'innovative' dynamics, in contrast to the 'imitative' dynamics (8.6)). Also the monotonicity axiom (8.7) is not satisfied in the strict sense required there: only the best strategy increases, the frequencies of all the other strategies decrease at the same rate -1.

This calls for a common extension of these game dynamics. A natural requirement is that the population should move towards a better reply against the present state, i.e. $\mathbf{x}(t + h) \cdot A\mathbf{x}(t) > \mathbf{x}(t) \cdot A\mathbf{x}(t)$ for small $h > 0$, whenever possible. Taking the limit $h \to 0$, this leads to the following concept.

A dynamics $\dot{\mathbf{x}} = \mathbf{f}(\mathbf{x})$ on S_n is said to be an *adjustment dynamics*, if $\dot{\mathbf{x}} A\mathbf{x} \geq 0$, with strict inequality whenever $\mathbf{x}$ is not a Nash equilibrium (or a rest point of the replicator equation).

Exercise 8.5.1 Show that payoff monotonic dynamics, the best-response dynamics and the adaptive dynamics of section 9.6 are all adjustment dynamics.

Exercise 8.5.2 Show that if the matrix A is symmetric, the average payoff $\mathbf{x} \cdot A\mathbf{x}$ is monotonically increasing for every adjustment dynamics.

Exercise 8.5.3 Show that strict Nash equilibria are asymptotically stable for all adjustment dynamics.

Exercise 8.5.4 Does each adjustment dynamics eliminate strategies that are strictly dominated by other pure strategies (as the payoff monotone dynamics do)?

Exercise 8.5.5 Construct examples of games with an interior ESS and smooth adjustment dynamics, for which the ESS is not asymptotically stable.

Exercise 8.5.6 Let $\mathbf{p}$ be an interior equilibrium. Show that heading straight towards the equilibrium, i.e. $\dot{\mathbf{x}} = \mathbf{p} - \mathbf{x}$, is an adjustment dynamics if and only if $\mathbf{p}$ is an ESS.

Exercise 8.5.7 Suppose the game A has a unique Nash equilibrium $\mathbf{p}$. Construct an adjustment dynamics for which $\mathbf{p}$ is globally stable.

8.6 A universally cycling game

In exercise 8.2.5 we have presented a rock–scissors–paper game for which any payoff monotonic dynamics leads to oscillations close to the boundary cycle from most initial conditions. We now present an example af a game that has this property for the considerably larger class of adjustment dynamics. Let

$$A = \begin{bmatrix} 0 & 0 & -1 & \varepsilon \\ \varepsilon & 0 & 0 & -1 \\ -1 & \varepsilon & 0 & 0 \\ 0 & -1 & \varepsilon & 0 \end{bmatrix}. \tag{8.24}$$

For $\varepsilon \neq 0$, this game has a unique Nash equilibrium: $\mathbf{p} = \left(\frac{1}{4}, \frac{1}{4}, \frac{1}{4}, \frac{1}{4}\right)$. Besides $\mathbf{p}$, and the four corners of the simplex, there are two more rest points for the replicator dynamics: $\mathbf{F}_{13} = (\frac{1}{2}, 0, \frac{1}{2}, 0)$ and $\mathbf{F}_{24} = (0, \frac{1}{2}, 0, \frac{1}{2})$. For $\varepsilon > 0$ this game has a best-response cycle among the pure strategies $1 \rightarrow 2 \rightarrow 3 \rightarrow 4 \rightarrow 1$.

For $\varepsilon = 0$, this game is a partnership game: $A = A^T$. The mean payoff is $P(\mathbf{x}) = \mathbf{x} \cdot A\mathbf{x} = -2(x_1 x_3 + x_2 x_4)$. The minimum value of P is $-\frac{1}{2}$ which is attained at the two points $\mathbf{F}_{13}$ and $\mathbf{F}_{24}$. At the interior equilibrium, $P(\mathbf{p}) = -\frac{1}{4}$. It is easy to see that P attains its maximum value 0 on the set Γ, consisting of the four edges $\mathbf{e}_1 \rightarrow \mathbf{e}_2 \rightarrow \mathbf{e}_3 \rightarrow \mathbf{e}_4 \rightarrow \mathbf{e}_1$. Each point in Γ is a Nash equilibrium.

$P(\mathbf{x}(t))$ is monotonically increasing along any solution of any adjustment dynamics, and it is strictly increasing at every point except the Nash equilibria in Γ and $\mathbf{P}$, and possibly $\mathbf{F}_{13}, \mathbf{F}_{24}$. In particular, Γ is an asymptotically stable attractor for every adjustment dynamics, and its basin of attraction contains the set $\{\mathbf{x} : P(\mathbf{x}) > -\frac{1}{4}\}$. (Usually, the unstable rest point $\mathbf{p}$ is a saddle point with a one-dimensional stable manifold. If the dynamics respects the cyclic symmetry of the game, then this stable manifold coincides with the line segment L bounded by $\mathbf{F}_{13}$ and $\mathbf{F}_{24}$. All other initial conditions lead to the set Γ.)

Theorem 8.6.1 *For every adjustment dynamics that depends continuously on the payoffs, and every $\delta > 0$, there exists an $\varepsilon_0 > 0$, such that for all choices ε in (8.24) with $|\varepsilon| < \varepsilon_0$, and all initial conditions $\mathbf{x}$ with $P(\mathbf{x}) > -\frac{1}{4} + \delta$, solutions do not converge to the interior Nash equilibrium $\mathbf{p}$, but enter the set $\{\mathbf{x} : P(\mathbf{x}) > -\delta\}$ and stay there for ever.*

Proof This follows since P is still a Lyapunov function for the perturbed dynamics outside some neighborhoods of Γ and $\mathbf{p}$. $\square$

Exercise 8.6.2 Show that the function $P(\mathbf{x})$ defined above is still monotonically increasing along solutions of the replicator dynamics for $-1 < \varepsilon < 0$.

Exercise 8.6.3 Compute the attractor for the best-response dynamics.

On the other hand, by exercise 8.5.7, one can construct for each (fixed) $\varepsilon \neq 0$ a smooth adjustment dynamics for which $\mathbf{p}$ is globally stable. However, those adjustment dynamics cannot be put together in a continuous way.

The example shows that equilibrium analysis is not enough for game dynamics: cycling behaviour is unavoidable. There is no reasonable dynamic process that leads to an equilibrium in every game.

8.7 Notes

Imitation dynamics have been studied by Helbing (1992), Björnerstedt and Weibull (1996), Weibull (1995) and Schlag (1994), who showed that the 'proportional imitation rule', which leads to the replicator dynamics, is in some sense optimal. Here, we have followed Hofbauer (1995b). Another economic motivation of the replicator dynamics through learning is described by Börgers and Sarin (1993). Payoff monotonic systems have been studied in Nachbar (1990), Friedman (1991), Samuelson and Zhang (1992), and Ritzberger and Weibull (1995), who also investigated aggregate monotonicity (see exercise 8.3.4). The observation in exercise 8.1.4 is due to Fudenberg and Levine (1998). The characterization of selection dynamics eliminating dominated strategies is from Hofbauer and Weibull (1996). The game (8.16) is due to Dekel and Scotchmer (1992) and serves as an almost universal counterexample (see e.g. exercise 8.3.6). Fictitious play was introduced in Brown (1951) and Robinson (1951). The best-response dynamics (8.20) appeared first in Matsui (1992). It has been thoroughly investigated in Hofbauer (1995a) which contains in particular theorem 8.4.4. Examples of nonconvergence are due to Shapley (1964) and Foster and Young (1995). For more on Shapley polygons and a general principle linking best-response dynamics with the time averages of the replicator equation, we refer to Gaunersdorfer and Hofbauer (1995). Exercise 8.4.6 is from Berger (1997b). Adjustment dynamics were introduced by Swinkels (1993). The universally cycling games in section 8.6 and exercise 8.2.5 are from Hofbauer and Swinkels (1996). For stochastic versions of game dynamics see Foster and Young (1990), Young (1993), Cabrales (1993), Kaniovsky and Young (1995), Fudenberg and Levine (1998), Dawid (1996), Kandori (1996), and Posch (1997).

9

Adaptive dynamics

If a population adopts an evolutionarily stable strategy, a selection–mutation regime cannot lead away from it; but it need not lead towards such a strategy, as we shall see by an example based on the Prisoner's Dilemma game. The evolutionary path in the space of strategies can be described by an adaptive dynamics which at first glance seems quite different from the replicator dynamics, but in fact generalizes it and leads to a further validation of evolutionary stability.

9.1 The repeated Prisoner's Dilemma

The Prisoner's Dilemma (or PD) game is a game between two players, each having the choice between two options: to cooperate (to play **C**) or to defect (to play **D**). The payoff matrix is given by

$$A = \begin{bmatrix} R & S \\ T & P \end{bmatrix}. \tag{9.1}$$

If both players cooperate, they receive as payoff R (the reward), which is assumed to be larger than the payoff P (the punishment) obtained if they both defect. But if one player plays **D** while the other plays **C**, then the defector receives a payoff T (the temptation) which is even higher than R, whereas the cooperator receives only the sucker's payoff S which is lower than P. In addition to this rank ordering of the payoff values

$$T > R > P > S, \tag{9.2}$$

one also assumes

$$2R > T + S, \tag{9.3}$$

which means that the total payoff for the two players is larger if both cooperate than if one cooperates and the other defects. (This is to prevent them from taking turns at defection and then sharing the payoffs.)

We see immediately that strategy **D** dominates. Two rational players will both defect, and end up with a payoff P instead of the higher payoff R for mutual cooperation. In a population of **C**- and **D**-players, the replicator dynamics leads inexorably to the extinction of cooperators.

Nevertheless, one finds a lot of spontaneous cooperation in economic and biological interactions. In particular, if the game is frequently repeated, cooperation becomes plausible. One will think twice about defecting if this makes one's co-player decide to defect on the next occasion — and *if* such an occasion is likely to occur.

Let us assume that with some fixed probability w, a further round can occur. Then w^n is the probability that the n-th round takes place, and $\sum w^n = (1-w)^{-1}$ is the expected length of the game. If A_n denotes the payoff in the n-th round, then the total payoff is given by

$$A = \sum A_n w^n. \qquad (9.4)$$

For $w = 1$, the limiting case where the game is repeated for infinitely many rounds, this series diverges. In this case one uses as payoff the limit of the mean

$$A = \lim \frac{A_0 + A_1 + \cdots + A_n}{n+1}. \qquad (9.5)$$

if it exists.

A strategy for the *repeated* PD is a program which tells one what to do in each round. In contrast to the one-round PD, there now exists a bewildering wealth of strategies, none of which serves as a best reply against all comers. If the co-player, for instance, plays *AllC* (the strategy which always cooperates), then it is best to play *AllD* - the total payoff is $(1-w)^{-1}T$. But if your co-player decides to cooperate until you defect and from then on never to cooperate again, it is best not to spoil the partnership by defecting: the gain in one round (T instead of R) is more than offset by the loss in the subsequent game, where one can attain at best P per round and hence cannot hope for more than $(1-w)^{-1}P$ altogether. Clearly, one would do better by steadily cooperating, provided

$$w > \frac{T-R}{T-P}. \qquad (9.6)$$

This absence of a best strategy is crucial. There is no hard and fast recipe for playing the iterated PD. In several well-publicized computer experiments (round-robin tournaments organised by Robert Axelrod) the highest payoff was obtained by the simplest strategy submitted, namely *Tit For Tat* (or TFT), which consists in playing **C** in the first move and then always repeating the co-player's previous move. Remarkably, a TFT player is never ahead of his co-player, but can nevertheless win the whole tournament. (This would not be possible for zero-sum games.) Of course TFT will not win every conceivable round-robin tournament: there is, as we have seen, no such strategy. But TFT does well in a wide variety of strategic environments.

On the other hand, the interaction between two TFT players is vulnerable to errors. One inadvertent defection starts a run of alternating defections. The total payoff, then, drops drastically. Obviously, the two TFT players ought to forgive occasionally — not on a predictable pattern, since this would make them exploitable, but randomly. This suggests considering strategies which are *stochastic*.

9.2 Stochastic strategies for the Prisoner's Dilemma

Let us consider strategies given by triples $(y, p, q) \in [0, 1]^3$, where y is the probability of playing **C** in the first round, and p and q are the conditional probabilities of playing **C** after an opponent's **C** (resp. **D**) in the previous round. Admittedly, these strategies are only a small subset in the space of all strategies for the repeated PD, but they contain important strategies like TFT (given by $(1, 1, 0)$) or *AllD* (given by $(0, 0, 0)$). Of course these two strategies are not properly stochastic. Since we are mostly interested in noisy interactions (caused, for instance, by errors in implementing a move), we shall usually consider only strategies with $0 < y, p, q < 1$. The other cases are always easy to deal with separately.

The **C**-*level* c_n is a player's probability of playing **C** in the n-th round. It is determined by the opponent's **C**-level in the previous round, and hence by the player's own **C**-level from two rounds before. For a player using strategy $E = (y, p, q)$ matched against an opponent using strategy $E' = (y', p', q')$, this 'echo effect' yields $c_n = px + q(1 - x) = q + rx$, where x is the opponent's **C**-level in the previous round, and $r = p - q$. Hence

$$c_{n+2} = q + r(q' + r'c_n). \tag{9.7}$$

Therefore the **C**-level c_n converges at the geometric rate $|rr'|^{\frac{1}{2}}$ to the *asymptotic* **C**-*level* of strategy E against E', given by

$$c = \frac{q + rq'}{1 - rr'}. \tag{9.8}$$

Clearly, we have $c = q + rc'$, where c' is the asymptotic C-level of E' against E. The asymptotic C-level of strategy E against itself is given by

$$s = \frac{q}{1-r}. \tag{9.9}$$

All strategies on the vertical plane through E and the TFT corner $(1,1,0)$ have the same asymptotic C-levels against themselves and consequently against each other.

Exercise 9.2.1 Show that the differences $c - c'$, $c - s'$ and $s - s'$ always have the same sign. (Here s' is the asymptotic C-level of E' against itself.)

For $w = 1$, the payoff obtained by an E-strategist against an E'-opponent is given by

$$A(E, E') = Rcc' + Sc(1 - c') + T(1 - c)c' + P(1 - c)(1 - c'). \tag{9.10}$$

This expression is independent of y, the probability of cooperating in the n-th round. Indeed, the value of y only affects the initial stage of the interaction, which plays no role in the long term.

Exercise 9.2.2 Compute explicitly the payoff $A(E, E')$ for $w < 1$.

9.3 Adaptive Dynamics for the Prisoner's Dilemma

If we consider only the case $w = 1$, we can neglect the initial probability y of cooperating. When can a population of players using strategy $E = (p, q)$ be invaded by a small minority using strategy $E' = (p', q')$? Obviously if and only if

$$A(E', E) > A(E, E). \tag{9.11}$$

In order to facilitate computations, we normalize the payoffs and set

$$R - P = 1, \qquad P - S = \beta, \qquad T - R = \gamma. \tag{9.12}$$

Using $c - s = r(c' - s)$, we see from (9.10) that $A(E', E) - A(E, E)$ is given by

$$(\beta - \gamma)(rc'^2 + qc' - s^2) + [r(1 + \gamma) - \beta](c' - s). \tag{9.13}$$

Let us consider first the special case $\beta = \gamma$. In this case

$$A(E', E) - A(E, E) = (c' - s)((\gamma + 1)r - \gamma). \tag{9.14}$$

The line $r = (\gamma + 1)^{-1}\gamma$ divides the square $[0, 1]^2$ of (p, q)-values into two parts (see fig. 9.1). If E lies in the south-east corner (a neighbourhood of the TFT-strategy $(1, 0)$), then precisely those strategies E' can invade which

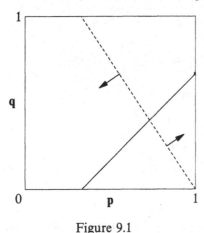

Figure 9.1

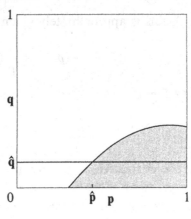

Figure 9.2

are more cooperative than E: indeed, if $s' > s$, then $c' > s$ and (9.11) holds. We define this region as the *cooperation-rewarding zone*. Conversely, if the population adopts a strategy E which does not lie in this zone, then every strategy which is *less* cooperative can invade. A strategy is more, or less, cooperative than E depending on whether it is above, or below, the line from E to the TFT-corner $(1,0)$. If E lies on the boundary of the cooperation-rewarding zone, i.e. satisfies $r = (\gamma + 1)^{-1}\gamma$, then all strategies E' do exactly as well, against E, as E does against itself.

The condition $\beta = \gamma$ holds if providing help to a co-player (i.e. playing **C**) costs c points to the donor and yields b points to the recipient (with $c < b$), whereas playing **D** costs nothing and yields nothing. In that case

$$R = b - c, \qquad S = -c, \qquad T = b, \qquad P = 0. \qquad (9.15)$$

The usual replicator dynamics describes the evolution under natural selection of the frequencies of a finite set of types in a heterogeneous population. We shall now propose a different dynamics which assumes that the population remains essentially homogeneous, but evolves under a selection-mutation regime. More precisely, we shall assume that the mutational jumps are small, and occur only rarely, so that a mutant will have either vanished or taken over before the next mutation arises. Of course, the mutational steps are random: but rather than turn to a stochastic process, we introduce a deterministic dynamics *in trait space* pointing into the direction which is most promising from the myopic point of view.

Let us investigate this first for the repeated Prisoner's Dilemma with $w = 1$. If the population is in state $E = (p,q)$, the payoff for a mutant strategy $E' = (p',q')$ is given by $A(E',E) = A((p',q'),(p,q))$. For (p',q') close to (p, q), this is approximately given by

$$A(E, E) + (p' - p)\frac{\partial A}{\partial p'}(E, E) + (q' - q)\frac{\partial A}{\partial q'}(E, E).$$

The *adaptive dynamics* is accordingly defined as

$$\dot{p} = \frac{\partial A}{\partial p'}(E', E), \quad \dot{q} = \frac{\partial A}{\partial q'}(E', E) \tag{9.16}$$

where the derivatives are evaluated for $E' = E$. We emphasize that (9.16) is *not* the gradient of the function $E \to A(E, E)$: the vector field does not point in the direction which is best for the whole (homogeneous) population, but in the direction which is best from the individual dissident's (or mutant's) point of view.

Using (9.13), the adaptive dynamics for the PD is easily seen to be

$$\dot{p} = \frac{q}{(1 - r)(1 - r^2)}F(p, q) \tag{9.17}$$

$$\dot{q} = \frac{1 - p}{(1 - r)(1 - r^2)}F(p, q) \tag{9.18}$$

where

$$F(p, q) = (\beta - \gamma)q\frac{1 + r}{1 - r} + r(1 + \gamma) - \beta. \tag{9.19}$$

The vector $(\dot{p}, \dot{q})$ at the point $E = (p, q)$ is orthogonal to the line from E to the TFT strategy $(1, 0)$ (see fig. 9.1).

It points upwards (towards larger p- and q-values) iff E is in the cooperation-rewarding zone, i.e. in the set where $F(p, q) > 0$. We note that this zone includes a neighborhood of TFT, i.e. of the south-east corner of the unit square. The boundary of this zone, where $F(p, q) = 0$, is given by the line $r = (\gamma + 1)^{-1}\gamma$ if $\beta = \gamma$ (see fig. 9.1), and by

$$q = \frac{\beta - r(1 + \gamma)}{\beta - \gamma}\frac{1 - r}{1 + r} \tag{9.20}$$

otherwise. We have sketched this in fig. 9.2 for the case $\beta < \gamma$ (where, it should be noted, $F(p, q) = 0$ implies $r > 0$, so that the cooperation-rewarding zone is below the diagonal). There is no evolutionary tendency towards TFT: this strategy is a pivot, rather than the aim, of the evolution.

Exercise 9.3.1 Analyse the adaptive dynamics if the game which is repeated is not the PD, but the Hawk-Dove game.

Exercise 9.3.2 Show that the adaptive dynamics for the PD with $\gamma = \beta$ and $w < 1$ is given by

$$\dot{y} = d \qquad \dot{p} = d\frac{w}{1-w}\frac{e}{1-wr} \qquad \dot{q} = d\frac{w}{1-w}\left(1 - \frac{e}{1-wr}\right) \qquad (9.21)$$

where

$$d = \frac{(\gamma+1)wr - \gamma}{1 - w^2r^2}, \qquad e = (1-w)y + wq \qquad (9.22)$$

Since $0 < e < 1 - wr$, the components of the dynamics all have the same sign, given by $(\gamma+1)wr - \gamma$. Again, we can speak of a cooperation-rewarding zone. Show that

$$\dot{y} = \frac{1-w}{1-w^2r^2}[A(AllC, E) - A(AllD, E)],$$

$$\dot{p} = \frac{w}{1-w^2r^2}[A(E, E) - A(AllD, E)]$$

$$\dot{q} = \frac{w}{1-w^2r^2}[A(AllC, E) - A(E, E)].$$

9.4 An ESS may be unattainable

The nonlinearity of the payoff $A(E', E)$ implies an interesting phenomenon. From (9.13) we can easily see that

$$A(E', E) - A(E, E) = (c' - s)[(\beta - \gamma)(s + rc') + r(1 + \gamma) - \beta]. \qquad (9.23)$$

Let us consider the case $\beta < \gamma$, and assume that some constraint keeps the q-value fixed throughout the population, at some value $\hat{q}$. The strategies are therefore restricted to the horizontal line $q = \hat{q}$, and we shall consider the case that this line intersects the line $F(p, q) = 0$ at a point $\hat{E} = (\hat{p}, \hat{q})$. Writing (9.23) as $A(E', E) - A(E, E) = (c' - s)G$ and using $G - F = r(\beta - \gamma)(c' - s)$, it follows (since $F = 0$ implies $r > 0$) that

$$A(E', \hat{E}) - A(\hat{E}, \hat{E}) = r(\beta - \gamma)(c' - s)^2 < 0 \qquad (9.24)$$

for all $E' \neq \hat{E}$. In our restricted strategy space, $\hat{p}$ is an ESS, and even a strict Nash equilibrium: indeed, for any $p \in]0, 1[$ with $p \neq \hat{p}$, we see that $E = (p, \hat{q})$ satisfies $A(E, \hat{E}) < A(\hat{E}, \hat{E})$. Thus a population where all members use $\hat{p}$ cannot be invaded. On the other hand, such a population is *unattainable*: indeed, if all members of the population use a strategy $p > \hat{p}$, this population can only be invaded by dissidents using a still larger value $p' > p$, since E is in the cooperation-rewarding zone; and if all members use a strategy with $p < \hat{p}$, then only values $p' < p$ can invade. This shows that the evolution is *away* from the ESS $\hat{p}$. Even if, by an extraordinary fluke, the population is originally in the 'Garden of Eden'-configuration $\hat{p}$, the slightest deviation – caused, for instance, by a change in environment – will start an evolution leading inexorably away from it. Such an unattainable ESS is only possible, of course, if there are infinitely many admissible strategies.

Exercise 9.4.1 Show that if, conversely, p is fixed at some value $\hat{p}$, the corresponding ESS $\hat{q}$ is attracting in the sense that if the homogeneous population uses a strategy q different from $\hat{q}$, then q' can invade whenever it lies in the same direction as $\hat{q}$.

Exercise 9.4.2 Investigate the invasion dynamics for $\beta \neq \gamma$, and give a geometric interpretation of the set of strategies E' in (p, q)-space which can invade a given E-population.

Exercise 9.4.3 Find three strategies E_1, E_2 and E_3 such that the resulting replicator dynamics is of 'rock–scissors–paper' type.

Exercise 9.4.4 Assume some minimal noise-level ε, so that the strategy space reduces to $\varepsilon \leq p, q \leq 1 - \varepsilon$. Find the strategy which maximizes the payoff of the population, subject to being invasion-proof against any strategy with a smaller p- or q-value. Show that it converges, for $\varepsilon \to 0$, to

$$p = 1 \qquad q = \min \left[1 - \frac{\gamma}{\beta + 1}, \frac{1}{\gamma + 1} \right] \tag{9.25}$$

This strategy is called *Generous TFT*.

9.5 A closer look at adaptive dynamics

Let us now leave the PD and consider, more generally, strategic traits determined by *continuous* variables (like size, sex ratio or the probability of escalating a fight). We assume that the population is essentially homogeneous: all members use the same strategy $\mathbf{x}$ except for an occasional dissident using a strategy $\mathbf{y} = \mathbf{x} + \mathbf{h}$ which is nearby. The payoff for such an individual is denoted by $A(\mathbf{y}, \mathbf{x})$, its relative advantage $A(\mathbf{y}, \mathbf{x}) - A(\mathbf{x}, \mathbf{x})$ by $W(\mathbf{h}, \mathbf{x})$. The adaptive dynamics (first version) is

$$\dot{x}_i = \frac{\partial}{\partial y_i} A(\mathbf{y}, \mathbf{x}) \tag{9.26}$$

for $i = 1, \ldots, n$, where the derivatives are evaluated at $\mathbf{y} = \mathbf{x}$. This vector, the gradient of $\mathbf{y} \to A(\mathbf{y}, \mathbf{x})$, i.e. its derivative $D_{\mathbf{y}}A(\mathbf{y}, \mathbf{x})|_{\mathbf{y}=\mathbf{x}}$, points in the direction of the maximal increase of the mutant's advantage.

The rationale is that a few dissidents or mutants test alternatives by using strategies close to $\mathbf{x}$, and that the whole population evolves in the most promising direction. This is supposed to mimic an evolution favouring individual fitness. It can also be used for learning models, under the assumption that trials are myopic, i.e. explore only the immediate vicinity of the current strategy. Of course we cannot really assume that a random mutation in the best direction will actually occur: but the adaptive dynamics points in the most favourable direction, and therefore defines (if it does not vanish) a half-space with the property that a mutant $\mathbf{y}$ close to $\mathbf{x}$ will invade if and only if it lies in that half-space.

If A is independent of what the others do, i.e. a function of $\mathbf{y}$ only, then we get the usual hill-climbing dynamics leading to local optimization.

Exercise 9.5.1 In the one-dimensional case, (9.26) reduces to

$$\dot{x} = \frac{\partial W}{\partial h}(0, x). \qquad (9.27)$$

A state $\hat{x}$ is locally evolutionarily stable (see section 6.5) if $W(h, \hat{x}) < 0$ for all small $h \neq 0$, and convergence-stable if $W(h, x)$ has the sign of $h(\hat{x} - x)$ for small h. Show that if the population is in a state x close to the convergence-stable state $\hat{x}$, then mutations in the direction towards $\hat{x}$ will succeed. The point $\hat{x}$ is a fixed point of (9.27) if it is locally evolutionarily stable or if it is convergence-stable. Generically, an equilibrium is convergence-stable if and only if $\frac{\partial^2 W}{\partial x \partial h} < 0$ and locally evolutionarily stable if and only if $\frac{\partial^2 W}{\partial h^2} < 0$.

9.6 Adaptive dynamics and gradients

So far, we have tacitly assumed that $\mathbf{x}$ varies in an open subset of $\mathbb{R}^n$ equipped with the Euclidean metric: fluctuations in every direction are equally likely. It may happen that another metric is more appropriate, for instance if genetic, developmental or social constraints render variations in one direction more likely than in another. It may also happen that $\mathbf{x}$ is restricted to some subset of $\mathbb{R}^n$ (for example to the simplex S_n, if the x_i are the probabilities of some strategies summing to 1).

Again, the prevalent state or strategy $\mathbf{x}$ will tend in the direction of the maximal local increase: $\dot{\mathbf{x}}$ will be proportional to the unit vector $\mathbf{h}$ maximizing $A(\mathbf{x} + \varepsilon\mathbf{h}, \mathbf{x}) - A(\mathbf{x}, \mathbf{x})$, in the limit $\varepsilon \to 0$. Obviously, this notion of unit vector depends on the Riemannian metric of the state space.

Let G be a general Riemannian metric which associates (in a smooth way) with each $\mathbf{x}$ a symmetric positive definite matrix $G(\mathbf{x}) = (g_{ij}(\mathbf{x}))$, such that the inner product in the tangent space at $\mathbf{x}$ is given by

$$\langle \boldsymbol{\eta}, \boldsymbol{\xi} \rangle_{\mathbf{x}} = \sum g_{ij}(\mathbf{x}) \eta_i \xi_j = \boldsymbol{\eta} \cdot G(\mathbf{x}) \boldsymbol{\xi}. \tag{9.28}$$

As in section 7.8 we define for a given Riemannian metric G the adaptive dynamics $\dot{\mathbf{x}}$ by requiring

$$\langle \boldsymbol{\eta}, \dot{\mathbf{x}} \rangle_{\mathbf{x}} = \sum g_{ij}(\mathbf{x}) \eta_i \dot{x}_j = D_{\mathbf{y}} A(\mathbf{y}, \mathbf{x})|_{\mathbf{y}=\mathbf{x}}(\boldsymbol{\eta}) \tag{9.29}$$

to hold for all $\boldsymbol{\eta}$ in the tangent space at $\mathbf{x}$.

Let us now consider the most important special case, where $\mathbf{x} \in S_n$ and A is linear in $\mathbf{y}$, i.e.

$$A(\mathbf{y}, \mathbf{x}) = \mathbf{y} \cdot \mathbf{f}(\mathbf{x}).$$

Then (9.29), i.e. $G\dot{\mathbf{x}} \cdot \boldsymbol{\eta} = \mathbf{f} \cdot \boldsymbol{\eta}$, must hold for all $\boldsymbol{\eta}$ in the tangent space $\mathbb{R}_0^n$ (i.e. the set of $\boldsymbol{\eta} \in \mathbb{R}^n$ satisfying $\sum_i \eta_i = 0$) and we must have

$$G(\mathbf{x})\dot{\mathbf{x}} - \mathbf{f}(\mathbf{x}) = \psi(\mathbf{x})\mathbf{1} \tag{9.30}$$

for some function ψ. Since the matrix G is invertible, we can transform (9.30), i.e. $G\dot{\mathbf{x}} = \mathbf{f} + \psi\mathbf{1}$, into

$$\dot{\mathbf{x}} = G^{-1}\mathbf{f}(\mathbf{x}) + \mathbf{g}\psi(\mathbf{x}), \tag{9.31}$$

with $\mathbf{g} = G^{-1}\mathbf{1}$. Then $\mathbf{1} \cdot \dot{\mathbf{x}} = 0$ implies $\psi(\mathbf{x}) = -(\mathbf{1} \cdot \mathbf{g})^{-1}\mathbf{g} \cdot \mathbf{f}$. Hence (9.31) is equivalent to the explicit form

$$\dot{x}_i = \sum_j c_{ij}(\mathbf{x})f_j(\mathbf{x}) \tag{9.32}$$

with

$$C = G^{-1} - (\mathbf{g} \cdot \mathbf{1})^{-1}\mathbf{g}\mathbf{g}^t. \tag{9.33}$$

where $\mathbf{g}^t$ is the transpose of $\mathbf{g}$. We illustrate this by two examples:

(a) The *Euclidean metric* on S_n. Here $g_{ij}(\mathbf{x}) = \delta_{ij}$, hence $\mathbf{g} = G^{-1}\mathbf{1} = \mathbf{1}$ and $c_{ij} = \delta_{ij} - \frac{1}{n}$, so that the adaptive dynamics reads

$$\dot{x}_i = f_i(\mathbf{x}) - \frac{1}{n}\sum f_k(\mathbf{x}). \tag{9.34}$$

(b) The *Shahshahani metric* on S_n, defined by $g_{ij}(\mathbf{x}) = \delta_{ij}\frac{1}{x_i}$ (see section 7.8). Here $\mathbf{g} = G^{-1}\mathbf{1} = \mathbf{x}$, so $c_{ij} = x_i\delta_{ij} - x_ix_j$, and the adaptive dynamics reads

$$\dot{x}_i = x_i(f_i(\mathbf{x}) - \bar{f}(\mathbf{x})), \tag{9.35}$$

with $\bar{f}(\mathbf{x}) = \sum x_k f_k(\mathbf{x})$. This is just the replicator equation (7.1).

Theorem 9.6.1 *If $\hat{\mathbf{x}} \in \text{int } S_n$ is a local ESS for a payoff function $A(\mathbf{y}, \mathbf{x}) = \mathbf{y}\mathbf{f}(\mathbf{x})$ which is linear in $\mathbf{y}$, then $\hat{\mathbf{x}}$ is asymptotically stable for each adaptive dynamics* (9.31).

Proof By definition (see section 6.5), we have

$$A(\hat{\mathbf{x}}, \mathbf{x}) > A(\mathbf{x}, \mathbf{x}) \tag{9.36}$$

for all $\mathbf{x}$ close to $\hat{\mathbf{x}}$ with $\mathbf{x} \neq \hat{\mathbf{x}}$. Let

$$V(\mathbf{x}) = \langle \mathbf{x} - \hat{\mathbf{x}}, \mathbf{x} - \hat{\mathbf{x}} \rangle_{\hat{\mathbf{x}}} = (\mathbf{x} - \hat{\mathbf{x}}) \cdot G(\hat{\mathbf{x}})(\mathbf{x} - \hat{\mathbf{x}})$$

(which, incidentally, is an approximation for the geodesic distance from $\mathbf{x}$ to $\hat{\mathbf{x}}$ in the Riemannian G-metric). Then

$$\dot{V}(\mathbf{x}) = 2(\mathbf{x} - \hat{\mathbf{x}}) \cdot G(\hat{\mathbf{x}})\dot{\mathbf{x}}, \tag{9.37}$$

which by (9.30) is approximately equal to

$$2(\mathbf{x} - \hat{\mathbf{x}}) \cdot G(\mathbf{x})\dot{\mathbf{x}} = 2(\mathbf{x} - \hat{\mathbf{x}}) \cdot (\mathbf{f}(\mathbf{x}) + \psi\mathbf{1}) = 2(\mathbf{x} - \hat{\mathbf{x}}) \cdot \mathbf{f}(\mathbf{x}) \tag{9.38}$$

which is strictly negative by (9.36). Hence V is a local Lyapunov function, and $\hat{\mathbf{x}}$ is asymptotically stable. $\qquad\square$

The proof shows that for each adaptive dynamics, the geodesic distance decreases monotonically near an ESS.

Exercise 9.6.2 Let C be an $n \times n$ positive semi-definite matrix whose kernel is spanned by $\mathbf{1}$, the vector orthogonal to $\mathbb{R}_0^n$. If $\boldsymbol{\xi} \cdot A\boldsymbol{\xi} < 0$ for all $\boldsymbol{\xi} \in \mathbb{R}_0^n$ with $\boldsymbol{\xi} \neq \mathbf{0}$, then show that CA is a stable matrix when restricted to $\mathbb{R}_0^n$, i.e. its eigenvalues have negative real part. (Hint: start with the eigenvalue equation $CA\mathbf{x} = \lambda\mathbf{x}, \mathbf{x} \in \mathbb{R}_0^n, \lambda \in \mathbb{R}$, find a $\mathbf{y} \in \mathbb{R}_0^n$ with $C\mathbf{y} = \mathbf{x}$ and conclude from $\mathbf{x} \cdot A\mathbf{x} = \lambda\mathbf{y} \cdot C\mathbf{y}$ that $\lambda < 0$. Extend this idea to complex eigenvalues.)

Exercise 9.6.3 Write (7.3) as

$$\dot{\mathbf{x}} = C(\mathbf{x})A\mathbf{x} \tag{9.39}$$

with $c_{ii}(\mathbf{x}) = x_i(1 - x_i)$ and $c_{ij} = -x_i x_j$ $(i \neq j)$. Since

$$\boldsymbol{\xi} \cdot C(\mathbf{x})\boldsymbol{\xi} = \sum x_i \xi_i^2 - \left(\sum x_i \xi_i\right)^2 = \sum x_i (\xi - \bar{\xi})^2 \geq 0, \tag{9.40}$$

the restriction of $C(\mathbf{x})$ to $\mathbb{R}_0^n$ is positive definite if $\mathbf{x} \in \text{int } S_n$. The linearization of (9.39) at a rest point $\mathbf{p}$ is given by

$$\dot{\boldsymbol{\xi}} = C(\mathbf{p})A\boldsymbol{\xi} \tag{9.41}$$

for $\xi \in \mathbb{R}_0^n$. Show that exercise 9.6.2 then implies that an interior ESS $\mathbf{p}$ is asymptotically stable.

Exercise 9.6.4 For a partnership game $A = A^T$ show that the adaptive dynamics (9.32) is a gradient system with respect to the Riemannian metric (9.28) and potential function $\frac{1}{2}\mathbf{x} \cdot A\mathbf{x}$.

9.7 Notes

For background information on the Prisoner's Dilemma game, see Axelrod (1984), Lindgren (1991), Sigmund (1993) and Nowak *et al.*(1995b). In the first few sections of this chapter, we follow Nowak and Sigmund (1990), see also Molander (1985), and Nowak and Sigmund (1994, 1995). The example of an unattainable ESS is due to Nowak (1990). Other such examples have been discovered by Eshel and Motro (1981) and Eshel (1983). The notion of convergence-stability (exercise 9.5.1) is due to Taylor (1989), see also Eshel (1996) and Lessard (1990). The last section on adaptive dynamics follows Hofbauer and Sigmund (1990). Hopkins (1995) arrives at (9.32) from a learning model. Exercises 9.6.2 and 9.6.3 are from Hines (1980). For other approaches to adaptive dynamics and trait substitution sequences, we refer to Christiansen (1991), Brown and Vincent (1987), Vincent *et al.* (1993), Dieckmann and Law (1996), and in particular to the monumental Metz *et al.* (1996a).

10

Asymmetric games

In asymmetric games, players in different positions have different strategy sets and payoff matrices. We discuss the replicator dyamics for such games, both for players who are forever tied to their positions and for players who can find themselves sometimes in one and sometimes in the other position.

10.1 Bimatrix games

So far we have always considered situations where the players are in symmetric positions: same set of strategies, same payoffs. However, there are many conflicts which are asymmetric. Thus food is more important for a starving animal than for a replete one, while the risk of injury is smaller for a stronger contestant, etc. In fact, asymmetries are not only incidental, but quite often essential features of the game: for example, in conflicts between males and females, between parents and offspring, between the owner of a habitat and an intruder, or between different species. If we restrict ourselves again to conflicts settled in pairwise encounters, and finite numbers of pure strategies, we are led to *bimatrix* games.

Thus let us distinguish between players in position I and in position II (for instance, White and Black in chess). In position I, a player has n strategies, and in position II he has m strategies. The payoffs are given by the matrices A for I and B for II. Thus a player in position I using strategy i against a player in position II using strategy j obtains the payoff a_{ij}, and the opponent obtains b_{ji}. The mixed strategies for player I are denoted by $\mathbf{p} \in S_n$, those for player II by $\mathbf{q} \in S_m$, and the corresponding payoffs are given by $\mathbf{p} \cdot A\mathbf{q}$ and $\mathbf{q} \cdot B\mathbf{p}$, respectively.

The pair $(\hat{\mathbf{p}}, \hat{\mathbf{q}}) \in S_n \times S_m$ is said to be a *Nash equilibrium* if $\hat{\mathbf{p}}$ is a best reply to $\hat{\mathbf{q}}$ and $\hat{\mathbf{q}}$ a best reply to $\hat{\mathbf{p}}$, i.e. if

$$\mathbf{p} \cdot A\hat{\mathbf{q}} \leq \hat{\mathbf{p}} \cdot A\hat{\mathbf{q}} \tag{10.1}$$

for all $\mathbf{p} \in S_n$ and

$$\mathbf{q} \cdot B\hat{\mathbf{p}} \leq \hat{\mathbf{q}} \cdot B\hat{\mathbf{p}} \tag{10.2}$$

for all $\mathbf{q} \in S_m$. The set of Nash equilibria for bimatrix games is always nonempty.

The games considered in the previous sections have all been *symmetric*, in the sense that both players have the same strategies at their disposal and that the payoff does not depend on the position: a player in position I using strategy *i* against a player in position II using strategy *j* obtains the same payoff as a player in position II using *i* against a player in position I using *j*. Thus symmetric games are those for which $A = B$. A Nash equilibrium $(\mathbf{p}, \mathbf{q})$ is said to be *symmetric* if $\mathbf{p} = \mathbf{q}$. A symmetric game can have asymmetric Nash equilibria: for instance, in the hawk–dove game, to escalate when in position I and not to escalate in position II. However, as long as the two players have no means of knowing their position, we shall only consider symmetric Nash equilibria for symmetric games.

The definition of strict Nash equilibria is analogous to (6.8). The pair $(\hat{\mathbf{p}}, \hat{\mathbf{q}})$ is a strict Nash equilibrium if strict inequalities hold in (10.1) and (10.2) whenever $\mathbf{p} \neq \hat{\mathbf{p}}$ resp. $\mathbf{q} \neq \hat{\mathbf{q}}$.

Exercise 10.1.1 Show that a strict Nash equilibrium $(\hat{\mathbf{p}}, \hat{\mathbf{q}})$ must consist of pure strategies, i.e. $\hat{\mathbf{p}} = \mathbf{e}_i$ and $\hat{\mathbf{q}} = \mathbf{f}_j$ for some corners $\mathbf{e}_i$ of S_n and $\mathbf{f}_j$ of S_m.

There is no obvious extension of the notion of evolutionary stability to asymmetric games. Indeed, an evolutionarily stable pair $(\hat{\mathbf{p}}, \hat{\mathbf{q}})$ must, first of all, be a Nash equilibrium, so that $\hat{\mathbf{p}}$ is a best reply to $\hat{\mathbf{q}}$. But what if $\mathbf{p}$ is another best reply? There is nothing to prevent it from invading, no reasonable condition analogous to the stability condition (6.12). So we have to exclude the existence of alternative best replies, which brings us back to strict Nash equilibria.

10.2 The Battle of the Sexes

As an example of an asymmetric game, we shall discuss a conflict between males and females concerning their respective shares in *parental investment*. In many species, it requires a considerable amount of time and energy to raise the offspring. Each parent might attempt to reduce its own share at the expense of the other. The outcome might depend on which sex is in a position to desert first. Whenever fertilization is internal, for example, females risk being deserted even before giving birth to the offspring. At an even more fundamental level, the game is rigged against females by the fact that they produce relatively few, large gametes, whereas males produce many small gametes. Females are thereby much more committed and can less afford to lose a child. Thus males are in many cases in a better position to desert. They can invest the corresponding gain in time and energy into increasing their offspring with the help of new mates.

The female counterstrategy is 'coyness', i.e. the insistence upon a long engagement period before copulation. Rather than undergoing a second costly engagement (for which it might be too late in the mating season), males would do better to stay faithfully home and help raise their offspring. Roughly speaking, in a population of coy females, males would have to be faithful. Among faithful males, however, it would not pay a female to be coy: the long engagement period is an unnecessary cost. Thus the proportion of 'fast' females would grow. But then 'philandering' males will have their chance and spread. Females, in that case, will do well to be coy. The argument thus runs full circle. In order to model this game-theoretically, let us assume that there are two types in the male population, namely E_1 (philandering) and E_2 (faithful) with frequencies x_1 and x_2, and two types in the female population, namely F_1 (coy) and F_2 (fast) with frequencies y_1 and y_2. Let us suppose that the successful raising of an offspring increases the fitness of both parents by G. The parental investment $-C$ will be entirely borne by the female if the male deserts. Otherwise, it is shared equally by both parents. A long engagement period represents a cost of $-E$ to both partners.

If a faithful male mates with a coy female, the payoff is $G - C/2 - E$ for both. A faithful male and a fast female skip the engagement cost and their payoff is $G - C/2$. But a philandering male meeting a fast female makes off with G, while her payoff is $G - C$. Finally if a philandering male encounters a coy female, nothing much happens and the payoff for both is 0.

With males in position I and females in position II, the payoff matrices are

$$A = \begin{bmatrix} 0 & G \\ G - \frac{C}{2} - E & G - \frac{C}{2} \end{bmatrix} \qquad B = \begin{bmatrix} 0 & G - \frac{C}{2} - E \\ G - C & G - \frac{C}{2} \end{bmatrix}. \qquad (10.3)$$

We shall assume $0 < E < G < C < 2(G - E)$. There exists no strict Nash equilibrium in this case (we have only to check the pure strategies), but a unique mixed Nash equilibrium $(\hat{\mathbf{p}}, \hat{\mathbf{q}})$. It is given by the solution of

$$a_{11}q_1 + a_{12}q_2 = a_{21}q_1 + a_{22}q_2 \qquad (q_2 = 1 - q_1)$$
$$b_{11}p_1 + b_{12}p_2 = b_{21}p_1 + b_{22}p_2 \qquad (p_2 = 1 - p_1)$$

i.e. by

$$\hat{p}_1 = \frac{E}{C - G + E} \qquad \hat{q}_1 = \frac{C}{2(G - E)} . \tag{10.4}$$

This equilibrium is not stable: if a fluctuation decreases, say, the amount of philandering males, then the payoff for the males will not change: each type has the same payoff, which depends only on the state of the female population. One cannot expect the frequency of philanderers to return to p_1. As to the female population, their payoff will even increase: but fast females gain more than coy females, since their risk of being deserted decreases. It is only when the amount of fast females increases that the male payoffs change. Again, they increase: but philanderers gain more than the faithful males; hence more philanderers, hence more coy females, hence fewer philanderers, and so on. This looks like an oscillating system. A static approach cannot deal with this situation.

10.3 A differential equation for asymmetric games

Let $\mathbf{x} \in S_n$ and $\mathbf{y} \in S_m$ denote the frequencies of the strategies for the players in position I resp. II. As in section 7.1, we may associate a differential equation with the game by assuming that the rate of increase $\dot{x}_i/x_i$ of strategy i is equal to the difference between its payoff $(A\mathbf{y})_i$ and the average payoff $\mathbf{x} \cdot A\mathbf{y}$ in the population X. This, and the corresponding assumption concerning type F_j, leads to the differential equation

$$\dot{x}_i = x_i((A\mathbf{y})_i - \mathbf{x} \cdot A\mathbf{y}) \qquad i = 1, \ldots, n \tag{10.5}$$
$$\dot{y}_j = y_j((B\mathbf{x})_j - \mathbf{y} \cdot B\mathbf{x}) \qquad j = 1, \ldots, m \tag{10.6}$$

on the invariant space $S_n \times S_m$.

Again, one checks easily that the boundary faces of $S_n \times S_m$ (i.e. the products of a face of S_n with a face of S_m) are invariant, and that the restriction of (10.5–6) to such a boundary face yields an equation of similar form. One obtains the boundary faces by setting some x_i or y_j equal to 0. Each such face in turn can be decomposed into boundary and interior, the boundary consisting of faces again. It is enough, therefore, to consider the restriction of (10.5–6) to subsets of the following form:

(a) at least one population consists of only one phenotype;
(b) both populations consist of several phenotypes.

This means:

(a) $x_i \equiv 1$ or $y_j \equiv 1$ for some i or some j;
(b) $x_i > 0$ for several i, and $y_j > 0$ for several j.

There is no loss of generality in studying only the restrictions of (10.5–6) to sets of the following type:

(a') $S_n \times \{\mathbf{f}_1\}$ with $\mathbf{f}_1 = \{1, 0, \ldots, 0\} \in S_m$;
(b') int $S_n \times S_m$.

All other restrictions of type (a) or (b) are of the same form as (a') or (b').

Exercise 10.3.1 (a') leads to

$$\dot{x}_i = x_i(a_i - \sum a_j x_j) \quad i = 1, \ldots, n \tag{10.7}$$

for constants a_i. Analyse this equation. Show that $x_i \to 0$ whenever a_i is not maximal.

It remains to consider case (b'). The rest points of (10.5–6) in the interior of $S_n \times S_m$ are the strictly positive solutions of the equations

$$(A\mathbf{y})_1 = \cdots = (A\mathbf{y})_n \qquad \sum_{j=1}^{m} y_j = 1 \tag{10.8}$$

$$(B\mathbf{x})_1 = \cdots = (B\mathbf{x})_m \qquad \sum_{i=1}^{n} x_i = 1 . \tag{10.9}$$

For $n > m$, (10.8) has solutions only if the matrix A is degenerate, while the solutions of (10.9) form a linear manifold of dimension at least $n - m$. The set of rest points in int $S_n \times S_m$ is thus either empty — this is the generic case — or it contains an $(n - m)$-dimensional subset.

An isolated rest point can thus exist only for $n = m$. If it exists, it is unique. We shall presently see that the rest points of (10.5–6) in int $S_n \times S_m$ cannot be sources or sinks.

Indeed (10.8) implies $(A\mathbf{y})_i = \mathbf{x} \cdot A\mathbf{y}$ and hence

$$\frac{\partial \dot{x}_i}{\partial x_j} = \frac{\partial}{\partial x_j} x_i ((A\mathbf{y})_i - \mathbf{x} \cdot A\mathbf{y}) = x_i \left(-\frac{\partial}{\partial x_j} (\mathbf{x} \cdot A\mathbf{y}) \right) . \tag{10.10}$$

But for $1 \leq j \leq n - 1$ one has

$$
\begin{aligned}
\frac{\partial}{\partial x_j}(\mathbf{x} \cdot A\mathbf{y}) &= \frac{\partial}{\partial x_j}(x_1(A\mathbf{y})_1 + \cdots + x_{n-1}(A\mathbf{y})_{n-1} \\
&\quad + (1 - x_1 - \cdots - x_{n-1})(A\mathbf{y})_n) \\
&= (A\mathbf{y})_j - (A\mathbf{y})_n = 0
\end{aligned}
\tag{10.11}
$$

and hence $\partial \dot{x}_i / \partial x_j = 0$ for $1 \leq i, j \leq n - 1$. A similar relation holds for the y_j, so that the Jacobian of (10.5–6) at a rest point in int $S_n \times S_m$ takes (after elimination of x_n and y_m) the form

$$
J = \begin{bmatrix} 0 & C \\ D & 0 \end{bmatrix},
$$

where the two blocks of zeros in the diagonal are an $(n - 1) \times (n - 1)$ and an $(m - 1) \times (m - 1)$ matrix respectively. For the characteristic polynomial $p(\lambda) = \det(J - \lambda I)$ (where I is the identity matrix) one obtains

$$
p(\lambda) = (-1)^{n+m} p(-\lambda)
\tag{10.12}
$$

as can be seen by changing the signs of the first $n - 1$ columns and the last $m - 1$ rows of $J - \lambda I$. Thus if λ is an eigenvalue of J, then $-\lambda$ is also an eigenvalue. Sinks and sources are therefore excluded. In particular, (10.5–6) admits sinks only at the corners of $S_n \times S_m$. This corresponds to the fact that mixed evolutionarily stable states are excluded for asymmetric games. We shall see in section 11.3 that mixed strategies cannot be asymptotically stable.

Exercise 10.3.2 If the ω-limit of an orbit of (10.5–6) is in int $S_n \times S_m$, then show that the time average exists and corresponds to a rest point in int $S_n \times S_m$.

Exercise 10.3.3 Show that if there is no rest point in int $S_n \times S_m$, then all orbits converge to bd $S_n \times S_m$. (Hint: compare with exercise 7.6.2.)

Exercise 10.3.4 If there is a manifold of fixed points in int $S_n \times S_m$, then show that there exists a corresponding decomposition of the state space into invariant manifolds. (Hint: construct invariants of motion.)

Exercise 10.3.5 Show that the characteristic polynomial of J is given by

$$
p(\lambda) = (-\lambda)^{n-m} \det(\lambda^2 I - DC) .
\tag{10.13}
$$

Hence nonzero eigenvalues occur as real pairs, imaginary pairs, or complex quadruples.

Exercise 10.3.6 In a bimatrix game given by the payoff matrices A and B, the Nash equilibria are rest points of (10.5–6). Show that every Lyapunov stable fixed point is a Nash equilibrium, and so is every fixed point which is the ω-limit of an interior orbit. (Hint: see theorem 7.2.1.)

10.4 The case of two players and two strategies

Let us now take a closer look at the case $n = m = 2$, which occurs in the Battle of the Sexes. Since we may add, as in the symmetric case, a constant to every column of A and B (cf. exercise 7.1.2), we may assume without loss of generality that the diagonal terms are zero. Hence the matrices are

$$ A = \begin{bmatrix} 0 & a_{12} \\ a_{21} & 0 \end{bmatrix} \qquad B = \begin{bmatrix} 0 & b_{12} \\ b_{21} & 0 \end{bmatrix} . \qquad (10.14) $$

Since $x_2 = 1 - x_1$ and $y_2 = 1 - y_1$, it is enough to consider the variables x_1 and y_1, which we shall denote by x and y. (10.5–6) now becomes

$$ \dot{x} = x(1-x)(a_{12} - (a_{12} + a_{21})y) \qquad (10.15) $$
$$ \dot{y} = y(1-y)(b_{12} - (b_{12} + b_{21})x) \qquad (10.16) $$

on the square $Q = \{(x,y) : 0 \leq x, y \leq 1\} \cong S_2 \times S_2$.

If $a_{12}a_{21} \leq 0$, then $\dot{x}$ does not change its sign in Q. One of the two strategies of player I dominates the other. In this case x is either constant, or converges monotonically to 0 or 1. A similar result holds for $b_{12}b_{21} \leq 0$. Thus it only remains to investigate the case when $a_{12}a_{21} > 0$ and $b_{12}b_{21} > 0$. In this case (10.15–16) admits a unique rest point in int Q, namely

$$ \mathbf{F} = \left(\frac{b_{12}}{b_{12} + b_{21}}, \frac{a_{12}}{a_{12} + a_{21}} \right) . $$

The Jacobian at $\mathbf{F}$ is

$$ A = \begin{bmatrix} 0 & -(a_{12} + a_{21})\frac{b_{12}b_{21}}{(b_{12}+b_{21})^2} \\ -(b_{12} + b_{21})\frac{a_{12}a_{21}}{(a_{12}+a_{21})^2} & 0 \end{bmatrix} \qquad (10.17) $$

and the eigenvalues are $\pm\lambda$ with

$$ \lambda^2 = \frac{a_{12}a_{21}b_{12}b_{21}}{(a_{12} + a_{21})(b_{12} + b_{21})} . \qquad (10.18) $$

If $a_{12}b_{12} > 0$, then $\mathbf{F}$ is a saddle, and almost all orbits in int Q converge to one or the other of two opposite corners of Q. (See fig. 10.1.)

If $a_{12}b_{12} < 0$, then the eigenvalues are purely imaginary, and all orbits in int Q are periodic orbits surrounding $\mathbf{F}$ (see fig. 10.2), as can best be seen by

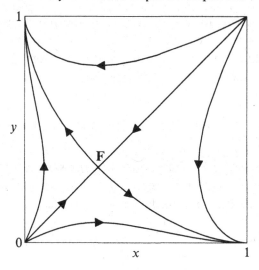

Fig. 10.1.

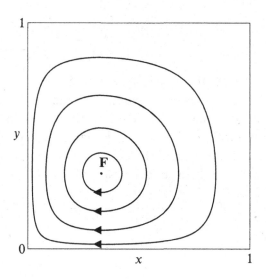

Fig. 10.2.

dividing the right hand side of (10.15–16) by the function $xy(1 - x)(1 - y)$, which is positive in int Q. This corresponds to a change in velocity and does not affect the orbits (see exercises 4.1.2 and 4.1.3). In int Q, we obtain

$$\dot{x} = \frac{a_{12} - (a_{12} + a_{21})y}{y(1 - y)} = \frac{a_{12}}{y} - \frac{a_{21}}{1 - y}, \tag{10.19}$$

$$\dot{y} = \frac{b_{12}}{x} - \frac{b_{21}}{1 - x}. \tag{10.20}$$

This system is *Hamiltonian* in int Q, i.e. it can be written as

$$\dot{x} = \frac{\partial H}{\partial y} \qquad \dot{y} = -\frac{\partial H}{\partial x} \qquad (10.21)$$

with

$$H(x, y) = a_{12} \log y + a_{21} \log(1 - y) - b_{12} \log x - b_{21} \log(1 - x) . \qquad (10.22)$$

H is a constant of motion for (10.15–16), since

$$\dot{H} = \frac{\partial H}{\partial x} \dot{x} + \frac{\partial H}{\partial y} \dot{y} \equiv 0 . \qquad (10.23)$$

Now if (10.15–16) admits a rest point $\mathbf{F}$ in int Q with imaginary eigenvalues, then the diverse conditions on the signs of the a_{ij} and b_{ji} imply that H attains its unique extremum in int Q at the point $\mathbf{F}$. All other orbits are periodic and surround $\mathbf{F}$.

Exercise 10.4.1 Show that for the time average along a periodic orbit with period T, one has

$$\left(\frac{1}{T} \int_0^T x(t)dt, \frac{1}{T} \int_0^T y(t)dt \right) = \mathbf{F} .$$

Let us return now to the Battle of the Sexes described in section 10.2. After the addition of appropriate constants, the matrices A and B from (10.3) turn into

$$A = \begin{bmatrix} 0 & \frac{C}{2} \\ G - \frac{C}{2} - E & 0 \end{bmatrix} \qquad B = \begin{bmatrix} 0 & -E \\ G - C & 0 \end{bmatrix} . \qquad (10.24)$$

The Nash equilibrium given by (10.4) is a centre surrounded by periodic orbits. Equation (10.15–16) takes the form

$$\dot{x} = x(1 - x)\left(\frac{C}{2} - (G - E)y \right) \qquad \dot{y} = y(1 - y)(-E + (C + E - G)x) . \qquad (10.25)$$

Up to the factors $(1 - x)$ and $(1 - y)$, this looks like the Lotka–Volterra equation (2.1). We conclude that the relations between the sexes resemble those between predators and prey and are subject to perpetual oscillations.

An interesting phenomenon occurs if the game is symmetric, i.e. if in (10.14) $a_{12} = b_{12} = a$ and $a_{21} = b_{21} = b$. In this case, the diagonal of Q (where $x = y$) is invariant and the dynamics, there, is that of the one-dimensional replicator equation $\dot{x} = x(1-x)(a-(a+b)x)$. Let us consider the generic case $ab \neq 0$. The interior fixed point, if it exists, is always a saddle.

If a pure strategy is a Nash equilibrium in the symmetric game, it will also be a Nash equilibrium of the corresponding bimatrix game, and there will be no asymmetric Nash equilibria. But if $a > 0$ and $b > 0$, the symmetric game has a unique *mixed* Nash equilibrium, and the corresponding point on the diagonal, which is a symmetric Nash equilibrium, will be a saddle. The orbits of the replicator equation flow away from the diagonal, which is a separatrix; orbits above the diagonal converge to $(0, 1)$ and orbits below the diagonal to $(1, 0)$ (see fig. 10.1). These off-diagonal corners of Q are additional Nash equilibria of the asymmetric game.

This occurs, for instance, in the hawk–dove game. As long as the players perceive each other in the same role, they are engaged in a symmetric game with a mixed ESS. But as soon as they perceive their roles as asymmetrical (for instance, if one is the owner of a contested territory and the other an intruder), this mixed ESS is replaced by two pure ESS: 'If owner, escalate; if intruder, retreat' or, alternatively, 'if owner, retreat; if intruder, escalate'. Which of the two strategies will actually be adopted in the population depends on the initial condition. In most real-life examples, the former ESS, which is usually termed Bourgeois, seems to prevail; but apparently, some spiders use the other, at first glance paradoxical ESS ('expropriate the expropriators'). In any case, escalated conflicts are avoided.

10.5 Role games

So far, we have assumed that the two populations are separate. This is appropriate for games between parasite and host, for instance, but hardly so for games between owner and intruder, or between parent and offspring, or between White and Black at chess, where one individual is sometimes in one position and sometimes in the other. It is also possible that for male–female or worker–queen conflicts, the genetic programs for the two roles are linked in the form of conditional strategies (for example: if male, be philanderer; if female, be coy).

We shall only consider the simplest possible case, where there are two strategies for each position: $\mathbf{e}_1$ and $\mathbf{e}_2$ for position I and $\mathbf{f}_1$ and $\mathbf{f}_2$ for II. Any individual will be in position I with probability p and in position II with probability $1 - p$. We assume that an individual's position is independent of the strategy adopted, and that individuals in one position interact only with those in the other position; interactions occur in random pairs.

The population will consist of four *behavioural types*: $\mathbf{G}_1 = \mathbf{e}_1\mathbf{f}_1$ (i.e. play $\mathbf{e}_1$ if in position I and $\mathbf{f}_1$ if in position II), $\mathbf{G}_2 = \mathbf{e}_2\mathbf{f}_1$, $\mathbf{G}_3 = \mathbf{e}_1\mathbf{f}_2$, and $\mathbf{G}_4 = \mathbf{e}_2\mathbf{f}_2$, with frequencies x_1 to x_4, respectively. The state of the population is given

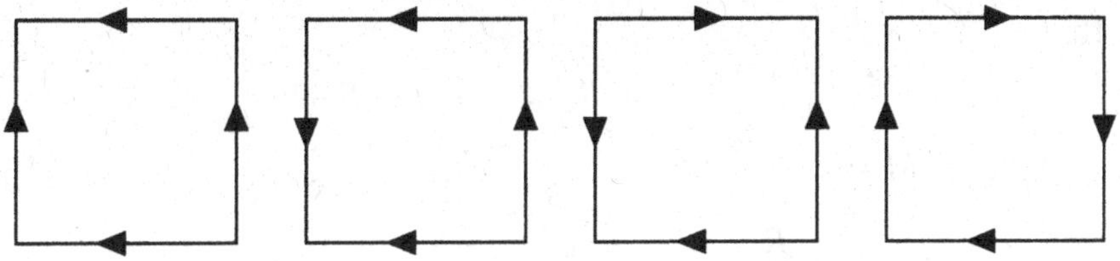

Fig. 10.3.

by a point $\mathbf{x} \in S_4$. The 2×2 payoff matrices for players in role I resp. II will be denoted by U resp. W.

Exercise 10.5.1 Let Γ be the sequence of four edges connecting the corners $\mathbf{G}_1$ to $\mathbf{G}_2$ to $\mathbf{G}_4$ to $\mathbf{G}_3$ and back to $\mathbf{G}_1$. Each edge connects two types using the same option in one position and different ones in the other. Generally, one of the alternatives is the better one, and we orient the edge accordingly. Show that this leads to the four possible orientations shown in fig. 10.3.

The payoff matrix is

$$M = \begin{bmatrix} A+a & A+b & B+a & B+b \\ C+a & C+b & D+a & D+b \\ A+c & A+d & B+c & B+d \\ C+c & C+d & D+c & D+d \end{bmatrix}, \qquad (10.26)$$

where

$$pU = \begin{bmatrix} A & B \\ C & D \end{bmatrix} \quad \text{and} \quad (1-p)W = \begin{bmatrix} a & b \\ c & d \end{bmatrix}. \qquad (10.27)$$

For example, a $\mathbf{G}_2$-individual is in position I with probability p: its strategy, then, is $\mathbf{e}_2$; it interacts only with individuals in position II of which x_j are of type $\mathbf{G}_j$. Thus its expected payoff is

$$\text{(payoff for } \mathbf{G}_2|\text{position I)} = x_1 u_{21} + x_2 u_{21} + x_3 u_{22} + x_4 u_{22}.$$

With probability $1-p$, the $\mathbf{G}_2$-individual is in role II, plays strategy $\mathbf{f}_1$, and obtains $x_1 w_{11} + x_2 w_{12} + x_3 w_{11} + x_4 w_{12}$. The expected payoff is therefore given by $\sum x_j M_{2j} = (M\mathbf{x})_2$.

We can now apply the usual replicator equation (7.3), i.e.

$$\dot{x}_i = x_i((M\mathbf{x})_i - \mathbf{x} \cdot M\mathbf{x}) \qquad (10.28)$$

and assume without loss of generality that the diagonal terms of M are all

0, so that

$$M = \begin{bmatrix} 0 & -R & -r & S+s \\ R & 0 & -S-r & s \\ r & -R-s & 0 & S \\ R+r & -s & -S & 0 \end{bmatrix}, \qquad (10.29)$$

where $R = C - A$, $S = B - D$, $r = c - a$ and $s = b - d$.

One checks immediately that the ratio $(x_2 x_3)^{-1} x_1 x_4$ is an invariant of motion for (10.28). Each equation $x_1 x_4 = K x_2 x_3$ (for $K > 0$) defines a saddle-shaped surface W_K in int S_4 which is bounded by Γ, the sequence of edges connecting $\mathbf{G}_1$ to $\mathbf{G}_2$ to $\mathbf{G}_4$ to $\mathbf{G}_3$ to $\mathbf{G}_1$.

Exercise 10.5.2 Show that the fixed points in int S_4 must satisfy

$$x_1 + x_2 = \frac{S}{R+S} \qquad x_1 + x_3 = \frac{s}{r+s}$$

if the denominators do not vanish. The corresponding line can be written as

$$x_i = m_i + \mu \quad (i = 1, 4) \quad \text{and} \quad x_i = m_i - \mu, \quad (i = 2, 3)$$

with μ as parameter and

$$\mathbf{m} = \frac{1}{(R+S)(r+s)}(Ss, Sr, Rs, Rr).$$

This line intersects int S_4 if and only if $\mathbf{m} \in W_1$, i.e. if and only if $RS > 0$ and $rs > 0$. Thus either every invariant surface W_K $(0 < K < \infty)$ contains a fixed point or none does. There are generically three possible types of dynamics:

(a) *no rest point in the interior.* Then there exists a corner of S_4 attracting all orbits in int S_4. This is the case of *global stability.* It corresponds to cases (1) and (2) in fig. 10.3.

(b) *a line of rest points in the interior, and $Rr > 0$.* Each rest point is a saddle on the corresponding invariant face W_K. This is the case of *bistability*: up to a set of measure 0, all initial conditions lead to one of two opposite corners (see case (3) in fig. 10.3).

(c) *a line of rest points in the interior, and $Rr < 0$.* This is the *cyclic* case (see case (4) in fig. 10.3): $\mathbf{G}_1$ beats $\mathbf{G}_2$, which beats $\mathbf{G}_4$, which beats $\mathbf{G}_3$, which in turn beats $\mathbf{G}_1$ (or the other way round). The Jacobian at the inner rest points has a pair of complex eigenvalues. On W_1, these are pure imaginary, and

$$S \log(x_1 + x_2) + R \log(x_3 + x_4) - s \log(x_1 + x_3) - r \log(x_2 + x_4)$$

is an invariant of motion, so that all orbits are periodic. For $K > 1$ the fixed point is a spiral sink, and for $K \in\ ]0, 1[$ a spiral source, or vice versa.

Exercise 10.5.3 Show all this. In case (c), check that off W_1, no periodic orbits exist: all orbits have the heteroclinic cycle Γ as their α- or ω-limit. Show that this case corresponds to the Battle of the Sexes.

Exercise 10.5.4 Show that a strategy for the game with matrix (10.26) is evolutionarily stable if and only if it corresponds to a strict Nash equilibrium; in particular (see Exercise 10.1.1) it must be one of the pure strategies G_1 to G_4.

Exercise 10.5.5 Analyse the case of an underlying symmetric game (i.e. $pU = (1 - p)W$) along the lines of the previous section. What happens to the symmetric Nash equilibria?

Exercise 10.5.6 Analyse the best-response dynamics for the game (10.26). Show that in the cyclic case (c) every best-response path outside the line of equilibria converges to **m**.

10.6 Notes

The Battle of the Sexes is due to Dawkins (1989). A similar game is described in Magurran and Nowak (1991), see also Sigmund (1993). Selten (1980) proved in a very general framework that evolutionarily stable strategies for asymmetric games are never mixed. For an application to the war of attrition see Hammerstein and Parker (1982). Equation (10.5–6) is analysed in Schuster *et al.* (1981b). Hofbauer (1996) derives its basic properties from a transformation to a 'bipartite system', and proves generic instability of interior equilibria in 3×3 games. The special case of the Battle of the Sexes was studied in Schuster and Sigmund (1981) and — under a discrete time dynamics — in Eshel and Akin (1983). A genetic model is studied in Maynard Smith and Hofbauer (1987). Game dynamics for asymmetric conflicts with self-interaction are studied in Taylor (1979), Schuster *et al.* (1981c) and Cressman *et al.* (1986). Cressman (1992, 1996) derives the correct notion of evolutionary stability for such games and proves dynamic stability. Section 10.5 is from Gaunersdorfer *et al.* (1991), see also Cressman *et al.* (1996). Exercise 10.5.6 is from Berger (1997a).

11

More on bimatrix games

There are alternative candidates for the dynamics of asymmetric games. We analyse, in particular, rescaled partnership games, which lead to gradient systems, and zero-sum games. The usual replicator dynamics is volume-preserving, whereas a modification is contracting. Finally, we discuss Nash–Pareto strategies, which provide an ersatz for mixed ESS in the asymmetric case.

11.1 Dynamics for bimatrix games

We have motivated equations (10.5–6), i.e. the replicator dynamics for an asymmetric game

$$\begin{aligned}
\dot{x}_i &= x_i((A\mathbf{y})_i - \mathbf{x} \cdot A\mathbf{y}) \\
\dot{y}_j &= y_j((B\mathbf{x})_j - \mathbf{y} \cdot B\mathbf{x})
\end{aligned} \tag{11.1}$$

by analogy with the replicator dynamics (7.3) for symmetric games.

A different dynamics including normalization by mean payoff, namely

$$\begin{aligned}
\dot{x}_i &= x_i \frac{(A\mathbf{y})_i - \mathbf{x} \cdot A\mathbf{y}}{\mathbf{x} \cdot A\mathbf{y}} \\
\dot{y}_j &= y_j \frac{(B\mathbf{x})_j - \mathbf{y} \cdot B\mathbf{x}}{\mathbf{y} \cdot B\mathbf{x}} ,
\end{aligned} \tag{11.2}$$

is motivated by the discrete time model. In discrete time, one assumes that the frequency x_i' of the E_i-players in the next generation is proportional to $x_i(A\mathbf{x})_i$. Since $\sum x_i' = 1$, the multiplication rate from one generation to the next has to be normalized by the mean payoff for the respective population. This gives

$$x_i' = x_i \frac{(A\mathbf{y})_i}{\mathbf{x} \cdot A\mathbf{y}} \qquad y_j' = y_j \frac{(B\mathbf{x})_j}{\mathbf{y} \cdot B\mathbf{x}} \quad . \tag{11.3}$$

The above differential equation (11.2) then follows from the straightforward approximation $x_i' - x_i \sim \dot{x}_i$. Since in general $\mathbf{x} \cdot A\mathbf{y} \neq \mathbf{y} \cdot B\mathbf{x}$ except for partnership games (i.e. for $A^t = B$), (11.2) cannot be reduced to (11.1) by a change of velocity. In fact we will observe soon that the replicator equation (11.1) and the modified replicator equation (11.2) show quite different qualitative behaviour.

Equations (11.3) and (11.2) are well defined on $S_n \times S_m$ provided the payoffs a_{ij} and b_{ji} are positive numbers, a restriction which is not needed for (11.1). The reason for this is the different meaning of payoffs: In the discrete model (11.3) and hence also in its offshoot (11.2), the a_{ij} are interpreted as *multiplication rates* from one generation to the next, whereas the a_{ij} in (11.1) measure only the *changes* in fitness caused by the game. In order to obtain a better understanding of the relation between (11.1) and the other two equations, we should therefore add some large constant C, measuring the common 'background fitness' of the individuals, to the payoffs a_{ij} and b_{ji} in (11.2) and (11.3). It is then easy to see that (11.1) is the limiting case of (11.2) as $C \to \infty$ (after a compression of the time-scale with factor C). This suggests that the dynamics of (11.1) should be somewhat simpler than that of the other two equations.

Exercise 11.1.1 Show that all three dynamics have the same fixed points.

Exercise 11.1.2 Compute the linearization of the modified replicator dynamics (11.2) and derive a result for the position of the eigenvalues similar to the one shown for (11.1) in Exercise 10.3.5. Prove that an isolated interior fixed point is always unstable for the discrete dynamics (11.3). Only pure states can be stable for (11.3).

Exercise 11.1.3 If no interior rest point exists, then all interior orbits of (11.2) converge to the boundary of $S_n \times S_m$. (For (11.1) cf. exercise 10.3.3.)

Exercise 11.1.4 A Nash equilibrium is a fixed point for any of the three dynamics, but the converse does not hold. Lyapunov stable fixed points and limit points of interior orbits are Nash equilibria.

11.2 Partnership games and zero-sum games

Equation (11.1) is not only simpler than the two other equations, it is also immune to rescalings of the payoffs by the addition of arbitrary constants to the columns of the payoff matrices A and B. More generally, a game

(A', B') is said to be a *rescaling* of (A, B) if there exist constants c_j, d_i and $\alpha > 0, \beta > 0$ such that

$$a'_{ij} = \alpha a_{ij} + c_j \qquad b'_{ji} = \beta b_{ji} + d_i \quad . \tag{11.4}$$

We then write $(A, B) \sim (A', B')$.

Exercise 11.2.1 Show that rescaling a game does not change its Nash equilibrium points.

Rescalings are of particular interest for partnership games, where the payoff is equally shared between the two players, and for zero-sum games, where the gain of one player is the loss of the other. Thus (A, B) is a *rescaled partnership game* if $(A, B) \sim (C, C^t)$ and a *rescaled zero-sum game* if $(A, B) \sim (C, -C^t)$ for some suitable $n \times m$ matrix C. More precisely, we shall say that (A, B) is a *c-partnership game* (with $c > 0$) or a *c-zero-sum game* (with $c < 0$) if there exist suitable c_{ij}, c_j and d_i such that

$$a_{ij} = c_{ij} + c_j \quad b_{ji} = c c_{ij} + d_i \tag{11.5}$$

for all i and j.

We shall first show that c-partnership games correspond exactly to certain gradient systems (see section 7.8). The tangent space at a point $(\mathbf{x}, \mathbf{y}) \in \text{int } S_n \times S_m$ consists of the vectors $(\boldsymbol{\xi}, \boldsymbol{\eta})$ with $\boldsymbol{\xi} \in \mathbb{R}_0^n$ and $\boldsymbol{\eta} \in \mathbb{R}_0^m$ (i.e. $\sum \xi_i = \sum \eta_j = 0$). As inner product in the tangent space, we define

$$\langle (\boldsymbol{\xi}, \boldsymbol{\eta}), (\boldsymbol{\xi}', \boldsymbol{\eta}') \rangle_{(\mathbf{x}, \mathbf{y})} = \sum \frac{\xi_i \xi'_i}{x_i} + \frac{1}{c} \sum \frac{\eta_j \eta'_j}{y_j} \quad . \tag{11.6}$$

The corresponding gradient is said to be a *c-gradient*.

Theorem 11.2.2 *The following conditions are equivalent:*
(RP1) *(A, B) is a c-partnership game;*
(RP2) *(11.1) is a c-gradient;*
(RP3) *$c\boldsymbol{\xi} \cdot A\boldsymbol{\eta} = \boldsymbol{\eta} \cdot B\boldsymbol{\xi}$ for all $\boldsymbol{\xi} \in \mathbb{R}_0^n$ and $\boldsymbol{\eta} \in \mathbb{R}_0^m$;*
(RP4) *For all $i, k \in \{1, \ldots, n\}$ and $j, l \in \{1, \ldots, m\}$*

$$c \left(a_{ij} - a_{il} - a_{kj} + a_{kl} \right) = b_{ji} - b_{li} - b_{jk} + b_{lk} \quad ; \tag{11.7}$$

(RP5) *There exist u_i, v_j such that $Q = cA - B^t$ satisfies $q_{ij} = u_i + v_j$ for all i and j.*

The potential of (11.1) is then given by $\mathbf{x} \cdot C\mathbf{y}$. If (A, B) is a partnership game, then this is just the average payoff, which therefore increases along every orbit.

Proof (RP1) $\Rightarrow$ (RP2). For every $(\boldsymbol{\xi}, \boldsymbol{\eta}) \in \mathbb{R}_0^n \times \mathbb{R}_0^m$ we have, according to (11.6):

$$\langle (\dot{\mathbf{x}}, \dot{\mathbf{y}}), (\boldsymbol{\xi}, \boldsymbol{\eta}) \rangle_{(\mathbf{x},\mathbf{y})} = \sum \xi_i [(A\mathbf{y})_i - \mathbf{x} \cdot A\mathbf{y}] + \frac{1}{c} \sum \eta_j [(B\mathbf{x})_j - \mathbf{y} \cdot B\mathbf{x}]$$

$$= \boldsymbol{\xi} \cdot C\mathbf{y} + \boldsymbol{\eta} \cdot C^t \mathbf{x} = \boldsymbol{\xi} \cdot C\mathbf{y} + \mathbf{x} \cdot C\boldsymbol{\eta} \quad .$$

This last expression is just $D_{(\mathbf{x},\mathbf{y})}V(\boldsymbol{\xi}, \boldsymbol{\eta})$, where V is the function $(\mathbf{x}, \mathbf{y}) \to \mathbf{x}C\mathbf{y}$.

(RP2) $\Rightarrow$ (RP3). If

$$\langle (\dot{\mathbf{x}}, \dot{\mathbf{y}}), (\boldsymbol{\xi}, \boldsymbol{\eta}) \rangle_{(\mathbf{x},\mathbf{y})} = D_{(\mathbf{x},\mathbf{y})}V(\boldsymbol{\xi}, \boldsymbol{\eta})$$

for some function V defined in a neighbourhood of int $S_n \times S_m$, then

$$\sum \xi_i [(A\mathbf{y})_i - \mathbf{x} \cdot A\mathbf{y}] + \frac{1}{c} \sum \eta_j [(B\mathbf{x})_j - \mathbf{y} \cdot B\mathbf{x}] = \sum \frac{\partial V}{\partial x_i} \xi_i + \sum \frac{\partial V}{\partial y_j} \eta_j$$

for all $\boldsymbol{\xi} \in \mathbb{R}_0^n, \boldsymbol{\eta} \in \mathbb{R}_0^m$. Hence

$$\sum \xi_i (A\mathbf{y})_i = \sum \frac{\partial V}{\partial x_i} \xi_i$$

for all $\boldsymbol{\xi} \in \mathbb{R}_0^n$, which implies

$$\frac{\partial V}{\partial x_i}(\mathbf{x}, \mathbf{y}) = (A\mathbf{y})_i + g(\mathbf{x}, \mathbf{y})$$

for a suitable function g which does not depend on i, and similarly

$$\frac{\partial V}{\partial y_j}(\mathbf{x}, \mathbf{y}) = \frac{1}{c}(B\mathbf{x})_j + h(\mathbf{x}, \mathbf{y}) \quad .$$

Differentiating again, we obtain

$$a_{ij} = \frac{\partial^2 V}{\partial x_i \partial y_j} - \frac{\partial g}{\partial y_j} \quad \text{and} \quad \frac{1}{c} b_{ji} = \frac{\partial^2 V}{\partial y_j \partial x_i} - \frac{\partial h}{\partial x_i} \quad .$$

This implies that $Q = cA - B^t$ satisfies $\boldsymbol{\xi} \cdot Q\boldsymbol{\eta} = 0$ for all $\boldsymbol{\xi} \in \mathbb{R}_0^n$ and $\boldsymbol{\eta} \in \mathbb{R}_0^m$.

(RP3) $\Rightarrow$ (RP4). This follows by choosing $\boldsymbol{\xi} = \mathbf{e}_i - \mathbf{e}_k$ (where $\mathbf{e}_i, \mathbf{e}_k$ are from the standard basis in $\mathbb{R}^n$) and $\boldsymbol{\eta} = \mathbf{f}_j - \mathbf{f}_l$ (with $\mathbf{f}_j, \mathbf{f}_l$ from the standard basis in $\mathbb{R}^m$).

(RP4) $\Rightarrow$ (RP5). By (11.7) we have for $Q = cA - B^t$

$$q_{ij} - q_{il} - q_{kj} + q_{kl} = 0 \quad .$$

With $k = n, l = m$ this implies

$$q_{ij} = q_{im} + q_{nj} - q_{nm} = u_i + v_j$$

where $u_i = q_{im}$ and $v_j = q_{nj} - q_{nm}$.

(RP5) $\Rightarrow$ (RP1). Since $ca_{ij} - b_{ji} = u_i + v_j$, we have only to set $c_{ij} = c^{-1}(b_{ji} + u_i)$ to obtain (11.5). $\qquad\qquad\qquad\qquad\qquad\qquad\qquad\qquad\square$

Exercise 11.2.3 Show that a strict maximum of $\mathbf{x} \cdot C \mathbf{y}$ on $S_n \times S_m$ corresponds to a pure strategy. Deduce that every rescaled partnership game has a pure Nash equilibrium.

Exercise 11.2.4 If (A, B) is a c-partnership game, then (11.2) is a gradient. (Hint: the definition of the inner product has to be changed, by replacing, in the right hand side of (11.6), x_i by $x_i(\mathbf{x} \cdot A\mathbf{y})^{-1}$ and y_j by $y_j(\mathbf{y} \cdot B\mathbf{x})^{-1}$.)

Theorem 11.2.5 *If* (11.5) *is valid for some* $c \in \mathbb{R}$, *and if* (A, B) *has an interior Nash equilibrium* $(\mathbf{p}, \mathbf{q})$, *then the function*

$$H(\mathbf{x}, \mathbf{y}) = c \sum p_i \log x_i - \sum q_j \log y_j \qquad (11.8)$$

is a constant of motion for (11.1).

Proof

$$
\begin{aligned}
\dot{H}(\mathbf{x}, \mathbf{y}) &= c \sum p_i \frac{\dot{x}_i}{x_i} - \sum q_j \frac{\dot{y}_j}{y_j} \\
&= c(\mathbf{p} - \mathbf{x}) \cdot A\mathbf{y} - (\mathbf{q} - \mathbf{y}) \cdot B\mathbf{x} \\
&= c(\mathbf{p} - \mathbf{x}) \cdot A(\mathbf{y} - \mathbf{q}) - (\mathbf{q} - \mathbf{y}) \cdot B(\mathbf{x} - \mathbf{p}) = 0 \quad.
\end{aligned}
$$

$\qquad\qquad\qquad\qquad\qquad\qquad\qquad\qquad\qquad\qquad\qquad\qquad\qquad\square$

   In particular, an interior rest point $(\mathbf{p}, \mathbf{q})$ for a c-zero-sum game is always stable (but not asymptotically stable).

Exercise 11.2.6 Prove that in a rescaled zero-sum game, the time average of every interior orbit of (11.1) converges to $(\mathbf{p}, \mathbf{q})$.

Exercise 11.2.7 Show that the Battle of the Sexes from section (10.2) is a rescaled zero-sum game.

Exercise 11.2.8 Show that all eigenvalues at an interior rest point of a zero-sum game are purely imaginary.

Exercise 11.2.9 Show that a game (A, B) is a c-zero-sum game if and only if (11.7) holds for all i, j, k, l. (Here, $c < 0$.)

Exercise 11.2.10 Analyse the behaviour of (11.1) for a rescaled zero-sum game without interior equilibrium, showing that (11.8) is a Lyapunov function, if $(\mathbf{p}, \mathbf{q})$ is a Nash equilibrium on the boundary.

Condition (11.7) can be interpreted in terms of populations of players playing mixed strategies. Let us assume that N mixed strategies $\mathbf{p}^i \in S_n$ are used in position I and occur with frequencies x_i ($i = 1, \ldots, N$); similarly, that M mixed strategies $\mathbf{q}^j \in S_m$ are used in position II and occur with frequencies y_j ($j = 1, \ldots, M$). The mean strategy in position I is given by $\mathbf{p} = \sum x_i \mathbf{p}^i$ and that in position II by $\mathbf{q} = \sum y_j \mathbf{q}^j$. Since the payoff for the mixed strategy $\mathbf{p}^i$ is given by $\mathbf{p}^i \cdot A\mathbf{q}$ and that for $\mathbf{q}^j$ by $\mathbf{q}^j \cdot B\mathbf{p}$, the replicator equation for the *mixed strategist game* is given by

$$
\begin{aligned}
\dot{x}_i &= x_i((\bar{A}\mathbf{y})_i - \mathbf{x} \cdot \bar{A}\mathbf{y}) \\
\dot{y}_j &= y_j((\bar{B}\mathbf{x})_j - \mathbf{y} \cdot \bar{B}\mathbf{x})
\end{aligned}
$$

where $\bar{a}_{ij} = \mathbf{p}^i \cdot A\mathbf{q}^j$ and $\bar{b}_{ji} = \mathbf{q}^j \cdot B\mathbf{p}^i$. Thus $\bar{A} = PAQ^t$ and $\bar{B} = QBP^t$, where P is the stochastic $N \times n$ matrix with rows $\mathbf{p}^i$ and Q is the stochastic $M \times m$ matrix with rows $\mathbf{q}^j$. (A nonnegative matrix is *stochastic* if each row sums to 1.)

Exercise 11.2.11 If (A', B') is a rescaling of (A, B), then $(\bar{A}', \bar{B}')$ is a rescaling of $(\bar{A}, \bar{B})$. If (A, B) is a c-partnership game (or a c-zero sum game), then so are all mixed strategist games (PAQ^t, QBP^t), for all stochastic $N \times n$ matrices P and $M \times m$ matrices Q.

Exercise 11.2.12 Show that (A, B) is a c-partnership game (or a c-zero-sum game) if and only if the mixed strategist games (PAQ^t, QBP^t) are c-partnership games (resp. c-zero sum games), for all $2 \times n$ matrices P having as rows the basis vectors $\mathbf{e}_i$ and $\mathbf{e}_k$ from $\mathbb{R}^n$, and all $2 \times m$ matrices Q having as rows the basis vectors $\mathbf{f}_j$ and $\mathbf{f}_l$ from $\mathbb{R}^m$. (Hint: in this case

$$
\bar{A} = \begin{bmatrix} a_{ij} & a_{il} \\ a_{kj} & a_{kl} \end{bmatrix} \qquad \bar{B} = \begin{bmatrix} b_{ji} & b_{jk} \\ b_{li} & b_{lk} \end{bmatrix}
$$

and $(\bar{A}, \bar{B})$ is a c-partnership game (resp. a c-zero-sum game) if and only if (11.7) is valid for some $c > 0$ (resp. $c < 0$.)

In particular, the following condition is equivalent to (RP1)–(RP5):

(RP6) *The mixed strategist games* (PAQ^t, QBP^t) *are c-partnership games, for all stochastic $N \times n$ matrices P and all stochastic $M \times m$ matrices Q.)*

The dynamic behaviour of the modified replicator equation (11.2) for general rescaled zero-sum games is not known. The results in section 11.4 suggest that a Nash equilibrium is asymptotically stable. A partial result in this direction is obtained in the following exercise.

Exercise 11.2.13 Show for $\beta a_{ij} + \alpha b_{ji} = $ const, that (11.8) is, for suitable $c > 0$, a Lyapunov function for (11.2).

Exercise 11.2.14 Show that generically a 2×2 game is either a rescaled zero-sum game or a rescaled partnership game.

Exercise 11.2.15 As a typical example for a game which can be rescaled neither to a zero-sum nor to a partnership game, consider the 3×3 game (A, B) with

$$A = \begin{pmatrix} 1 & 0 & 0 \\ 0 & 1 & 0 \\ 0 & 0 & 1 \end{pmatrix} \quad \text{and} \quad B = \begin{pmatrix} a & b & c \\ c & a & b \\ b & c & a \end{pmatrix}$$

Under what conditions is (A, B) a rescaled zero-sum game? When does (11.1) have an attractor on the boundary? When does a Hopf bifurcation occur?

11.3 Conservation of volume

In higher dimensions, the replicator equation (11.1) need not admit a constant of motion, but it will still preserve some (properly modified) volume.

Let us first recall the *Liouville formula*: If $\dot{\mathbf{x}} = \mathbf{f}(\mathbf{x})$ is defined on the open set U in $\mathbb{R}^n$ and if $G \subset U$ has volume V, then the volume $V(t)$ of $G(t) = \{\mathbf{y} = \mathbf{x}(t) : \mathbf{x} \in G\}$ satisfies

$$\dot{V}(t) = \int_{G(t)} \operatorname{div} \mathbf{f}(\mathbf{x}) d(x_1, \ldots, x_n) \tag{11.9}$$

Indeed, the substitution rule for integrals implies that if $\mathbf{g} : G \to \mathbf{g}(G)$ is invertible and $\det D_{\mathbf{x}}\mathbf{g} > 0$ for all $\mathbf{x} \in G$, then the volume of $\mathbf{g}(G)$ is given by

$$\int_G \det D_{\mathbf{x}}\mathbf{g} d(x_1, \ldots, x_n).$$

If one sets $\mathbf{g} : \mathbf{x} \mapsto \mathbf{x}(T)$, lets T converge to 0 and uses the fact that the derivative of $\det D_{\mathbf{x}}\mathbf{g}$ is the trace of $D_{\mathbf{x}}\mathbf{f}$, this yields (11.9).

Now let us divide the vector fields (11.1) and (11.2) by the positive function $P = \prod_{i=1}^n x_i \prod_{j=1}^m y_j$. We call these modified vector fields (I) and (II) and compute their divergence. The divergence of (I) in $\operatorname{int}\mathbb{R}^n \times \mathbb{R}^m$ is given by

$$
\begin{aligned}
\operatorname{div}(\mathrm{I}) &= \sum_{i=1}^n \partial\dot{x}_i/\partial x_i + \sum_{j=1}^m \partial\dot{y}_j/\partial y_j \\
&= \tfrac{1}{P}\left(-\sum_{i=1}^n x_i(A\mathbf{y})_i - \sum_{j=1}^m y_j(B\mathbf{x})_j\right) = -\tfrac{1}{P}\left(\mathbf{x}\cdot A\mathbf{y} + \mathbf{y}\cdot B\mathbf{x}\right) .
\end{aligned}
\tag{11.10}
$$

In order to obtain the divergence div_0 within the state space $\operatorname{int} S_n \times S_m$, we have to subtract the eigenvalues of the Jacobian which are orthogonal to S_n and S_m. Since

$$
\left(\sum x_i\right)^{\cdot} = \left(1 - \sum x_i\right)\mathbf{x}\cdot A\mathbf{y},
$$

for (11.1) these two superfluous eigenvalues are $-\mathbf{x}\cdot A\mathbf{y}$ and $-\mathbf{y}\cdot B\mathbf{x}$. Taking into account the factor $1/P$, we see from (11.10) that the replicator dynamics (11.1) is divergence free:

$$
\operatorname{div}_0(\mathrm{I}) \equiv 0 .
\tag{11.11}
$$

By Liouville's formula (11.9), (I) preserves volume. This yields the following result.

Theorem 11.3.1 *Up to a change in velocity the flow (11.1) in* $\operatorname{int} S_n \times S_m$ *is incompressible.*

Exercise 11.3.2 Show that (11.1) cannot have attractors in $\operatorname{int} S_n \times S_m$. In particular, an interior rest point cannot be asymptotically stable for (11.1). A rest point of (11.1) is asymptotically stable if and only if it is a strict Nash equilibrium.

By a similar calculation we obtain for the modification (II) of (11.2):

$$
\operatorname{div}(\mathrm{II}) = -\frac{1}{P}\left(\sum_{i=1}^n x_i\frac{(A\mathbf{y})_i^2}{(\mathbf{x}\cdot A\mathbf{y})^2} + \sum_{j=1}^m y_j\frac{(B\mathbf{x})_j^2}{(\mathbf{y}\cdot B\mathbf{x})^2}\right) .
$$

The eigenvalues orthogonal to S_n and S_m are easily seen to be both -1. Thus

$$
\begin{aligned}
\operatorname{div}_0(\mathrm{II}) &= \operatorname{div}(\mathrm{II}) + \frac{2}{P} \\
&= -\frac{1}{P}\left(\sum_{i=1}^n x_i\left[\frac{(A\mathbf{y})_i - \mathbf{x}\cdot A\mathbf{y}}{\mathbf{x}\cdot A\mathbf{y}}\right]^2 + \sum_{j=1}^m y_j\left[\frac{(B\mathbf{x})_j - \mathbf{y}\cdot B\mathbf{x}}{\mathbf{y}\cdot B\mathbf{x}}\right]^2\right) .
\end{aligned}
\tag{11.12}
$$

Thus $\operatorname{div}_0(\mathrm{II}) \le 0$ on $\operatorname{int} S_n \times S_m$ and by (11.9), (11.2) is volume-contracting. The rate of contraction corresponds to the variance of the vector field and tends to zero near an interior rest point point. This reflects the fact that

(11.1) and (11.2) have essentially the same linearization near fixed points (see Exercise 11.1.2).

Exercise 11.3.3 Consider an aggregate monotonic dynamics

$$\dot{x}_i = x_i[(Ay)_i - \mathbf{x} \cdot A\mathbf{y}]\phi(\mathbf{x} \cdot A\mathbf{y})$$
$$\dot{y}_j = y_j[(Bx)_j - \mathbf{y} \cdot B\mathbf{x}]\phi(\mathbf{y} \cdot B\mathbf{x})$$

for a monotonically increasing function ϕ and apply the above to show that volume increases. Conclude that there are no attractors in the interior, and that interior rest points are unstable. Only pure equilibria can be stable.

Let us now apply this result to complete the discussion of two-strategy games ($n = m = 2$). What is still lacking is the behaviour of the modified replicator equation (11.2) for rescaled zero-sum games.

Consider the general Battle of the Sexes

$$A = \begin{bmatrix} \alpha & b + \beta \\ a + \alpha & \beta \end{bmatrix} \qquad B = \begin{bmatrix} c + \gamma & \delta \\ \gamma & d + \delta \end{bmatrix} \qquad (11.13)$$

with $a, b, c, d > 0$ and $\alpha, \beta, \gamma, \delta \geq 0$.

Exercise 11.3.4 Show that if $n = m = 2$ this is the general form of a rescaled zero-sum game with a unique interior fixed point.

For (11.1) we may discard the constants $\alpha, \beta, \gamma, \delta$ and compute the interior Nash equilibrium as

$$\bar{p} = \frac{d}{c + d} \qquad \bar{q} = \frac{b}{a + b} \qquad . \qquad (11.14)$$

For (11.2), formula (11.12) shows that P^{-1} is a Dulac function and that periodic orbits are impossible (see section 4.1). Since the area is decreasing we conclude that *the interior rest point* (11.14) *is globally asymptotically stable.* Although the mixed strategy $(\bar{p}, \bar{q})$ is not an ESS, it is an attractor for (11.2).

Exercise 11.3.5 Give another proof of this by finding a global Lyapunov function.

Exercise 11.3.6 Analyse the behaviour of the difference equation (11.3) for (11.13).

11.4 Nash–Pareto pairs

We have seen in the last section that in the Battle of the Sexes endowed with dynamics (11.2), the mixed Nash equilibrium is asymptotically stable. Hence mixed strategies are of relevance in evolutionary terms, although they cannot be evolutionarily stable. This raises the problem of relaxing the notion of ESS, which is rather narrow for asymmetric contests, so that it can include mixed strategies.

Suppose that the two populations are in state $(\mathbf{p}, \mathbf{q}) \in S_n \times S_m$. This state will certainly *not* be stable in any evolutionary sense if there exists a state $(\mathbf{x}, \mathbf{y})$ near $(\mathbf{p}, \mathbf{q})$ such that *both* populations can increase their mean payoff by switching to it. More formally, we say that a pair of strategies $(\mathbf{p}, \mathbf{q})$ is a *Nash–Pareto pair* for an asymmetric game with payoff matrices A and B if the following two conditions hold:

(i) *Equilibrium condition:* $\mathbf{p} \cdot A\mathbf{q} \geq \mathbf{x} \cdot A\mathbf{q}$ and $\mathbf{q} \cdot B\mathbf{p} \geq \mathbf{y} \cdot B\mathbf{p}$ for all
   $(\mathbf{x}, \mathbf{y}) \in S_n \times S_m$.

(ii) *Stability condition:* For all states $(\mathbf{x}, \mathbf{y}) \in S_n \times S_m$ for which equality holds in (i) we have

$$\text{if } \mathbf{x} \cdot A\mathbf{y} > \mathbf{p} \cdot A\mathbf{y} \quad \text{then} \quad \mathbf{y} \cdot B\mathbf{x} < \mathbf{q} \cdot B\mathbf{x}$$

and

$$\text{if } \mathbf{y} \cdot B\mathbf{x} > \mathbf{q} \cdot B\mathbf{x} \quad \text{then} \quad \mathbf{x} \cdot A\mathbf{y} < \mathbf{p} \cdot A\mathbf{y} \quad .$$

The first condition just says that $(\mathbf{p}, \mathbf{q})$ is a Nash equilibrium pair. The second condition says that it is impossible for *both* players to simultaneously take advantage of a deviation from the equilibrium $(\mathbf{p}, \mathbf{q})$: at least one of them gets penalized. This corresponds to the concept of *Pareto optimality* or *efficiency* in classical game theory. Note that in condition (ii) it is essential to allow both populations to deviate simultaneously. If $\mathbf{p}$ is replaced by $\mathbf{x}$ and $\mathbf{y}$ remains equal to $\mathbf{q}$ then condition (ii) is void.

Exercise 11.4.1 Show that the Nash equilibrium of the Battle of Sexes game (11.13) is a Nash–Pareto pair. Find other examples of Nash–Pareto pairs. Show that for $A = B = 0$ every state is Nash–Pareto.

We now characterize Nash–Pareto pairs.

Theorem 11.4.2 $(\mathbf{p}, \mathbf{q}) \in \text{int } S_n \times S_m$ *is a Nash–Pareto pair of the bimatrix game* (A, B) *if and only if there exists a constant* $c > 0$ *such that*

$$(\mathbf{x} - \mathbf{p}) \cdot A\mathbf{y} + c(\mathbf{y} - \mathbf{q}) \cdot B\mathbf{x} = 0 \tag{11.15}$$

for all $(\mathbf{x}, \mathbf{y}) \in S_n \times S_m$, *i.e. if and only if the game is a rescaled zero-sum game. Such a Nash–Pareto pair is stable for the replicator dynamics* (11.1).

This shows that the harmless-looking conditions (i) and (ii) (which are purely qualitative assertions, involving only inequalities) imply an identity between the payoff matrices.

Proof It is straightforward to show that (11.15) implies the Nash–Pareto conditions. For the converse we note that for interior $(\mathbf{p}, \mathbf{q})$, we have equality in (i) and the second condition (ii) becomes effective. Consider the bilinear form $F : \mathbb{R}_0^n \times \mathbb{R}_0^m \to \mathbb{R}^2$ given by

$$F(\xi, \eta) = (\xi \cdot A\eta, \eta \cdot B\xi) \quad . \tag{11.16}$$

With $\xi = \mathbf{x} - \mathbf{p}$ and $\eta = \mathbf{y} - \mathbf{q}$, (ii) means that F does not take values in the positive quadrant $\mathbb{R}_+^2 \setminus \{\mathbf{o}\}$. This implies that the image of F is a line in $\mathbb{R}^2$ with negative slope, i.e. there exists $c > 0$ such that

$$\xi \cdot A\eta = -c\eta \cdot B\xi \qquad \text{for all} \qquad (\xi, \eta) \in \mathbb{R}_0^n \times \mathbb{R}_0^m. \tag{11.17}$$

This shows (11.15).

If we choose $\xi = \mathbf{e}_i - \mathbf{e}_k$ and $\eta = \mathbf{f}_j - \mathbf{f}_l$ (where $\mathbf{e}_i$ and $\mathbf{f}_j$ denote the unit vectors, i.e. the corners of S_n, S_m) in (11.17), we obtain

$$a_{ij} - a_{il} - a_{kj} + a_{kl} = -c\,(b_{ji} - b_{li} - b_{jk} + b_{lk}) \tag{11.18}$$

for all i, j, k, l. As shown in (11.8) and exercise 11.2.9, this means precisely that (A, B) is a rescaling of a zero-sum game.
The function

$$H = \sum p_i \log x_i + c \sum q_j \log y_j \tag{11.19}$$

satisfies $\dot{H} = 0$ which proves the stability of $(\mathbf{p}, \mathbf{q})$. $\square$

It is instructive to compare this with the notion of evolutionary stability for symmetric conflicts. As we have seen in section 7.2, $\mathbf{p} \in \text{int } S_n$ is evolutionarily stable if and only if $P(\mathbf{x}) = \prod x_i^{p_i}$ is a strict global Lyapunov function: $\dot{P}(\mathbf{x}) > 0$ for $\mathbf{x} \neq \mathbf{p}$. For asymmetric conflicts the above shows that $(\mathbf{p}, \mathbf{q}) \in \text{int } S_n \times S_m$ is a Nash–Pareto pair if and only if the power product

$$P(\mathbf{x}, \mathbf{y}) = \prod x_i^{p_i} \prod y_j^{cq_j} \tag{}$$

is a global Lyapunov function (and even a constant of motion) for the replicator equation (11.1): $\dot{P}(\mathbf{x}, \mathbf{y}) \geq 0$ for $(\mathbf{x}, \mathbf{y})$ near $(\mathbf{p}, \mathbf{q})$.

Exercise 11.4.3 Show that if $P(\mathbf{x}, \mathbf{y})$ is a strict Lyapunov function for (11.1), i.e. if $\dot{P}(\mathbf{x}, \mathbf{y}) > 0$ for all $(\mathbf{x}, \mathbf{y})$ near $(\mathbf{p}, \mathbf{q})$, then $(\mathbf{p}, \mathbf{q})$ is a strict Nash equilibrium, and hence both $\mathbf{p}$ and $\mathbf{q}$ are pure strategies.

In symmetric games, an evolutionarily stable state is in general robust against small perturbations in the payoffs. In asymmetric conflicts the situation is different. Indeed, let $(\mathbf{p}, \mathbf{q})$ be a Nash–Pareto pair for a bimatrix game (A, B). It is said to be *robust* if every game $\left(\tilde{A}, \tilde{B} \right)$ in a suitable neighbourhood of the given (A, B) has a Nash–Pareto pair $(\tilde{\mathbf{p}}, \tilde{\mathbf{q}})$ near $(\mathbf{p}, \mathbf{q})$.

Exercise 11.4.4 Show that a robust Nash–Pareto pair is a mixture of *at most two* strategies in each population. Hint: if n or $m \geq 3$, then a slight perturbation can destroy the identity (11.15).

Hence if $(\mathbf{p}, \mathbf{q})$ is totally mixed and robust, then $n, m \leq 2$. Thus there are only two possibilities for a robust Nash–Pareto pair: either (a) it consists of pure strategies (and is then a strict Nash equilibrium) or (b) it is the Nash equilibrium of a subgame of type (11.13).

Conjecture 11.4.5 *If an interior equilibrium is isolated and stable for* (11.1), *then it is a Nash–Pareto pair.*

11.5 Game dynamics and Nash–Pareto pairs

In section 7.3 we interpreted the ESS conditions in game-dynamical terms. We shall now do this for Nash–Pareto pairs too, and ask in which sense they are immune against mutant strategies.

Suppose that the two populations are homogeneous, and consist of $\mathbf{p}$- and $\mathbf{q}$-players, respectively. Let small numbers of $\mathbf{x}$- resp. $\mathbf{y}$-players invade. We shall denote the frequency of $\mathbf{x}$-players by x and that of $\mathbf{y}$-players by y. The dynamics (11.1) of this game reads

$$\begin{aligned} \dot{x} &= x(1-x)(b-(a+b)y) \\ \dot{y} &= y(1-y)(d-(c+d)x) \end{aligned} \tag{11.20}$$

on $[0, 1] \times [0, 1]$, with $a = (\mathbf{p} - \mathbf{x}) \cdot A\mathbf{y}$, $b = (\mathbf{x} - \mathbf{p}) \cdot A\mathbf{q}$, $c = (\mathbf{q} - \mathbf{y}) \cdot B\mathbf{x}$, $d = (\mathbf{y} - \mathbf{q}) \cdot B\mathbf{q}$. The pair $(\mathbf{p}, \mathbf{q})$ is a Nash equilibrium if and only if $b \leq 0$ and $d \leq 0$ for all $\mathbf{x}$ and $\mathbf{y}$. It is stable against $(\mathbf{x}, \mathbf{y})$, in the sense that the corner $(\mathbf{p}, \mathbf{q})$ is an attractor (see fig. 11.1) if and only if $b < 0$ and $d < 0$, i.e. if and only if

$$\mathbf{p} \cdot A\mathbf{q} > \mathbf{x} \cdot A\mathbf{q} \text{ and } \mathbf{q} \cdot B\mathbf{p} > \mathbf{y} \cdot B\mathbf{p} \text{ for all } \mathbf{x} \neq \mathbf{p} \text{ and } \mathbf{y} \neq \mathbf{q}. \tag{11.21}$$

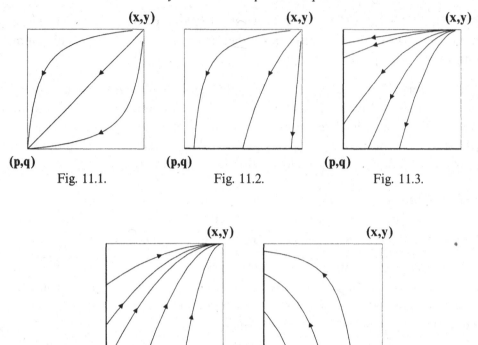

Fig. 11.1.　　　　Fig. 11.2.　　　　Fig. 11.3.

Fig. 11.4.　　　　Fig. 11.5.

Thus we recover exactly the definition of a strict Nash equilibrium for an asymmetric game. There is no 'second condition' in the sense of (6.12), since if $b = 0$, i.e. if

$$\mathbf{p} \cdot A\mathbf{q} = \mathbf{x} \cdot A\mathbf{q} \quad \text{for some} \quad \mathbf{x} \neq \mathbf{p} \quad , \tag{11.22}$$

then a line of rest points arises and $\mathbf{p}$ can be invaded by $\mathbf{x}$-mutants due to stochastic fluctuations (see figs. 11.2–5). Thus in the strict sense, (11.22) is incompatible with evolutionary stability.

On the other hand, the introduction of *any* $\mathbf{y}$-mutant satisfying $a > 0$ and $d < 0$, i.e. $\mathbf{p} \cdot A\mathbf{y} > \mathbf{x} \cdot A\mathbf{y}$ and $\mathbf{q} \cdot B\mathbf{p} > \mathbf{y} \cdot B\mathbf{p}$, into the population of players in position II would lead back to a point closer to $(\mathbf{p}, \mathbf{q})$. Hence the following concept of a 'weak ESS' could still guarantee some evolutionary stability: $b \leq 0$, with $a > 0$ if $b = 0$; and $d \leq 0$, with $c > 0$ if $d = 0$. More explicitly,

$$
\begin{aligned}
&\mathbf{p} \cdot A\mathbf{q} \geq \mathbf{x} \cdot A\mathbf{q} \quad \text{for all} \quad \mathbf{x}; \\
&\text{if} \quad \mathbf{p} \cdot A\mathbf{q} = \mathbf{x} \cdot A\mathbf{q}, \quad \text{then} \quad \mathbf{p} \cdot A\mathbf{y} > \mathbf{x} \cdot A\mathbf{y} \quad \text{for all} \quad \mathbf{y}; \\
&\mathbf{q} \cdot B\mathbf{p} \geq \mathbf{y} \cdot B\mathbf{p} \quad \text{for all} \quad \mathbf{y}; \\
&\text{if} \quad \mathbf{q} \cdot B\mathbf{p} = \mathbf{y} \cdot B\mathbf{p}, \quad \text{then} \quad \mathbf{q} \cdot B\mathbf{x} > \mathbf{y} \cdot B\mathbf{x} \quad \text{for all} \quad \mathbf{x}.
\end{aligned}
\tag{11.23}
$$

Exercise 11.5.1 Show that (11.23) implies that the phase portrait of the reduced game endowed with any of the three dynamics (11.1–3) looks as shown in fig. 11.2–3. Show that (11.23) implies (again) that both **p** and **q** are pure strategies.

For mixed strategies **p** and **q**, the equalities $b = 0$ and $d = 0$, i.e. $\mathbf{p}A\mathbf{q} = \mathbf{x}A\mathbf{q}$ and $\mathbf{q} \cdot B\mathbf{p} = \mathbf{y} \cdot B\mathbf{p}$, hold for many $\mathbf{x}, \mathbf{y}$, and thus by (11.22) fixed-point edges are inevitable. What we have to preclude is that both $a < 0$ and $c < 0$, i.e. $\mathbf{x} \cdot A\mathbf{y} > \mathbf{p} \cdot A\mathbf{y}$ and $\mathbf{y} \cdot B\mathbf{x} > \mathbf{q} \cdot B\mathbf{x}$, hold for some $(\mathbf{x}, \mathbf{y})$, for then $(\mathbf{x}, \mathbf{y})$ would certainly invade $(\mathbf{p}, \mathbf{q})$.

Due to the asymmetry of the game, however, there remains one intermediate possibility, which has no analogue in the symmetric case. This is exactly our concept of Nash–Pareto pair: $b \leq 0$ and $d \leq 0$; if $b = 0$ and $d = 0$, then $ac < 0$.

Exercise 11.5.2 Show that for a Nash–Pareto pair $(\mathbf{p}, \mathbf{q})$ and for $(\mathbf{x}, \mathbf{y})$ having support contained in that of $(\mathbf{p}, \mathbf{q})$, the phase portrait of the reduced game, with any dynamics, looks like fig. 11.5. All orbits travel from some fixed point on one line to some fixed point on the other line.

Hence evolution will not further increase the 'distance' of the deviation from a Nash–Pareto pair $(\mathbf{p}, \mathbf{q})$, caused by some mutation. It then depends on the actual dynamics and on the strength of stochastic effects whether this distance will decay to zero and the mutant will be eliminated. This may be conjectured at least for the modified replicator dynamics (11.2). In any case, a Nash–Pareto point $(\mathbf{p}, \mathbf{q})$ will be physically relevant as the time average of the orbits.

11.6 Notes

The model (11.1) was introduced by Taylor (1979), see also Schuster *et al.* (1981b). The model (11.2) was proposed by Maynard Smith (1982, Appendix J). Cressman *et al.* (1986) study yet another dynamics for bimatrix games. An 'alternating moves' discrete time dynamics for bimatrix games which is different from (11.3) but shares many of the conservative properties of (11.1) was suggested in Hofbauer (1996). We owe the idea of section 11.3 to Ethan Akin (personal communication), who proved (in a different way) that (11.1) is volume-preserving (see also Eshel and Akin (1983)). For generalizations to asymmetric *n*-person games, see Weibull (1995). The result in exercise 11.3.3 is due to Ritzberger and Weibull (1995). Partnership games are also known as 'games with identical interests' or 'potential games', see Monderer and Shapley (1996).

Part three

Permanence and Stability

12

Catalytic hypercycles

The hypercycle equation is a particular replicator equation whose interior rest point is unstable for higher dimensions. Nevertheless, the boundary is a repellor, so that the frequencies are bounded away from 0. We prove this permanence with the help of an average Lyapunov function.

12.1 The hypercycle equation

Let us consider the *hypercycle equation*

$$\dot{x}_i = x_i \left(k_i x_{i-1} - \sum_{j=1}^{n} k_j x_j x_{j-1} \right) \qquad (12.1)$$

where the indices $i = 1, 2, \ldots, n$ are counted mod n (so that $x_0 = x_n$). This is just the replicator equation $\dot{x}_i = x_i((A\mathbf{x})_i - \mathbf{x} \cdot A\mathbf{x})$ on S_n, with

$$A = \begin{bmatrix} 0 & 0 & 0 & . & . & . & k_1 \\ k_2 & \cdot 0 & 0 & . & . & . & 0 \\ 0 & k_3 & 0 & . & . & . & 0 \\ . & . & . & . & . & . & . \\ 0 & 0 & 0 & . & . & k_n & 0 \end{bmatrix} . \qquad (12.2)$$

This equation plays a crucial role in a theory of Manfred Eigen and Peter Schuster on prebiotic evolution: the x_i are the relative frequencies of RNA molecules that catalyse each other's replication in a closed feedback loop which provides a possible mechanism for the 'ganging up' of molecular information carriers.

The hypercycle equation (12.1) admits a unique rest point $\mathbf{p}$ in int S_n, given

143

by $k_1 x_n = k_2 x_1 = \cdots = k_n x_{n-1}$ and $x_1 + x_2 + \cdots + x_n = 1$. This yields

$$p_i = k_{i+1}^{-1}\left(\sum_j k_{j+1}^{-1}\right)^{-1} \qquad i = 1,\ldots,n. \qquad (12.3)$$

The stability analysis of **p** is considerably simplified by a change in coordinates. Thus we set

$$y_i = \frac{k_{i+1}x_i}{\sum_{j=1}^n k_{j+1}x_j}. \qquad (12.4)$$

The projective transformation $\mathbf{x} \to \mathbf{y}$ of S_n onto itself transforms (12.1) into

$$\dot{y}_i = y_i\left(y_{i-1} - \sum y_j y_{j-1}\right)\left(\sum_s k_{s+1}^{-1}y_s\right)^{-1}. \qquad (12.5)$$

We may drop the last factor (see exercises 4.1.2, 4.1.3) and obtain

$$\dot{x}_i = x_i\left(x_{i-1} - \sum_j x_j x_{j-1}\right) \qquad (12.6)$$

which is just (12.1) with $k_1 = k_2 = \cdots = k_n = 1$. Thus a change in coordinates followed by a change in velocity allows one to get rid of the coefficients k_i. The Jacobian of (12.6) at the unique rest point $\mathbf{m} = \frac{1}{n}\mathbf{1}$ in int S_n is a circulant matrix (see section 5.5). The elements in the first row are

$$-\frac{2}{n^2}, \quad -\frac{2}{n^2}, \quad \ldots, \quad -\frac{2}{n^2}, \quad \frac{1}{n} - \frac{2}{n^2}$$

and the other rows are obtained by cyclic permutation. The eigenvalues are given by (5.20). This yields $\gamma_0 = -1/n$ and, for $j = 1,\ldots,n-1$,

$$\gamma_j = \sum_{k=0}^{n-1}\left(-\frac{2}{n^2}\right)\lambda^{kj} + \frac{1}{n}\lambda^{(n-1)j} = \frac{\lambda^{-j}}{n} \qquad (12.7)$$

with $\lambda = e^{2\pi i/n}$. The eigenvalue γ_0 corresponds to the eigenvector **1** which is orthogonal to the simplex S_n. Since we are only interested in the restriction of (12.6) to S_n, we need not consider it any further. We see that for $n \geq 5$, the rest point **m** is unstable.

The function

$$P(\mathbf{x}) = x_1 x_2 \ldots x_n \qquad (12.8)$$

vanishes on bd S_n (where $x_i = 0$ for at least one i) and is strictly positive in int S_n. It attains its maximal value (in S_n) at the point **m**. The time derivative of $t \to \log P(\mathbf{x}(t))$ is given by

$$\left(\log P\right)^{\cdot} = \sum_{i=1}^{n} \frac{\dot{x}_i}{x_i} = 1 - n \sum_{j=1}^{n} x_j x_{j-1} \,. \tag{12.9}$$

For $n = 2$ and $n = 3$, this function is positive and vanishes only at $\mathbf{m}$.

Exercise 12.1.1 Show that for $n = 4$, the function $\dot{P}/P$ is still nonnegative, and vanishes on the set $\{\mathbf{x} \in S_4 : x_1 + x_3 = x_2 + x_4\}$, which contains only one invariant set, consisting of $\mathbf{m}$ only.

The following theorem is an immediate consequence.

Theorem 12.1.2 *For short hypercycles (i.e. $n = 2,3,4$) the inner rest point is globally stable.*

For $n \geq 5$, however, the interior rest point is unstable, and one is left with the question whether a permanent coexistence of the different types of polynucleotides is possible. This question will be addressed in the next section.

Exercise 12.1.3 An alternative way of proving the global stability of $\mathbf{m}$ for $n = 3$ would be to use other Lyapunov functions than (12.8). Try e.g. $x_1^{-p} + x_2^{-p} + x_3^{-p}$.

Exercise 12.1.4 Write down a global Lyapunov function for (12.1) if $n \leq 4$.

Exercise 12.1.5 Consider the hypercycle equation (12.1) with $n = 3$. For which values of k_i is the unique interior rest point evolutionarily stable? Show that for $n = 4$, the interior rest point (which, as we know, is globally stable) is never an ESS.

12.2 Permanence

As we have just seen, hypercycles do not reach a stable rest point for $n \geq 5$. They may nevertheless serve their purpose, which is to allow the coexistence of several types of self-reproducing macromolecules: whether the concentrations converge or oscillate in a regular or irregular way is of secondary importance. The main thing is that they do not vanish: more precisely, that there exists a *threshold value* $\delta > 0$ such that every solution of (12.6) in int S_n satisfies $x_i(t) > \delta$ for $i = 1, \ldots, n$ whenever t is large enough. This implies that if initially all species are present, even if only in very small quantities, then after some time some sizeable amount of each will be present. No perturbation which is smaller than δ could wipe out a molecular species.

This property, which is important in many other contexts, deserves a name. A dynamical system defined on S_n is said to be *permanent* if there exists a $\delta > 0$ such that $x_i(0) > 0$ for $i = 1, \ldots, n$ implies

$$\liminf_{t \to +\infty} x_i(t) > \delta \tag{12.10}$$

for $i = 1, \ldots, n$.

Let us stress that δ does *not* depend on the initial values $x_i(0)$. Permanence means more than just that no component will vanish. If every state is a rest point, for example, the system is not permanent. Even if initially the concentrations were abundant, a sequence of tiny perturbations could lead to the extinction of a species. For a permanent system, on the other hand, perturbations which are sufficiently small and rare cannot lead to extinction. The boundary of the state space S_n acts as a *repellor*.

The proof that the hypercycle is permanent is not obvious. We shall start with a more general theorem giving conditions for permanence which will also be useful in many other situations.

Theorem 12.2.1 *Let us consider a dynamical system on S_n leaving the boundary invariant. Let $P : S_n \to \mathbb{R}$ be a differentiable function vanishing on $\mathrm{bd}\, S_n$ and strictly positive in $\mathrm{int}\, S_n$. If there exists a continuous function Ψ on S_n such that the following two conditions hold:*

$$\textit{for } \mathbf{x} \in \mathrm{int}\, S_n, \qquad \frac{\dot{P}(\mathbf{x})}{P(\mathbf{x})} = \Psi(\mathbf{x}) \tag{12.11}$$

$$\textit{for } \mathbf{x} \in \mathrm{bd}\, S_n, \quad \int_0^T \Psi(\mathbf{x}(t))dt > 0 \textit{ for some } T > 0 , \tag{12.12}$$

then the dynamical system is permanent.

The value $P(\mathbf{x})$ measures the distance from $\mathbf{x}$ to the boundary. If one had $\Psi > 0$ on $\mathrm{bd}\, S_n$ — a condition implying (12.12) — then $\dot{P}(\mathbf{x}) > 0$ for any $\mathbf{x} \in \mathrm{int}\, S_n$ near the boundary, and so P would increase, i.e. the orbit would be repelled from $\mathrm{bd}\, S_n$. In that case, P would act like a Lyapunov function. Quite often, however, one cannot find a function P of this type. The weaker version defined above is said to be an *average Lyapunov function*: its time average acts like a Lyapunov function.

Proof The $T > 0$ in (12.12) can obviously be chosen as a locally continuous function $T(\mathbf{x})$. Its infimum τ is positive, since bd S_n is compact. For $h > 0$, we define

$$U_h = \left\{ \mathbf{x} \in S_n : \text{there is a } T > \tau \text{ such that } \frac{1}{T} \int_0^T \Psi(\mathbf{x}(t)) dt > h \right\}.$$

For $\mathbf{x} \in U_h$ we set

$$T_h(\mathbf{x}) = \inf \left\{ T > \tau : \frac{1}{T} \int_0^T \Psi(\mathbf{x}(t)) dt > h \right\}.$$

We show first that U_h is open and T_h upper semicontinuous: in other words, if $\mathbf{x} \in U_h$ and $\alpha > 0$ are given, then for $\mathbf{y} \in S_n$ sufficiently close to $\mathbf{x}$, one has

$$\mathbf{y} \in U_h \quad \text{and} \quad T_h(\mathbf{y}) < T_h(\mathbf{x}) + \alpha. \tag{12.13}$$

Indeed, for α and $\mathbf{x}$ there is a $T \in [\tau, T_h(\mathbf{x}) + \alpha[$ such that

$$\varepsilon = \frac{1}{T} \int_0^T \Psi(\mathbf{x}(t)) dt - h > 0.$$

Since the solutions of ordinary differential equations depend continuously on the initial values, $\mathbf{x}(t)$ and $\mathbf{y}(t)$ are near each other, for all $t \in [0, T]$, if $\mathbf{x}$ and $\mathbf{y}$ are sufficiently close. The uniform continuity of Ψ implies $|\Psi(\mathbf{x}(t)) - \Psi(\mathbf{y}(t))| < \varepsilon$ for all times $t \in [0, T]$, and hence

$$\frac{1}{T} \int_0^T \Psi(\mathbf{y}(t)) dt > \frac{1}{T} \int_0^T \Psi(\mathbf{x}(t)) dt - \varepsilon = h$$

from which (12.13) follows.

By (12.12) the family of nested sets U_h (with $h > 0$) is an open covering of the compact set bd S_n. There exists one $h > 0$, then, such that U_h is an open neighbourhood of bd S_n (in S_n). Since $S_n \backslash U_h$ is also compact, P attains its minimum on this set. If we choose $p > 0$ smaller than this minimum, then the set

$$I(p) = \{\mathbf{x} \in S_n : 0 < P(\mathbf{x}) \leq p\}$$

is contained in U_h. $I(p)$ is a 'boundary layer' which is very thin if p is small. We shall show that if $\mathbf{x} \in I(p)$, then there exists a $t > 0$ such that $\mathbf{x}(t) \notin I(p)$. Indeed, otherwise $\mathbf{x}(t)$ would have to be in U_h for all $t > 0$. In this case, there exists a $T \geq \tau$ such that

$$\frac{1}{T} \int_t^{T+t} \Psi(\mathbf{x}(s))ds > h \ .$$

But since $\Psi = (\log P)^{\cdot}$ holds in int S_n, this implies

$$h < \frac{1}{T} \int_t^{T+t} (\log P)^{\cdot}(\mathbf{x}(s))ds = \frac{1}{T}\left[\log P(\mathbf{x}(T+t)) - \log P(\mathbf{x}(t))\right] \ ,$$

that is,

$$P(\mathbf{x}(T+t)) > P(\mathbf{x}(t))e^{hT} \geq P(\mathbf{x}(t))e^{h\tau} \ .$$

Hence there would exist a sequence t_n for which $P(\mathbf{x}(t_n))$ tends to $+\infty$, in contradiction to the boundedness of P.

Let us denote by $\bar{I}(p)$ the union of $I(p)$ with bd S_n. All that remains to be shown is that there exists a $q \in \,]0, p[$ such that $\mathbf{x}(0) \notin \bar{I}(p)$ implies $\mathbf{x}(t) \notin I(q)$ for all $t \geq 0$.

The upper semicontinuous function T_h admits an upper bound $\bar{T}$ on the compact set $\bar{I}(p)$. Let t_0 be the first time when $\mathbf{x}(t)$ reaches $\bar{I}(p)$, i.e.

$$t_0 = \min\{t > 0 : \mathbf{x}(t) \in \bar{I}(p)\},$$

and let $\mathbf{x}(t_0) = \mathbf{y}$. Obviously $P(\mathbf{y}) = p$. Let m be the minimum of Ψ on S_n. For $m \geq 0$, all is clear, since P never decreases. In case $m < 0$ we set $q = pe^{m\bar{T}}$. For $t \in \,]0, \bar{T}[$,

$$\frac{1}{t} \int_0^t \Psi(\mathbf{y}(s))ds \geq m,$$

and hence, just as above

$$P(\mathbf{y}(t)) \geq P(\mathbf{y})e^{mt} > pe^{m\bar{T}} = q \ .$$

Hence the solution does not reach $I(q)$ for $t \in \,]0, \bar{T}[$. Furthermore, since $\mathbf{y} \in I(p)$, there is a time $T \in [\tau, \bar{T}[$ such that

$$P(\mathbf{y}(T)) \geq pe^{h\tau} \geq p \ .$$

At time $t + T$, thus, the orbit of $\mathbf{x}$ has left $I(p)$ without having reached $I(q)$. Repeating this argument, one sees that the orbit can never reach $I(q)$. $\quad\square$

Theorem 12.2.2 *It is sufficient to verify* (12.12) *for all* $\mathbf{x}$ *in the ω-limits of orbits on the boundary of S_n.*

Proof There exists, as before, an $h > 0$ such that U_h is an open neighbourhood of $\omega(\mathbf{x})$ (in S_n) for all $\mathbf{x} \in \text{bd } S_n$. Since $\mathbf{x}(t)$ converges to $\omega(\mathbf{x})$, there is a t_1 such that $\mathbf{x}(t) \in U_h$ for all $t \geq t_1$. There is a $t_2 \geq t_1 + \tau$ such that

$$\frac{1}{t_2 - t_1} \int_{t_1}^{t_2} \Psi(\mathbf{x}(t)) dt > h \,,$$

similarly a $t_3 \geq t_2 + \tau$ such that

$$\frac{1}{t_3 - t_2} \int_{t_2}^{t_3} \Psi(\mathbf{x}(t)) dt > h,$$

etc. We obtain a sequence $t_1, t_2, t_3, \ldots$ satisfying

$$\frac{1}{t_k - t_1} \int_{t_1}^{t_k} \Psi(\mathbf{x}(t)) dt > h \,.$$

If k is sufficiently large, the time average

$$\frac{1}{t_k} \int_{0}^{t_k} \Psi(\mathbf{x}(t)) dt$$

is close to the previous expression and hence positive. □

Exercise 12.2.3 Show that theorems 12.2.1 and 12.2.2 are valid if Ψ is only assumed to be lower semicontinuous. This can always be achieved by defining, for $\mathbf{x} \in \text{bd } S_n$,

$$\Psi(\mathbf{x}) = \liminf_{\mathbf{y} \to \mathbf{x}} \frac{\dot{P}}{P}(\mathbf{y})$$

where $\mathbf{y}$ belongs to int S_n.

12.3 The permanence of the hypercycle

Theorem 12.3.1 *The hypercycle* (12.1) *is permanent.*

Proof We shall use theorem 12.2.1. As the average Lyapunov function $P(\mathbf{x})$ we choose the product $x_1 x_2 \ldots x_n$ which has already stood us in good stead in section 12.1. We have $\dot{P} = P\Psi$ with $\Psi = \sum k_i x_{i-1} - n\bar{f}$, where $\bar{f} = \sum k_j x_j x_{j-1}$. In order to show that P is an average Lyapunov function, it remains to verify condition (12.12). Thus we have to show that for every $\mathbf{x} \in \operatorname{bd} S_n$, there exists a $T > 0$ such that

$$\frac{1}{T} \int_0^T \sum_{i=1}^n (k_i x_{i-1} - n\bar{f}) dt > 0 \tag{12.14}$$

i.e. such that

$$\frac{1}{T} \int_0^T \bar{f}(\mathbf{x}(t)) dt < \frac{1}{nT} \int_0^T \sum k_i x_{i-1} dt . \tag{12.15}$$

Since $k := \min k_i > 0$ and $\sum k_i x_{i-1} \geq k$ for all $\mathbf{x} \in S_n$, the right hand side of (12.15) is not smaller than k/n. It is enough, therefore, to show that there is no $\mathbf{x} \in \operatorname{bd} S_n$ such that for all $T > 0$

$$\frac{1}{T} \int_0^T \bar{f}(\mathbf{x}(t)) dt \geq \frac{k}{n} . \tag{12.16}$$

Let us proceed indirectly and assume that there is such an $\mathbf{x} \in S_n$. We shall show by induction that

$$\lim_{t \to +\infty} x_i(t) = 0 \tag{12.17}$$

for $i = 1, \ldots, n$. Since $\mathbf{x} \in \operatorname{bd} S_n$, there exists an index i_0 such that $x_{i_0}(t) \equiv 0$. Now if $x_i(t)$ converges to 0, then so does x_{i+1}; indeed, if $x_{i+1}(t) > 0$, one obtains from

$$\left(\log x_{i+1} \right)^{\cdot} = \frac{\dot{x}_{i+1}}{x_{i+1}} = k_{i+1} x_i - \bar{f}$$

by integrating from 0 to T and dividing by T

$$\frac{\log x_{i+1}(T) - \log x_{i+1}(0)}{T} = \frac{1}{T} \int_0^T \left(\log x_{i+1} \right)^{\cdot} dt$$

$$= \frac{1}{T} \int_0^T k_{i+1} x_i(t) dt - \frac{1}{T} \int_0^T \bar{f}(\mathbf{x}(t)) dt .$$

From $x_i(t) \to 0$ it follows that

$$\frac{1}{T} \int_0^T k_{i+1} x_i(t) dt < \frac{k}{2n}$$

for all sufficiently large T. Together with (12.16) this implies

$$\log x_{i+1}(T) - \log x_{i+1}(0) < -\frac{kT}{2n}$$

or

$$x_{i+1}(T) < x_{i+1}(0) \exp\left(-\frac{kT}{2n}\right). \tag{12.18}$$

Hence $x_{i+1}(t) \to 0$ and (12.17) must hold. This contradicts the relation $\sum x_i = 1$. Hence P is an average Lyapunov function and (12.1) is permanent.

$\square$

Let us mention that the same proof shows that the considerably more general equation obtained by replacing the constants $k_i > 0$ in (12.1) by positive functions $F_i(\mathbf{x})$ is also permanent. Such equations describe the reaction kinetics for more realistic hypercycle models.

Exercise 12.3.2 Analyse the discrete time dynamics (7.23) for the hypercycle. Prove permanence for $C > 0$. Analyse the local stability for the interior fixed point $\mathbf{p}$. Prove that $\mathbf{p}$ is globally stable if $n \le 3$ and $C > 0$. Analyse the asymptotic behaviour for $C = 0$.

12.4 The competition of disjoint hypercycles

Let us assume now that the 'primordial soup' contains n types of RNA molecules organized into *several* disjoint hypercycles. This can be described by a permutation π of the set $\{1,\ldots,n\}$. Every permutation can be decomposed into *elementary cycles* $\Gamma_1,\ldots\Gamma_s$: these correspond to hypercycles. The dynamics is given by

$$\dot{x}_i = x_i \left(k_i x_{\pi(i)} - \sum k_j x_j x_{\pi(j)} \right) \tag{12.19}$$

with $k_i > 0$ for $i = 1,\ldots,n$.

If π consists of a unique cycle, we obtain — up to a reordering of the indices — the familiar hypercycle equation (12.1). If the elementary cycle Γ_j consists of a unique element i (a fixed point of the permutation π), then M_i is an autocatalytic molecular type. As in section 12.1 we may perform a transformation

$$y_i = \frac{k_{\tau(i)} x_i}{\sum_j k_{\tau(j)} x_j} \tag{12.20}$$

(with $\tau = \pi^{-1}$) and get rid of the k_i. Again there is a unique rest point in int S_n, namely the centre $\mathbf{m} = \frac{1}{n}\mathbf{1}$.

Exercise 12.4.1 Show that short hypercycles go to equilibrium: if the length $|\Gamma_k|$ of Γ_k is smaller than 5 and if i and j belong to Γ_k, then

$$\frac{x_i}{x_j} \to 1 \tag{12.21}$$

for $t \to +\infty$. (Hint: if Γ_k consists of the indices $1,\ldots,m$ (with $m \le 4$) consider $V = PS^{-m}$ with $P = x_1 \ldots x_m$ and $S = x_1 + \cdots + x_m$.)

Exercise 12.4.2 Show that as soon as there is more than one hypercycle in the reactor, the interior rest point **m** is unstable and the system is not permanent.

Exercise 12.4.3 For the catalytic chain described by (12.1) with $k_1 = 0$, $k_2 > 0,\ldots,k_n > 0$, all x_i with $i \ne n$ will tend to extinction. (Hint: show by induction, for $1 \le i \le n-1$, that $x_i \to 0$ and x_i/x_{i+1} converges in an ultimately monotonic way to some limit. Alternatively, use Theorem 7.6.1.)

12.5 Notes

For the hypercycle theory see Eigen and Schuster (1979) and Eigen *et al.* (1981). Permanence was introduced in Schuster *et al.* (1979a). Permanence of a general 'hypercycle equation' was shown in Hofbauer *et al.* (1981). Theorem 12.2.1 on average Lyapunov functions and its application to hypercycles is from Hofbauer (1981b), while exercise 12.2.3 is due to Hutson (1984), who set it up in a considerably more general situation. A discrete time version of the hypercycle (see exercise 12.3.2) was analysed in Hofbauer (1984). In Hofbauer *et al.* (1991), it has been shown that (12.1) admits a stable periodic orbit for $n \ge 5$.

13

Criteria for permanence

Using topological arguments, we show that every permanent dynamical system admits an interior rest point. We introduce the notion of a saturated rest point, which corresponds to a Nash equilibrium. For replicator equations we derive a series of necessary and of sufficient conditions for permanence.

13.1 Permanence and persistence for replicator equations

For the replicator equation

$$\dot{x}_i = x_i(f_i(\mathbf{x}) - \bar{f}) \tag{13.1}$$

on S_n, with $\bar{f}(\mathbf{x}) = \sum x_j f_j(\mathbf{x})$, and for ecological equations of the type

$$\dot{x}_i = x_i f_i(\mathbf{x}) \tag{13.2}$$

on $\mathbb{R}^n_+$, we can rarely expect to obtain a full description of the attractors. For many purposes, however, the precise asymptotic behaviour is less important than the question of *extinction*, i.e. whether all species in the system can survive or not. There are several mathematical concepts dealing with this aspect. In particular, a system of type (13.1) or (13.2) is said to be *permanent* if there exists a compact set K in the interior of the state space such that all orbits in the interior end up in K. This means that the boundary is a *repellor* (for (13.2), we have to consider the points at infinity as part of the boundary of $\mathbb{R}^n_+$).

Equivalently, permanence means for (13.1) that there exists a $\delta > 0$ such that

$$\delta < \liminf_{t \to +\infty} x_i(t) \tag{13.3}$$

for all i, whenever $x_i(0) > 0$ for all i. (Thus we recover the definition of

153

section 12.2.) For (13.2), permanence requires in addition that there exists a D such that

$$\limsup_{t \to +\infty} x_i(t) \le D \tag{13.4}$$

for all i, whenever $\mathbf{x} \in \operatorname{int} \mathbb{R}^n_+$. If (13.4) holds for all $\mathbf{x} \in \mathbb{R}^n_+$, we shall say that the orbits of (13.2) are *uniformly bounded* (obviously a minimal concession to reality). Condition (13.3) means that if all types are initially present, the dynamics will not lead to extinction. The threshold δ is a uniform one, independent of the initial condition. Thus permanence is a more stringent property than *strong persistence* (which requires (13.3) with $\delta = 0$) and *persistence* (which requires

$$\limsup_{t \to +\infty} x_i(t) > 0 \tag{13.5}$$

for all orbits in the interior of the state space).

One general way to prove permanence is to construct an *average Lyapunov function*. For (13.1) this was proved in section 12.2. The same proof applies for (13.2) if its orbits are uniformly bounded.

Exercise 13.1.1 Check this. Why does one need the uniform boundedness condition?

For most examples, the terms f_i in (13.1) and (13.2) are linear. This yields the first-order replicator equation

$$\dot{x}_i = x_i\big((A\mathbf{x})_i - \mathbf{x} \cdot A\mathbf{x}\big) \tag{13.6}$$

on S_n, resp. the Lotka–Volterra equation

$$\dot{x}_i = x_i\big(r_i + (A\mathbf{x})_i\big) \tag{13.7}$$

on $\mathbb{R}^n_+$. In sections 13.5 and 13.6, we shall present useful necessary and sufficient conditions for the permanence of such systems.

Exercise 13.1.2 Show that (13.6) with the modified rock–scissors–paper matrix

$$A = \begin{bmatrix} 0 & 1+\varepsilon & -1 \\ -1 & 0 & 1+\varepsilon \\ 1+\varepsilon & -1 & 0 \end{bmatrix} \tag{13.8}$$

is persistent for all $\varepsilon > -1$, but not strongly persistent for $\varepsilon < 0$. It is strongly persistent but not permanent for $\varepsilon = 0$.

13.2 Brouwer's degree and Poincaré's index

Let U be a bounded open subset of $\mathbb{R}^n$ and $\mathbf{f}$ a vector field defined on a neighbourhood of its closure $\bar{U}$. A point $\mathbf{x} \in U$ is said to be *regular* if $\det D_{\mathbf{x}}\mathbf{f} \neq 0$. A point $\mathbf{y} \in \mathbb{R}^n$ is said to be a *regular value* if all $\mathbf{x} \in \bar{U}$ with $\mathbf{f}(\mathbf{x}) = \mathbf{y}$ are regular.

As a consequence of the *implicit function theorem*, $\mathbf{f}$ is locally invertible around every such $\mathbf{x}$, and hence the roots of $\mathbf{f}(\mathbf{x}) = \mathbf{y}$ are isolated. *Sard's theorem* states that almost every $\mathbf{y} \in \mathbb{R}^n$ is a regular value, and hence that regular values are dense in $\mathbb{R}^n$.

Now suppose $\mathbf{y} \notin \mathbf{f}(\mathrm{bd}\, U)$. Then the *Brouwer degree* of $\mathbf{f}$ at the value $\mathbf{y} \in \mathbb{R}^n$ is defined by

$$\deg(\mathbf{f}, \mathbf{y}) := \sum_{\mathbf{f}(\mathbf{x})=\mathbf{y}} \mathrm{sgn}\, \det D_{\mathbf{x}}\mathbf{f} \tag{13.9}$$

if $\mathbf{y}$ is a regular value, and by

$$\deg(\mathbf{f}, \mathbf{y}) := \lim_{n \to \infty} \deg(\mathbf{f}, \mathbf{y}_n) \tag{13.10}$$

otherwise, where $\mathbf{y}_n$ is a sequence of regular values converging to $\mathbf{y}$ (it can be shown that this limit is well defined). For a regular value $\mathbf{y}$, the degree counts the difference between the number of orientation-preserving and orientation-reversing roots of $\mathbf{f}(\mathbf{x}) = \mathbf{y}$.

The basic property of the degree is its *homotopy invariance*: if $\mathbf{f}_t$ (with $t \in [0,1]$) is a continuous family of mappings from $\bar{U}$ to $\mathbb{R}^n$, and if $\mathbf{y}$ does not belong to any $\mathbf{f}_t(\mathrm{bd}\, U)$, then $t \to \deg(\mathbf{f}_t, \mathbf{y})$ is constant and hence $\deg(\mathbf{f}_1, \mathbf{y}) = \deg(\mathbf{f}_0, \mathbf{y})$. It follows that if $\mathbf{f}$ and $\mathbf{g}$ are defined on $\bar{U}$ and coincide on $\mathrm{bd}\, U$, then $\deg(\mathbf{f}, \mathbf{y}) = \deg(\mathbf{g}, \mathbf{y})$ for all $\mathbf{y} \notin \mathbf{f}(\mathrm{bd}\, U)$. One just has to apply the previous result to the family $\mathbf{f}_t = t\mathbf{g} + (1-t)\mathbf{f}$, noting that $\mathbf{f}_0 = \mathbf{f}$ and $\mathbf{f}_1 = \mathbf{g}$.

Furthermore, if $\mathbf{y}_0$ and $\mathbf{y}_1$ belong to the same component of $\mathbb{R}^n \backslash \mathbf{f}(\mathrm{bd}\, U)$, i.e. if there is a continuous path $t \to \mathbf{y}_t$ not intersecting $\mathbf{f}(\mathrm{bd}\, U)$ for $t \in [0,1]$, then $\deg(\mathbf{f}, \mathbf{y}_0) = \deg(\mathbf{f}, \mathbf{y}_1)$. The degree $\deg(\mathbf{f}, \mathbf{y})$ depends only on the values of $\mathbf{f}$ on $\mathrm{bd}\, U$ and on the component of $\mathbb{R}^n \backslash \mathbf{f}(\mathrm{bd}\, U)$ containing $\mathbf{y}$.

Of particular interest, of course, are the zeros of the vector field $\mathbf{f}$, i.e. the rest points of the differential equation $\dot{\mathbf{x}} = \mathbf{f}(\mathbf{x})$. If the value $\mathbf{y}$ is not otherwise specified, the degree of $\mathbf{f}$ is understood to be $\deg(\mathbf{f}, \mathbf{0})$. If $\mathbf{f}$ and $\mathbf{g}$ are two vector fields on $\bar{U}$ such that the degree is defined (which means that they do not vanish on $\mathrm{bd}\, U$) and if they never point in opposite directions on $\mathrm{bd}\, U$ (i.e. if there are no $\mathbf{x} \in \mathrm{bd}\, U$ and $\lambda > 0$ such that $\mathbf{f}(\mathbf{x}) + \lambda\mathbf{g}(\mathbf{x}) = \mathbf{0}$) then $\deg(\mathbf{f}, \mathbf{0}) = \deg(\mathbf{g}, \mathbf{0})$. This can be seen by looking at the family $\mathbf{f}_t = t\mathbf{g} + (1-t)\mathbf{f}$, which has no zero on $\mathrm{bd}\, U$.

As a consequence, we obtain the celebrated *fixed point theorem* of Brouwer:

Theorem 13.2.1 *For any map* $\mathbf{h}$ *of the unit ball* $D = \{\mathbf{x} \in \mathbb{R}^n : \|\mathbf{x}\| \leq 1\}$ *into itself there exists an* $\mathbf{x} \in D$ *such that* $\mathbf{h}(\mathbf{x}) = \mathbf{x}$.

Indeed, if $\mathbf{h}(\mathbf{x}) \neq \mathbf{x}$ for all $\mathbf{x} \in D$, then the vector field $\mathbf{f}(\mathbf{x}) = \mathbf{h}(\mathbf{x}) - \mathbf{x}$ points inwards for all $\mathbf{x} \in \mathrm{bd}\, D$. On $\mathrm{bd}\, D$, therefore, it never points in the opposite direction of $\mathbf{g}(\mathbf{x}) = -\mathbf{x}$. But then $\deg(\mathbf{f}, \mathbf{0}) = \deg(\mathbf{g}, \mathbf{0}) = (-1)^n \neq 0$, and so there is some $\mathbf{x} \in D$ with $\mathbf{f}(\mathbf{x}) = \mathbf{0}$, which is a contradiction. (We have assumed here that the map $\mathbf{h}$ is differentiable. The theorem of Brouwer holds, in fact, for any continuous $\mathbf{h}$. This can be proved by approximating $\mathbf{h}$ uniformly by a sequence of differentiable vector fields $\mathbf{h}_n$.)

Exercise 13.2.2 Show that if $\dot{\mathbf{x}} = \mathbf{f}(\mathbf{x})$ is a differential equation for which the ball D (or any set homeomorphic to it) is forward invariant, then D contains a rest point.

Now let $\hat{\mathbf{x}}$ be an isolated rest point of the differential equation $\dot{\mathbf{x}} = \mathbf{f}(\mathbf{x})$ defined on the open set $U \subseteq \mathbb{R}^n$. The *Poincaré index* of $\hat{\mathbf{x}}$ with respect to the vector field $\mathbf{f}$ is defined as

$$i(\hat{\mathbf{x}}) = \deg(\mathbf{f}, \mathbf{0}) \qquad (13.11)$$

where $\mathbf{f}$, here, denotes the restriction to a closed ball $\bar{B} \subseteq U$ containing $\hat{\mathbf{x}}$ but no other rest point. It can easily be shown that this degree does not depend on the particular choice of B.

If $\hat{\mathbf{x}}$ is regular, then $i(\hat{\mathbf{x}})$ is just the sign of $\det D_{\hat{\mathbf{x}}}\mathbf{f}$. Hence

$$i(\hat{\mathbf{x}}) = (-1)^\sigma \qquad (13.12)$$

where σ is the number of real negative eigenvalues of the Jacobian $D_{\hat{\mathbf{x}}}\mathbf{f}$. For $n = 2$, for example, the index of a centre, a sink or a source is $+1$, while that of a saddle is -1.

There exists an important *topological characterization* of $i(\hat{\mathbf{x}})$, valid also for nonregular $\hat{\mathbf{x}}$. Let us consider the map

$$\mathbf{x} \to \mathbf{h}(\mathbf{x}) = \frac{\mathbf{f}(\mathbf{x})}{\|\mathbf{f}(\mathbf{x})\|} \qquad (13.13)$$

from $\mathrm{bd}\, B$ to the unit sphere in $\mathbb{R}^n$. The degree of such a map between two $(n-1)$-dimensional spheres can be defined just as before. Then

$$i(\hat{\mathbf{x}}) = \deg(\mathbf{h}, \mathbf{y}) \qquad (13.14)$$

where $\mathbf{y}$ is any regular value of $\mathbf{h}$ (it does not matter which one, since $\mathrm{bd}\, B$

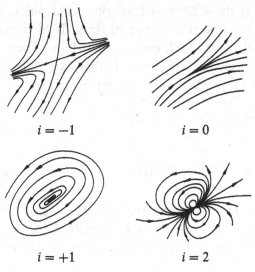

$i = -1$ $i = 0$

$i = +1$ $i = 2$

Fig. 13.1.

has an empty boundary). In fig. 13.1 we sketch a few examples for $n = 2$. If $\hat{x}$ is nonregular, the index $i(\hat{x})$ may take values different from ± 1.

Exercise 13.2.3 Show that for the differential equation $\dot{z} = z^n$ on $\mathbb{R}^2$ (written in complex notation), the origin has index n.

The amazing *Poincaré–Hopf theorem* states that if M is a compact manifold with boundary and **f** is a vector field on M pointing outward at the boundary points and having only finitely many fixed points, then

$$\sum i = \chi(M) \tag{13.15}$$

where the sum is over all rest points in M and where the right hand side, the so-called *Euler characteristic*, is a topological invariant depending only on M. The sum of the indices, thus, does not depend on the vector field.

We shall illustrate this with some two-dimensional examples. If B is the unit disk in $\mathbb{R}^2$, then $\chi(B) = 1$. Let us consider a planar vector field admitting a closed orbit γ and defined everywhere in its interior Γ. Along the boundary of Γ, i.e. along γ, it will point outwards (at least after a continuous transformation). It follows from the index theorem that Γ must contain a rest point (we know this also from the Poincaré–Bendixson theorem). If all rest points in Γ are regular, there must be an odd number $2n + 1$ of them, of which exactly n are saddles. In particular, if there is a unique rest point in Γ, it cannot be a saddle.

Next let us look at a vector field on the two-dimensional sphere M, for which $\chi(M) = 2$. It must have at least one fixed point. If all fixed points are regular, there must be at least two of them. If M is the two-dimensional surface of the solid torus in $\mathbb{R}^3$, on the other hand, there exist vector fields without rest points: indeed, $\chi(M) = 0$. If there are only regular rest points, they must be even in number, half of them being saddles.

Exercise 13.2.4 Draw such vector fields.

13.3 An index theorem for permanent systems

Theorem 13.3.1 *If (13.2) is permanent, then the degree of the vector field is* $(-1)^n$ *with respect to any bounded set* U *(with* $\bar{U} \subseteq \text{int}\,\mathbb{R}^n_+$*) containing all interior* ω*-limits. In particular, there exists a rest point in* $\text{int}\,\mathbb{R}^n_+$*.*

Proof Let $K \subseteq \text{int}\,\mathbb{R}^n_+$ be a compact set containing all ω-limits in its interior. For an orbit starting at $\mathbf{x} \in \text{int}\,\mathbb{R}^n_+$, let $\tau(\mathbf{x})$ be the time of first entry into $\text{int}\,K$:

$$\tau(\mathbf{x}) = \inf\{t \geq 0 : \mathbf{x}(t) \in \text{int}\,K\} \quad .$$

It is easy to see that τ is well defined on $\text{int}\,\mathbb{R}^n_+$, upper semicontinuous and therefore locally bounded. Let

$$T = \max_{\mathbf{x} \in K} \tau(\mathbf{x})$$

be the maximum time for orbits leaving K to return to K. The set

$$K^+ = \{\mathbf{x}(t) : \mathbf{x} \in K, 0 \leq t \leq T\}$$

is compact and forward invariant.
Now let $B \subseteq \text{int}\,\mathbb{R}^n_+$ be homeomorphic to a ball and contain K^+. The entry time $\tau(\mathbf{x})$ will again attain an upper bound $\hat{T}$ on B. Let $\mathbf{h}(\mathbf{x})$ denote the right hand side of (13.2), and consider the following homotopy:

$$\begin{aligned}
\mathbf{h}_t(\mathbf{x}) &= \mathbf{h}(\mathbf{x}) && \text{for } t = 0 \\
\mathbf{h}_t(\mathbf{x}) &= \tfrac{1}{t}(\mathbf{x}(t) - \mathbf{x}(0)) && \text{for } t > 0 \quad .
\end{aligned} \tag{13.16}$$

Clearly $\mathbf{h}_t(\mathbf{x}) \neq \mathbf{0}$ for all $\mathbf{x} \in \text{bd}\,B$ and $t \in \mathbb{R}_+$, since there are no fixed or periodic points on $\text{bd}\,B$. Thus the degree of the vector field $\mathbf{x} \to \mathbf{h}_t(\mathbf{x})$ with respect to B is defined for all $t \in [0, \hat{T}]$ and is independent of t. For $t = \hat{T}$, the vector field $\mathbf{h}_t(\mathbf{x})$ points inward along $\text{bd}\,B$, and hence its degree is $(-1)^n$. Thus the degree of $\mathbf{h}_0(\mathbf{x}) = \mathbf{h}(\mathbf{x})$ with respect to B (or any other open bounded set containing K) is $(-1)^n$. $\qquad\square$

The same result obviously holds for (13.1), except that the degree is $(-1)^{n-1}$, since the dimension is reduced by 1.

13.4 Saturated rest points and a general index theorem

Before proceeding further, it will be useful to define a rest point $\mathbf{p}$ of (13.2) (resp. (13.1)) as *saturated* if $f_i(\mathbf{p}) \leq 0$ (resp. $f_i(\mathbf{p}) \leq \bar{f}(\mathbf{p})$) whenever $p_i = 0$ (for $p_i > 0$, of course, the equality sign must hold). Every rest point in the interior of the state space is trivially saturated. For a rest point on the boundary, the condition means that the dynamics does not 'call for' the missing species.

To be saturated is an eigenvalue condition. Indeed, the Jacobian of (13.2) at $\mathbf{p}$ is of the form

$$\frac{\partial \dot{x}_i}{\partial x_j} = \delta_{ij} f_i(\mathbf{p}) + p_i \frac{\partial f_i}{\partial x_j}(\mathbf{p}) \ .$$

(where δ_{ij} is the Kronecker delta: $\delta_{ij} = 1$ if $i = j$, $\delta_{ij} = 0$ otherwise). If $p_i = 0$, this reduces to $f_i(\mathbf{p})$ for $i = j$ and to 0 for $i \neq j$, so that $f_i(\mathbf{p})$ is an eigenvalue for the (left) eigenvector $\mathbf{e}_i$. Hence we call it a *transversal eigenvalue*. It measures the rate of approach to the face $x_i = 0$ close to the rest point $\mathbf{p}$. Similarly the transversal eigenvalues of (13.1) are of the form $f_i(\mathbf{p}) - \bar{f}(\mathbf{p})$. The rest point $\mathbf{p}$ is saturated if and only if all its transversal eigenvalues are nonpositive. With this concept at hand we can state a general *index theorem for replicator equations.*

Theorem 13.4.1 *There exists at least one saturated rest point for* (13.1). *If all saturated rest points are regular, the sum of their indices is* $(-1)^{n-1}$, *and hence their number is odd.*

Proof Let us consider

$$\dot{x}_i = x_i \big(f_i(\mathbf{x}) - \bar{f}(\mathbf{x}) - n\varepsilon \big) + \varepsilon \tag{13.17}$$

as a perturbation of (13.1), with a small $\varepsilon > 0$ representing an influx term. Clearly (13.17) maintains the relation $\sum \dot{x}_i = 0$ on S_n. On the boundary, the flow now points *into* int S_n. The sum of the indices of the rest points of (13.17) in int S_n is therefore $(-1)^{n-1}$.

For every $\varepsilon > 0$, there must consequently exist at least one rest point $\mathbf{p}(\varepsilon)$ of (13.17) in int S_n. Then

$$f_i(\mathbf{p}(\varepsilon)) - \bar{f}(\mathbf{p}(\varepsilon)) - n\varepsilon = -\frac{\varepsilon}{p_i(\varepsilon)} < 0 \ . \tag{13.18}$$

Any accumulation point $\mathbf{p}$ of $\mathbf{p}(\varepsilon)$ for $\varepsilon \to 0$ satisfies $f_i(\mathbf{p}) - \bar{f}(\mathbf{p}) \leq 0$ by continuity and is therefore a saturated rest point of (13.1). This shows the first claim.

If every saturated rest point $\mathbf{p}$ of (13.1) is regular, then there are only finitely many of them (since S_n is compact), and each one continues, by the implicit function theorem, to a unique family $\mathbf{p}(\varepsilon)$ of fixed points of (13.17), as long as ε is sufficiently small. Clearly $\mathbf{p}(\varepsilon) \in \operatorname{int} S_n$ for $\varepsilon > 0$ since $f_i(\mathbf{p}) < \bar{f}(\mathbf{p})$ implies $p_i(\varepsilon) > 0$ by (13.18). But as we have seen, fixed points of (13.17) accumulate for $\varepsilon \to 0$ only at *saturated* rest points of (13.1). Hence every rest point of (13.17) lies on one of these families $\mathbf{p}(\varepsilon)$, for ε sufficiently small. As the index does not change when ε reaches $\mathbf{0}$ (by regularity), the above index relation for $\varepsilon > 0$ holds also for $\varepsilon = 0$. □

Exercise 13.4.2 Give a heuristic argument (by drawing a picture) that regular saturated rest points move into the interior and nonsaturated ones into the exterior of S_n, if an influx term $\varepsilon > 0$ is introduced as in (13.17).

Exercise 13.4.3 Show that a 'degenerate' saturated rest point (with at least one transversal eigenvalue being equal to 0) can move inwards, outwards, disappear, or split up into several rest points, if an immigration term is introduced. (As a typical example, consider a one- or two-dimensional Lotka–Volterra equation with $\mathbf{0}$ as a degenerate saturated rest point, i.e. $r_i = \mathbf{0}$ for all i.)

Exercise 13.4.4 Prove the *index theorem for ecological equations*: if (13.2) has uniformly bounded orbits, then it has a saturated fixed point, and if all saturated rest points are regular, then the sum of their indices is $(-1)^n$.

Exercise 13.4.5 Let $\hat{\mathbf{x}} \in \mathbb{R}^n_+$ be an isolated fixed point of (13.2) and let U be the intersection of an isolating neighbourhood of $\hat{\mathbf{x}}$ with $\operatorname{int} \mathbb{R}^n_+$. Let $-\varepsilon$ be the vector with components $-\varepsilon_i < 0$ and $\mathbf{h}$ the vector field with components $x_i f_i(\mathbf{x})$ on U. Show that bd-ind $(\hat{\mathbf{x}}) := \lim_{\varepsilon \to \mathbf{0}} \deg(\mathbf{h}, -\varepsilon)$ is well defined. Show that bd-ind $(\hat{\mathbf{x}}) = i(\hat{\mathbf{x}})$ if $\hat{\mathbf{x}}$ is saturated and regular, but that bd-ind $(\hat{\mathbf{x}}) = 0$ if $\hat{\mathbf{x}}$ is not saturated, and that theorem 13.4.1 extends to isolated fixed points for this 'boundary-index'. If bd-ind $(\hat{\mathbf{x}}) \neq \mathbf{0}$, then there exist points $\mathbf{x} \in \mathbb{R}^n_+$ arbitrarily close to $\hat{\mathbf{x}}$ such that $f_i(\hat{\mathbf{x}}) < 0$ for all i with $\hat{x}_i = 0$. Prove the equivalent statements for (13.1).

For equations (13.6) and (13.7), the rest point $\mathbf{p}$ is saturated if $(A\mathbf{p})_i \leq \mathbf{p} \cdot A\mathbf{p}$ resp. $r_i + (A\mathbf{p})_i \leq 0$ whenever $p_i = 0$. Thus *the saturated rest points $\mathbf{p}$ of* (13.6) *are the (symmetric) Nash equilibria for the game with payoff matrix A:*

$$\mathbf{x} \cdot A\mathbf{p} \le \mathbf{p} \cdot A\mathbf{p} \quad \text{for all} \quad \mathbf{x} \in S_n .$$

As immediate corollaries of theorem 13.4.1, one obtains therefore *the existence of a Nash equilibrium* and the *odd number theorem for regular Nash equilibria,* both classical results in game theory.

Exercise 13.4.6 Show that the same 'dynamical' proof also works for bimatrix games (see chapter 10).

Exercise 13.4.7 Show that a persistent system (13.1) always has a rest point $\mathbf{x} \in S_n$ satisfying $f_i(\mathbf{x}) = \bar{f}(\mathbf{x})$ for all i (but $\mathbf{x}$ may lie on the boundary).

We shall call (13.1) *robustly persistent* if it remains persistent under small perturbations of the f_i. (Here 'small' means that the values of the $f_i(\mathbf{x})$ *and* of the partial derivatives $\frac{\partial f_i}{\partial x_j}$ change only a little.) Similarly (13.6) will be called robustly persistent if it remains persistent under small perturbations of the a_{ij}.

We can now extend the index theorem 13.4.1 to robustly persistent systems.

Theorem 13.4.8 *If (13.1) is robustly persistent, then it has an interior fixed point. Moreover the degree of (13.1) is $(-1)^{n-1}$ with respect to any open set U with $\bar{U} \subseteq \operatorname{int} S_n$ which contains all interior fixed points. In particular for (13.6) the interior fixed point is unique and has index $(-1)^{n-1}$.*

Proof We note first that there are no saturated rest points on the boundary. Indeed, if $\mathbf{p}$ were such a rest point with $f_i(\mathbf{p}) \le \bar{f}(\mathbf{p})$ for all i with $p_i = 0$, then a suitable perturbation would turn $\mathbf{p}$ into a hyperbolic fixed point with all transversal eigenvalues negative. But then the stable manifold of $\mathbf{p}$ meets $\operatorname{int} S_n$. Thus there are interior orbits converging to $\mathbf{p}$, contradicting persistence.

Therefore the set S of saturated fixed points (which is always closed) is a compact subset of $\operatorname{int} S_n$. Let U be a convex neighbourhood of S with $\bar{U} \subseteq \operatorname{int} S_n$. For every $\mathbf{x} \in S_n \setminus U$ we have $\max_i (f_i(\mathbf{x}) - \bar{f}(\mathbf{x})) > 0$ since there is no saturated fixed point outside U. By compactness we obtain a $c > 0$ such that

$$\max_i \left[f_i(\mathbf{x}) - \bar{f}(\mathbf{x}) \right] \ge c > 0 \text{ for } \mathbf{x} \in S_n \setminus U.$$

Inserting this into (13.18) we observe that the perturbed equation (13.17) cannot have a fixed point in $S_n \setminus U$ either, as long as $\varepsilon < c/n$. By homotopy the degree of (13.1) with respect to U coincides with the degree of (13.17) with respect to U (or with respect to int S_n), which is $(-1)^{n-1}$. For (13.6) this implies that the interior fixed point exists and is unique and therefore regular. Hence its index is $(-1)^{n-1}$. $\qquad\square$

Exercise 13.4.9 Prove a similar result for robustly persistent systems (13.2) with uniformly bounded orbits.

13.5 Necessary conditions for permanence

Theorem 13.5.1 *If* (13.6) *or* (13.7) *is permanent, then there exists a unique interior rest point* $\hat{\mathbf{x}}$ *and, for each* $\mathbf{x}$ *in the interior of the state space,*

$$\lim_{T \to +\infty} \frac{1}{T} \int_0^T \mathbf{x}(t) dt = \hat{\mathbf{x}}.$$

Proof Theorem 13.3.1 implies that there exists at least one interior rest point. If there were two of them, the line l joining them would consist only of rest points, and intersect the boundary of the state space. But since the boundary is a repellor, there cannot be fixed points arbitrarily close by. The convergence of the time averages, finally, follows from theorems 5.2.3 and 7.6.4, and the fact that by permanence, if $\mathbf{x}$ is in the interior of the state space, then so is $\omega(\mathbf{x})$. $\qquad\square$

Theorem 13.5.2 *Let* (13.7) *be permanent and denote the Jacobian at the unique interior rest point* $\hat{\mathbf{x}}$ *by* D. *Then*

$$(-1)^n \det D \; > \; 0 \; , \tag{13.19}$$
$$\operatorname{tr} D \; < \; 0 \; , \tag{13.20}$$
$$(-1)^n \det A \; > \; 0 \; . \tag{13.21}$$

Proof The conditions on the signs of the determinants are simple consequences of the index theorem 13.3.1. Indeed, since $\hat{\mathbf{x}}$ is the unique solution of $-\mathbf{r} = A\mathbf{x}$, the matrix A is nonsingular. Also, theorem 13.3.1 implies that $i(\hat{\mathbf{x}}) = (-1)^n$. Now

$$d_{ij} = \hat{x}_i a_{ij} \tag{13.22}$$

and so D is also nonsingular. Thus $i(\hat{\mathbf{x}})$ is just the sign of $\det D$. This establishes (13.19) and via (13.22) also (13.21).
In order to prove the trace condition (13.20), we shall have to use the

Liouville formula (see section 11.3). Let us multiply the right hand side of (13.7) by the positive function

$$B(\mathbf{x}) = \prod x_i^{s_i - 1} \tag{13.23}$$

where the s_i will be specified later. The resulting equation $\dot{x}_i = h_i(\mathbf{x})$ with

$$h_i(\mathbf{x}) = x_i \left(r_i + \sum a_{ij} x_j \right) B(\mathbf{x}) \tag{13.24}$$

differs from (13.7) just by a change in velocity and is therefore also permanent. Since

$$\frac{\partial h_i}{\partial x_i} = B(\mathbf{x})\left[s_i(r_i + (A\mathbf{x})_i) + x_i a_{ii} \right] \quad, \tag{13.25}$$

we obtain

$$\operatorname{div} \mathbf{h}(\mathbf{x}) = B(\mathbf{x}) \left[\sum s_i(r_i + (A\mathbf{x})_i) + \sum x_i a_{ii} \right] \tag{13.26}$$

which at the rest point $\hat{\mathbf{x}}$ reduces to

$$\operatorname{div} \mathbf{h}(\hat{\mathbf{x}}) = B(\hat{\mathbf{x}}) \sum \hat{x}_i a_{ii} = B(\hat{\mathbf{x}}) \operatorname{tr} D \quad . \tag{13.27}$$

Hence

$$\begin{aligned}
\operatorname{div} \mathbf{h}(\mathbf{x}) &= B(\mathbf{x})\left[\sum_i s_i \sum_j a_{ij}(x_j - \hat{x}_j) + \sum_i a_{ii}(x_i - \hat{x}_i) + \operatorname{tr} D \right] \\
&= B(\mathbf{x})\left[\sum_j (x_j - \hat{x}_j)\left(\sum_i s_i a_{ij} + a_{jj} \right) + \operatorname{tr} D \right] . \tag{13.28}
\end{aligned}$$

Since A is nonsingular, we may choose s_i such that for all j

$$\sum_i s_i a_{ij} + a_{jj} = 0 \quad . \tag{13.29}$$

Then

$$\operatorname{div} \mathbf{h}(\mathbf{x}) = B(\mathbf{x}) \operatorname{tr} D \quad . \tag{13.30}$$

Since there exists a ball in int $\mathbb{R}^n_+$ which, in time T, gets shrunk (see the proof of the index theorem 13.3.1), Liouville's formula (11.9) together with (13.30) implies (13.20). □

The latter part of this proof shows that if tr$D$ is positive, negative or zero, then the flow of the Lotka–Volterra equation is (essentially) volume expanding, volume contracting, or volume preserving, respectively. Thus the Lotka–Volterra equation (13.7) and its linearization affect volumes in the same way.

Theorem 13.5.3 *Let* (13.6) *with* $a_{ii} = 0$ *be permanent and* $\hat{\mathbf{x}}$ *be the interior fixed point. Then*

$$\hat{\mathbf{x}} \cdot A\hat{\mathbf{x}} > 0 \qquad\qquad (13.31)$$

and

$$(-1)^{n-1} \det A > 0 \quad . \qquad\qquad (13.32)$$

Proof We transform (13.6) into

$$\dot{y}_i = y_i(r_i + (A'\mathbf{y})_i) \quad i = 1, \ldots, n-1 \quad , \qquad\qquad (13.33)$$

as described in theorem 7.5.1, with $a_{ij}' = a_{ij} - a_{nj}$, and apply theorem 13.5.2. Computing the Jacobian of (13.6),

$$\frac{\partial \dot{x}_i}{\partial x_j} = \delta_{ij}\left((A\mathbf{x})_i - \mathbf{x} \cdot A\mathbf{x}\right) + x_i(a_{ij} - (A\mathbf{x})_j - (\mathbf{x}A)_j)$$

we see that its trace at $\hat{\mathbf{x}}$ is given by

$$\sum_{i=1}^{n} \frac{\partial \dot{x}_i}{\partial x_i}\Big|_{\hat{\mathbf{x}}} = \sum_i \hat{x}_i a_{ii} - 2\hat{\mathbf{x}} \cdot A\hat{\mathbf{x}} \quad .$$

Since we restrict (13.6) to the simplex we have to omit the one eigenvalue corresponding to the left eigenvector $\mathbf{1}$, which is $-\hat{\mathbf{x}} \cdot A\hat{\mathbf{x}}$. Thus the divergence of (13.6) *within* S_n at the point $\hat{\mathbf{x}}$ is given by

$$\sum \hat{x}_i a_{ii} - \hat{\mathbf{x}} \cdot A\hat{\mathbf{x}} \quad . \qquad\qquad (13.34)$$

Since we assume $a_{ii} = 0$, (13.20) implies (13.31).

Furthermore, applying Cramer's rule to the system $(A\hat{\mathbf{x}})_i = \hat{\mathbf{x}} \cdot A\hat{\mathbf{x}}$ yields

$$\hat{x}_n \det A = (\hat{\mathbf{x}} \cdot A\hat{\mathbf{x}}) \det A_n \qquad\qquad (13.35)$$

with

$$A_n = \begin{bmatrix} a_{11} & \cdots & a_{1,n-1} & 1 \\ \vdots & & \vdots & \vdots \\ a_{n1} & \cdots & a_{n,n-1} & 1 \end{bmatrix}.$$

Clearly

$$\det A_n = \begin{bmatrix} a_{11}' & \cdots & a_{1,n-1}' & 0 \\ \vdots & & \vdots & \vdots \\ a_{n-1,1}' & \cdots & a_{n-1,n-1}' & 0 \\ a_{n1} & \cdots & a_{n,n-1} & 1 \end{bmatrix} = \det A' \quad .$$

Since by (13.21) $(-1)^{n-1} \det A' > 0$, equations (13.31) and (13.35) imply (13.32). $\qquad\square$

Exercise 13.5.4 Subzero-sum games are games with $a_{ii} = 0$ and $a_{ij} + a_{ji} \leq 0$. An example is (13.8) with $\varepsilon \leq 0$. Show that a subzero-sum game cannot have attractors in int S_n or in the interior of a subface.

Exercise 13.5.5 If the replicator equation (13.6) admits a unique interior fixed point $\hat{\mathbf{x}}$, show that its index (within S_n) is given by

$$i(\hat{\mathbf{x}}) = \text{sgn} \frac{\det A}{\hat{\mathbf{x}} \cdot A\hat{\mathbf{x}}} \quad .$$

(Hint: use (13.35) and sgn det $A_n = i(\hat{\mathbf{x}})$. Note that $a_{ii} = 0$ is not required. As a consequence, $\hat{\mathbf{x}} \cdot A\hat{\mathbf{x}} = 0$ if and only if $\det A = 0$.)

Exercise 13.5.6 Show that (13.32) holds for permanent equations (13.6) whenever all $a_{ij} \geq 0$.

We conclude with another necessary condition for the permanence of Lotka–Volterra systems.

Theorem 13.5.7 *Assume that (13.7) is permanent or robustly persistent and admits a unique rest point* $\mathbf{y}$ *in the interior of the face* $x_k = 0$. *Then the minor* $A^{(k)}$ *obtained by deleting the k-th row and column of A satisfies*

$$(-1)^{n-1} \det A^{(k)} > 0 \quad .$$

Proof It is enough to show that the transversal eigenvalue at $\mathbf{y}$ is given by

$$- \hat{x}_k \frac{\det A}{\det A^{(k)}} \quad . \tag{13.36}$$

Since $A\hat{\mathbf{x}} + \mathbf{r} = \mathbf{0}$, Cramer's rule implies that this expression is equal to

$$\frac{1}{\det A^{(k)}} \det (a_{i1}, \ldots, r_i, \ldots, a_{in})$$

with $\mathbf{r} = (r_i)$ in the k-th column. Developing this along the k-th row yields

$$\frac{1}{\det A^{(k)}} \sum_j a'_{kj} (-1)^{k+j} \det^k_j (a_{i1}, \ldots, r_i, \ldots, a_{in})$$

where $a'_{kj} = a_{kj}$ if $k \neq j, a'_{kk} = r_k$ and $\det^k_j$ indicates that the k-th row and the j-th column of $\det(a_{i1}, \ldots, r_i, \ldots, a_{in})$ are deleted. But we also have $A^{(k)}\mathbf{y} + \mathbf{r}^{(k)} = \mathbf{0}$ (where $\mathbf{r}^{(k)}$ is obtained from $\mathbf{r}$ by removing the k-th component), so that the previous expression coincides with

$$r_k + \sum_{j \neq k} a_{kj} y_j \quad .$$

Now this is just the transversal eigenvalue at $\mathbf{y}$, which has to be positive. □

Exercise 13.5.8 Let (13.6) admit an interior fixed point $\hat{\mathbf{x}}$ and a fixed point $\mathbf{y}$ in the interior of the face $x_k = 0$. Show that the transversal eigenvalue at $\mathbf{y}$ has the same sign as

$$\frac{-(\mathbf{y}\cdot A\mathbf{y})\det A}{(\hat{\mathbf{x}}\cdot A\hat{\mathbf{x}})\det A^{(k)}} \quad .$$

13.6 Sufficient conditions for permanence

The following theorem is a very useful strengthening of theorem 12.2.2 on average Lyapunov functions.

Theorem 13.6.1 *The replicator system* (13.6) *is permanent if there exists a* $\mathbf{p} \in \text{int } S_n$ *such that*

$$\mathbf{p}\cdot A\mathbf{x} > \mathbf{x}\cdot A\mathbf{x} \tag{13.37}$$

for all rest points $\mathbf{x} \in \text{bd } S_n$.

It is remarkable that only rest points are involved: the ω-limit sets on the boundary may be considerably more complicated. The similarity to the ESS condition (6.16) is also intriguing.

Proof It is enough to show that

$$P(\mathbf{x}) = \prod_i x_i^{p_i} \tag{13.38}$$

is an average Lyapunov function. Clearly $P(\mathbf{x}) = 0$ for $\mathbf{x} \in \text{bd } S_n$ and $P(\mathbf{x}) > 0$ for $\mathbf{x} \in \text{int } S_n$. The function

$$\Psi(\mathbf{x}) = \mathbf{p}\cdot A\mathbf{x} - \mathbf{x}\cdot A\mathbf{x} \tag{13.39}$$

satisfies $\dot{P} = P\Psi$ in int S_n. It remains to show that for every $\mathbf{y} \in \text{bd } S_n$, there is a $T > 0$ such that

$$\int_0^T \Psi(\mathbf{y}(t))dt > 0 \quad . \tag{13.40}$$

The proof proceeds by induction on the number r of positive components of $\mathbf{y}$. For $r = 1$, (13.40) is an obvious consequence of (13.37), since $\mathbf{y}$ is then a corner of S_n and hence a boundary rest point. Assume now that (13.40) is valid for $r = 1, \ldots, m-1$ and let $I = \text{supp}(\mathbf{y})$ have cardinality m. We have to distinguish two cases:

(a) $\mathbf{y}(t)$ converges to the boundary of the simplex

$$S_n(I) = \{\mathbf{x} \in S_n : x_i = 0 \text{ for all } i \notin I\} \quad .$$

Then $\omega(\mathbf{y})$ is contained in a union of faces of dimension $\leq m - 1$. By our assumption, (13.40) holds for all $\mathbf{z} \in \omega(\mathbf{y})$. The argument in theorem 12.2.2 then shows that (13.40) holds for $\mathbf{y}$ as well.

(b) $\mathbf{y}(t)$ does not converge to bd $S_n(I)$.

In that case, there exists an $\varepsilon > 0$ and a sequence $T_k \to +\infty$ such that $y_i(T_k) > \varepsilon$ for all $i \in I$ and $k = 1, 2, \ldots$ Let us write

$$\bar{y}_i(T) = \frac{1}{T} \int_0^T y_i(t)dt \quad \text{and} \quad a(T) = \frac{1}{T} \int_0^T \mathbf{y}(t) \cdot A\mathbf{y}(t)dt \quad .$$

Since the sequences $\bar{y}_i(T_k)$ and $a(T_k)$ are obviously bounded, we may obtain a subsequence — which we shall again denote by T_k — such that $\bar{y}_i(T_k)$ and $a(T_k)$ converge: the limits will be denoted by $\bar{x}_i$ and $\bar{a}$. For $i \in I$, we have

$$(\log y_i)' = (A\mathbf{y})_i - \mathbf{y} \cdot A\mathbf{y} \quad .$$

Integrating this from 0 to T_k and dividing by T_k, we obtain

$$\frac{1}{T_k}(\log y_i(T_k) - \log y_i(0)) = (A\bar{\mathbf{y}}(T_k))_i - a(T_k) \quad . \tag{13.41}$$

Since $\log y_i(T_k)$ is bounded, the left hand side converges to 0. Hence

$$(A\bar{\mathbf{x}})_i = \bar{a} \text{ for all } i \in I \quad . \tag{13.42}$$

Furthermore $\sum \bar{x}_i = 1, \bar{x}_i \geq 0$ and $\bar{x}_i = 0$ for $i \notin I$. Hence $\bar{a} = \bar{\mathbf{x}} \cdot A\bar{\mathbf{x}}$ and $\bar{\mathbf{x}}$ is a rest point in $S_n(I)$. Now

$$\frac{1}{T_k} \int_0^{T_k} \Psi(\mathbf{y}(t))dt = \sum_{i=1}^n p_i \frac{1}{T_k} \int_0^{T_k} [(A\mathbf{y})_i - \mathbf{y} \cdot A\mathbf{y}] \, dt$$

converges to

$$\sum_{i=1}^n p_i [(A\bar{\mathbf{x}})_i - \bar{\mathbf{x}} \cdot A\bar{\mathbf{x}}] = \mathbf{p} \cdot A\bar{\mathbf{x}} - \bar{\mathbf{x}} \cdot A\bar{\mathbf{x}}$$

which is positive by (13.37). This implies (13.40) and hence permanence. $\quad\square$

Thus in order to prove permanence, we have to find out whether there exists a positive solution $\mathbf{p}$ for the linear inequalities

$$\sum_{i:x_i=0} p_i [(A\mathbf{x})_i - \mathbf{x} \cdot A\mathbf{x}] > 0 \tag{13.43}$$

(where $\mathbf{x}$ runs through the boundary rest points). The coefficients $(A\mathbf{x})_i - \mathbf{x} \cdot A\mathbf{x}$ are the transversal eigenvalues of $\mathbf{x}$. The number of inequalities may be quite large, but the following exercise shows that the problem is finite. Applications of theorem 13.6.1 will be given in chapters 14 and 16.

Exercise 13.6.2 If there is a continuum of rest points on the boundary, show that it is sufficient to test the inequalities (13.43) on the extremal points of the continuum.

Exercise 13.6.3 Show that similarly a Lotka–Volterra equation (13.7) with uniformly bounded orbits is permanent if there exists a $\mathbf{p} \in \operatorname{int} \mathbb{R}^n_+$ such that

$$\mathbf{p} \cdot (\mathbf{r} + A\mathbf{x}) > 0 \tag{13.44}$$

for all fixed points $\mathbf{x} \in \operatorname{bd} \mathbb{R}^n_+$, i.e. if there exists a positive solution $\mathbf{p}$ for

$$\sum_{i:x_i=0} p_i \left[r_i + (A\mathbf{x})_i \right] > 0 \tag{13.45}$$

(where $\mathbf{x}$ runs through the boundary rest points).

Exercise 13.6.4 Show that the set of solutions $\mathbf{p}$ of (13.43) is an open convex subset of $\operatorname{int} S_n$. Show that in general, this subset need not contain the interior fixed point. But for the hypercycle, every $\mathbf{p} \in \operatorname{int} S_n$ works.

Exercise 13.6.5 Investigate the inhomogeneous hypercycle

$$\dot{x}_i = x_i(b_i + k_i x_{i-1} - \bar{f})$$

with $b_i, k_i > 0$. Show that it is permanent if and only if it admits an interior rest point. (Hint: either use theorem 13.6.1 with $p_i = 1/k_i$ or imitate the argument in section 12.3.)

A geometric condition for permanence is given by the following:

Theorem 13.6.6 *Consider a Lotka–Volterra equation (13.7) with uniformly bounded orbits. If the set*

$$D = \{\mathbf{x} \in \mathbb{R}^n_+ : \mathbf{r} + A\mathbf{x} \leq \mathbf{0}\} \tag{13.46}$$

(where no x_i increases) is disjoint from the convex hull C of all boundary fixed points, then (13.7) is permanent.

Proof A rest point $\mathbf{z}$ of (13.7) lies in D if and only if it is saturated. The assumption $C \cap D = \emptyset$ implies that there are no saturated rest points on the boundary. Theorem 13.4.1 then shows the existence of an interior rest point $\hat{x}$; this point is unique, since otherwise we would have a line of fixed points intersecting $\text{bd}\,\mathbb{R}_+^n$, i.e. a nonempty intersection of C and D. It follows that A is nonsingular. The convex set C can be separated from the convex set $\hat{D} = \{\mathbf{x} \in \mathbb{R}^n : \mathbf{r} + A\mathbf{x} \le \mathbf{0}\}$ by a hyperplane. Since A is nonsingular we can write the separating functional in the form $\mathbf{p}A$, with $\mathbf{p} \in \mathbb{R}^n$. Then

$$\mathbf{p}\cdot A\mathbf{z} > \mathbf{p}\cdot A\mathbf{x} \qquad\qquad (13.47)$$

for all $\mathbf{x} \in \hat{D}$ and all fixed points $\mathbf{z} \in \text{bd}\,\mathbb{R}_+^n$. Since the interior fixed point $\hat{x}$ lies in $\hat{D}$, we obtain in particular

$$\mathbf{p}\cdot A\mathbf{z} > \mathbf{p}\cdot A\hat{x} = -\mathbf{p}\cdot\mathbf{r} \quad . \qquad\qquad (13.48)$$

Thus $\mathbf{p}$ is a solution of (13.44).

It remains to show that $\mathbf{p}$ is positive. Let $\mathbf{v}$ be any vector in $\mathbb{R}^n$ with $A\mathbf{v} \le \mathbf{0}$. Then $\hat{x} + c\mathbf{v}$ belongs to $\hat{D}$ for every $c > 0$ since

$$\mathbf{r} + A(\hat{x} + c\mathbf{v}) = cA\mathbf{v} \le \mathbf{0} \quad .$$

Thus

$$\mathbf{p}\cdot A\mathbf{z} > \mathbf{p}\cdot A\hat{x} + c\mathbf{p}\cdot A\mathbf{v} \quad . \qquad\qquad (13.49)$$

This can hold for arbitrarily large c only if $\mathbf{p}\cdot A\mathbf{v} \le 0$ for all such $\mathbf{v}$. Therefore $\mathbf{p} \ge \mathbf{0}$. Since $\mathbf{z}$ varies in a compact set, the set of solutions $\mathbf{p}$ of (13.48) is open, and we can find a solution $\mathbf{p} > \mathbf{0}$. Hence (13.7) is permanent. $\square$

As a simple special case, let us consider two-species competition (see section 3.3). In the presence of an interior rest point there are two possible types of behaviour:

(a) *global stability* (and hence permanence) if the interior fixed point is above the line joining the two one-species rest points;
(b) *bistability* (and thus competitive exclusion) if it is below.

Exercise 13.6.7 Show that the geometric condition of theorem 13.6.6 is actually equivalent to the one given in exercise 13.6.3.

Exercise 13.6.8 Check the three-species system (5.18) from the point of view of theorem 13.6.6.

These examples might suggest that the geometric condition $C \cap D = \emptyset$ is not only sufficient, but also necessary for (at least a robust type of) permanence. This will be shown for $n = 3$ in section 16.2. For $n \ge 4$ this is not true, however. See exercise 16.4.6 for a counterexample.

13.7 Notes

The notion of permanence was introduced in Schuster *et al.* (1979a). Persistence (now usually called weak persistence) was introduced by Freedman and Waltman (1977). A completely different approach yielding sufficient conditions for persistence is due to Butler and Waltman (1986). Butler *et al.* (1986) give conditions under which persistence implies permanence. Freedman and Moson (1990) survey various definitions of persistence. The first to show that permanence implies the existence of an interior rest point were Hutson and Moran (1982), see also Hutson (1990). The present approach, in particular the index theorems 13.3.1 and 13.4.1 and their application to Lotka–Volterra equations, is due to Hofbauer (1988); for a generalization to semiflows see Hofbauer (1990). Theorem 13.6.1 is due to Jansen (1987). The importance of chain recurrent sets on the boundary for the study of permanence was pointed out in Garay (1989), Hofbauer (1989), and Hofbauer and So (1989). For a comprehensive survey including notions of permanence for other dynamical systems, see Hutson and Schmitt (1992). For shorter introductions see Anderson *et al.* (1992) and Sigmund (1995). For applications to food webs and ecological communities, see Cohen (1988), Pimm *et al.* (1991) and Yodzis (1990). For numerical experiments on community construction, we refer to Taylor (1988), Stadler and Happel (1993), Law and Blackford (1992), Law and Morton (1993), and the survey by Nee (1990).

14

Replicator networks

In the first part of this chapter we study replicator equations with cyclic symmetry and show that for n > 3, Hopf bifurcations can lead to periodic attractors. We then describe the relation between the graph of an autocatalytic network and its permanence. Finally, we characterize permanence for essentially hypercyclic networks, by using M-matrices.

14.1 A periodic attractor for $n = 4$

We have seen in section 4.2 that the two-dimensional Lotka–Volterra equation admits no limit cycle. The three-dimensional one does, however. This follows from the fact that *the replicator equation*

$$\dot{x}_i = x_i((A\mathbf{x})_i - \mathbf{x} \cdot A\mathbf{x}) \tag{14.1}$$

on S_n admits a (supercritical) Hopf bifurcation for $n = 4$.

We shall use the matrix

$$A = \begin{bmatrix} 0 & 0 & -\mu & 1 \\ 1 & 0 & 0 & -\mu \\ -\mu & 1 & 0 & 0 \\ 0 & -\mu & 1 & 0 \end{bmatrix} . \tag{14.2}$$

It is obvious that $\mathbf{m} = (\frac{1}{4}, \frac{1}{4}, \frac{1}{4}, \frac{1}{4})$ is a rest point for (14.1). Since A is circulant, the Jacobian D of (14.1) at $\mathbf{m}$ will also be circulant. A simple computation shows that the first row of D is given by

$$\frac{-1 + \mu}{8}, \frac{-1 + \mu}{8}, \frac{-1 - \mu}{8}, \frac{1 + \mu}{8} . \tag{14.3}$$

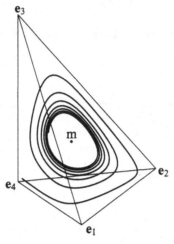

Fig. 14.1.

The eigenvalues can easily be computed by (5.20). The eigenvalue

$$\gamma_0 = \frac{1}{4}(-1 + \mu)$$

belongs to the eigenvector **1** which is orthogonal to S_4. Since we are only interested in the restriction of (14.1) to S_4, we may exclude it from further consideration. The other eigenvalues are

$$\gamma_1 = \frac{1}{4}(\mu - i) \quad \gamma_2 = \frac{1}{4}(-1 - \mu) \quad \gamma_3 = \frac{1}{4}(\mu + i). \qquad (14.4)$$

If μ varies from $-\frac{1}{2}$ to $+\frac{1}{2}$, then γ_2 will be negative, while the complex conjugate pair γ_1 and γ_3 crosses the imaginary axis, from left to right, for $\mu = 0$.

For $\mu < 0$, therefore, **m** is a sink; for $\mu > 0$, it is unstable; for $\mu = 0$, finally, the matrix A describes the hypercycle with $n = 4$: as we know from theorem 12.1.2, the point **m** is asymptotically stable in this case.

All conditions for the occurrence of a Hopf bifurcation are therefore satisfied (see section 4.5). For sufficiently small values $\mu > 0$, there will be a periodic attractor in the neighbourhood of **m**, see fig. 14.1.

Exercise 14.1.1 Show that for $-1 < \mu < 1$, the system is permanent. (Hint: take $P = \prod x_i$ as an average Lyapunov function and proceed as in section 13.6.) What happens for $\mu > 1$ and $\mu < -1$?

Exercise 14.1.2 Analyse the general four-species system (14.1) with cyclic symmetry

$$A = \begin{bmatrix} 0 & a_1 & a_2 & a_3 \\ a_3 & 0 & a_1 & a_2 \\ a_2 & a_3 & 0 & a_1 \\ a_1 & a_2 & a_3 & 0 \end{bmatrix}.$$

(a) If $a_1 + a_3 \geq a_2 > 0$, then $P = x_1 x_2 x_3 x_4$ is a global Lyapunov function.

(b) Show that a Hopf bifurcation occurs at $a_2 = 0$ if $a_1 \neq a_3$, and leads to periodic attractors for small $a_2 < 0$ if $a_1 + a_3 > 0$.

(c) The system is permanent if and only if

$$a_1 a_3 \leq 0 \quad \text{and} \quad -(a_1 + a_3) < a_2 < a_1 + a_3$$

or

$$a_1 a_3 > 0 \quad \text{and} \quad -\frac{(a_1 - a_3)^2}{a_1 + a_3} < a_2 < a_1 + a_3 .$$

(d) If $a_2 > a_1 + a_3 > 0$ then $V = x_1 x_3 + x_2 x_4$ is a global Lyapunov function.

14.2 Cyclic symmetry

We have seen in sections 5.5 and 14.1 that the assumption of *cyclic symmetry*, unrealistic as it may be, helps to display interesting properties of replicator equations. We shall investigate now why it leads to a supercritical Hopf bifurcation for $n \geq 4$, but not for $n = 3$.

We shall count the indices modulo n, denote the coefficients $a_{i,i+k}$ by a_k (since they do not depend on i) and assume without loss of generality that $a_0 = 0$. The replicator equation becomes

$$\dot{x}_i = x_i \left(\sum_{j=1}^{n} a_j x_{i+j} - \mathbf{x} \cdot A \mathbf{x} \right). \tag{14.5}$$

The central point $\mathbf{m} = \frac{1}{n} \mathbf{1}$ is an inner rest point for (14.5). The Jacobian at $\mathbf{m}$ is again circulant, its first row being $c_0, \ldots, c_{n-1}$ with

$$c_i = \frac{1}{n}(a_i - 2\bar{a}) \quad \text{and} \quad \bar{a} = \frac{1}{n} \sum_{j=1}^{n-1} a_j . \tag{14.6}$$

According to (5.20), its eigenvalues are $\gamma_0 = -\bar{a}$ (with eigenvector $\mathbf{1}$ orthogonal to S_n, and henceforth omitted) and

$$\gamma_j = \sum_{s=0}^{n-1} c_s e^{2\pi i j s/n} = \frac{1}{n} \sum_{s=1}^{n-1} a_s e^{2\pi i j s/n} \tag{14.7}$$

for $j = 1, \ldots, n-1$.

The exercises at the end of this section yield a proof of the following:

Theorem 14.2.1 *If* **m** *is a sink, then it is globally stable. If* **m** *is a source, all orbits in* int S_n *(except* **m***) converge to* bd S_n.

In the case $n = 3$, which corresponds to the rock–scissors–paper game from section 7.7, the eigenvalues are

$$\gamma_{1,2} = -\frac{1}{6}\left[(a_1 + a_2) \pm i\sqrt{3}(a_2 - a_1)\right] \quad . \tag{14.8}$$

The function $P = x_1 x_2 x_3$ satisfies

$$\dot{P} = \frac{3P}{2}(a_1 + a_2)\sum_{j=1}^{3}\left(x_j - \frac{1}{3}\right)^2 \quad . \tag{14.9}$$

Since there are no limit cycles for $n = 3$, we know that Hopf bifurcations leading to the emergence of periodic attractors cannot occur. The reason is that when $\operatorname{Re}\gamma_{1,2} = -\frac{1}{6}(a_1 + a_2) = 0$, the centre **m** is not asymptotically stable. P is a constant of motion and all orbits circle around **m**. The Hopf bifurcation is *degenerate* (see section 4.5).

For $n \geq 4$, however, **m** is asymptotically stable if one pair of complex eigenvalues lies on the imaginary axis and all other eigenvalues to its left (as shown in the next exercise). By Hopf's theorem (see section 4.5), a supercritical Hopf bifurcations leading to stable limit cycles occurs.

Exercise 14.2.2 Introduce complex coordinates

$$y_p = \sum_{j=1}^{n} x_j \exp(2\pi i j p/n) \qquad (p = 0,\ldots,n-1) \tag{14.10}$$

and check that $y_0 = 1, \bar{y}_p = y_{n-p}$ (for $p = 1,\ldots,n-1$) and

$$x_j = \frac{1}{n}\sum_{p=0}^{n-1} y_p \exp(-2\pi i j p/n) \qquad (j = 1,\ldots,n) \quad . \tag{14.11}$$

Exercise 14.2.3 (14.5) leads to

$$\dot{y}_p = \sum_{m} w_m \bar{y}_m y_{p+m} - y_p \sum_{m=0}^{n-1} \operatorname{Re} w_m |y_m|^2 \tag{14.12}$$

(where $w_0 = -\gamma_0$ and $w_m = \gamma_m$ for $m \neq 0$).

Exercise 14.2.4 Check that with $P = x_1 x_2 \ldots x_n$ one has

$$\dot{P} = -nP \sum_{m=1}^{n-1} \operatorname{Re} w_m |y_m|^2 \quad . \tag{14.13}$$

Theorem 14.2.1 follows from (14.13).

Exercise 14.2.5 If $n \geq 4$, $\operatorname{Re} w_1 = \operatorname{Re} w_{n-1} = 0$ and $\operatorname{Re} w_i < 0$ for $i = 2,\ldots,$ $n-2$, then $\dot{P} \geq 0$, so that every orbit in int S_n converges to the maximal invariant subset M of $\{\dot{P} = 0\}$, i.e. of $\{y_m = 0$ for $m = 2,\ldots,n-2\}$. Show that $y_1 = y_{m-1} = 0$ on M and hence that **m** is asymptotically stable.

Exercise 14.2.6 Show that (5.18) is equivalent to (14.5) with $n = 3$.

14.3 Permanence and irreducibility

With a replicator equation (14.1) with $a_{ij} \geq 0$ for all i and j, we associate a *directed graph* whose vertices correspond to the indices i: an arrow from j to i denotes that $a_{ij} > 0$, i.e. that x_j enhances the growth of x_i.

A directed graph is said to be *irreducible* if for any two vertices i and j there exists an oriented path leading from i to j.

Theorem 14.3.1 *If* (14.1) *with* $a_{ij} \geq 0$ *is permanent, then its graph is irreducible.*

Proof Let us assume that the system is not irreducible. There exists, then, a proper subset $D \subseteq \{1,\ldots,n\}$ which is *closed* in the sense that no arrow leads out of it, i.e. that $a_{ji} = 0$ for all $i \in D, j \notin D$. Let us define

$$M = \max_{1 \leq j \leq n} \sum_{s=1}^{n} a_{js} \qquad m = \min_{i \in D} \sum_{t \in D} a_{it} \quad . \tag{14.14}$$

Clearly some a_{js} are positive, since otherwise the system would have only fixed points, which does not agree with permanence. This implies $M > 0$. We shall show that we also have $m > 0$.

This can be seen indirectly. If $m = 0$, there exists an $i \in D$ such that $a_{it} = 0$ for all $t \in D$. In this case we replace D by the closed set $D \setminus \{i\}$. If we still have $m = 0$, we repeat the procedure. We end up either with $m > 0$ or with a set containing a single element k, which satisfies $a_{kk} = 0$. Clearly, $a_{jk} = 0$ for all j. In that case, x_k does not occur in the equations $(A\mathbf{x})_1 = \cdots = (A\mathbf{x})_n$ for the unique interior rest point guaranteed by permanence: this is a contradiction.

Now we may define

$$G = \{\mathbf{x} \in \text{int } S_n : Mx_j < mx_i \text{ for all } i \in D, j \notin D\} \quad . \tag{14.15}$$

The set G is forward invariant. Indeed, the quotient rule implies that for $\mathbf{x} \in G, i \in D$ and $j \notin D$ we have

$$\left(\frac{x_j}{x_i}\right)^{\cdot} = \frac{x_j}{x_i}\left(\sum_s a_{js}x_s - \sum_t a_{it}x_t\right) \leq \frac{x_j}{x_i}\left(\sum_{s \notin D} a_{js}x_s - \sum_{t \in D} a_{it}x_t\right)$$

$$\leq \frac{x_j}{x_i}\left(\max_{s \notin D} x_s \sum_{s \notin D} a_{js} - \min_{t \in D} x_t \sum_{t \in D} a_{it}\right)$$

$$\leq \frac{x_j}{x_i}\left(M \max_{s \notin D} x_s - m \min_{t \in D} x_t\right) < 0 . \tag{14.16}$$

Hence $\mathbf{x}(t)$ remains in G and converges to bd S_n for $t \to +\infty$, a contradiction to permanence. $\qquad\square$

14.4 Permanence of catalytic networks

By a *catalytic network*, we shall understand a replicator equation (14.1) with $a_{ij} \geq 0$ and $a_{ii} = 0$. Hypercycles are simple examples of catalytic networks: they guarantee permanence, but of course many other types of network can also do that job.

We have seen that the directed graph of a permanent catalytic network is irreducible. It is tempting to conjecture that it must even be *Hamiltonian*, i.e. contain a closed path visiting every vertex exactly once. This would mean that such a network had to contain a full hypercycle. It turns out, however, that this is valid only in low dimensions.

Theorem 14.4.1 *If $n \leq 5$, then the graph of a permanent catalytic network is Hamiltonian.*

Proof We know from (13.32) that the sign of det A is $(-1)^{n-1}$. Now

$$\det A = \sum_{\sigma}(\text{sgn}\,\sigma)a_{1\sigma(1)}a_{2\sigma(2)}\ldots a_{n\sigma(n)} \tag{14.17}$$

where the summation extends over all permutations σ of $\{1,\ldots,n\}$. Since $a_{ii} = 0$ we need only consider permutations without fixed elements. Every permutation σ can be split into (say) k elementary cycles, and $\text{sgn}\,\sigma = (-1)^{n-k}$. If $n \leq 5$, k can be 1 or 2. Any permutation which is the product of two smaller cycles has sign $(-1)^{n-2}$. In order that $\text{sgn}\det A = (-1)^{n-1}$, in accordance with (13.32), there must be at least one permutation σ consisting of a single cycle such that

$$a_{1\sigma(1)}a_{2\sigma(2)}\ldots a_{n\sigma(n)} > 0 \quad . \tag{14.18}$$

This corresponds to a closed feedback loop visiting every vertex precisely once. ☐

Exercise 14.4.2 For $n = 6$ and

$$A = \begin{bmatrix} 0 & 1 & 0 & 0 & 0 & 3 \\ 2 & 0 & 0 & 0 & 0 & 0 \\ 1 & 0 & 0 & 2 & 0 & 0 \\ 0 & 3 & 1 & 0 & 0 & 0 \\ 0 & 0 & 0 & 3 & 0 & 1 \\ 0 & 0 & 0 & 0 & 1 & 0 \end{bmatrix},$$

show that the replicator equation is permanent although the graph is not Hamiltonian. (Hint: apply theorem 13.6.1 with $\mathbf{p} = \frac{1}{49}(9, 1, 2, 13, 23, 1)$. The inequalities corresponding to boundary fixed points $\mathbf{x}$ with $\mathbf{x} \cdot A\mathbf{x} = 0$ are trivially satisfied, since A is irreducible. The only remaining fixed points are $(\frac{1}{2}, \frac{2}{3}, 0, 0, 0, 0)$, $(0, 0, \frac{1}{3}, \frac{2}{3}, 0, 0)$, $(0, 0, 0, 0, \frac{1}{2}, \frac{1}{2})$, $\frac{1}{28}(5, 0, 9, 2, 9, 3)$ and $\frac{1}{41}(9, 6, 0, 5, 18, 3)$.)

Exercise 14.4.3 Show that for $n = 3$ a catalytic network is permanent if and only if there exists a unique interior fixed point. List the graphs of all such networks and check that $\det A > 0$.

Theorem 14.4.4 *For a catalytic network with $n = 4$, a necessary and sufficient condition for permanence is that there exists an interior fixed point and $\det A < 0$.*

Necessity follows from theorems 13.3.1 and 13.5.3. We shall omit here a proof of sufficiency (which can be based on theorem 13.6.1).

14.5 Essentially hypercyclic networks

A large class of interaction matrices displays an *essentially hypercyclic struc-ture*: $a_{ij} > 0$ if $i = j + 1 \pmod{n}$ and $a_{ij} \leq 0$ otherwise (with $a_{ii} = 0$), or in a self-explanatory notation

$$A = \begin{bmatrix} 0 & - & - & . & . & . & - & + \\ + & 0 & - & . & . & . & - & - \\ - & + & 0 & . & . & . & - & - \\ . & . & . & . & . & . & . & . \\ . & . & . & . & . & . & . & . \\ . & . & . & . & . & . & . & . \\ - & - & - & . & . & . & + & 0 \end{bmatrix}. \tag{14.19}$$

As we shall presently see, permanence of such networks can be characterized in terms of M-matrices. An $n \times n$ matrix C with off-diagonal terms $c_{ij} \leq 0$ $(i \neq j)$ is said to be a (nonsingular) *M-matrix* if one of the following equivalent conditions is satisfied (the equivalence of these conditions will be shown in section 15.1):

(i) There is a $\mathbf{p} > 0$ such that $C\mathbf{p} > 0$;
(ii) all leading principal minors of C are positive;
(iii) all real eigenvalues of C are positive;
(iv) for all $\mathbf{x} > \mathbf{0}$ there is an i with $(C\mathbf{x})_i > 0$.
(v) C^{-1} exists and $C^{-1} \geq 0$.

We recall that a *principal submatrix* A_I of an $n \times n$ matrix A is a matrix (a_{ij}) with $i, j \in I \subseteq \{1, \ldots, n\}$. If I is of the form $\{1, 2, \ldots, k\}$, $k = 1, \ldots, n$, then A_I is said to be a *leading principal submatrix*. The *minors* are the corresponding determinants. We also recall that $\mathbf{p} > \mathbf{0}$ means that $p_i > 0$ for all i.

Theorem 14.5.1 *For an essentially hypercyclic matrix A, the following conditions are equivalent:*

(1) The replicator equation (14.1) is permanent;
(2) there is an inner rest point $\hat{\mathbf{x}}$ of (14.1) where the Jacobian D has negative trace;
(3) there is an inner rest point $\hat{\mathbf{x}}$ it of (14.1) with $\hat{\mathbf{x}} \cdot A\hat{\mathbf{x}} > 0$;
(4) the matrix C obtained by moving the top row of A to the bottom is an M-matrix;
(5) there is a $\mathbf{p} > \mathbf{0}$ such that $\mathbf{p} \cdot A > \mathbf{0}$.

Proof The implication from (1) to (2) has been shown in theorem 13.5.2. The equivalence of (2) and (3) follows from (13.34). (3) implies (4) since $(A\hat{\mathbf{x}})_i = \hat{\mathbf{x}} \cdot A\hat{\mathbf{x}} > 0$, hence $A\hat{\mathbf{x}} > \mathbf{0}$ and thus $C\hat{\mathbf{x}} > \mathbf{0}$. Condition (5) follows from (4) because there is a vector $\mathbf{p} > \mathbf{0}$ such that $(p_2, \ldots, p_n, p_1)C > \mathbf{0}$. The implication from (5) to (1), finally, follows from theorem 13.6.1 since $\mathbf{x} \cdot A\mathbf{x} < 0$ holds at each rest point $\mathbf{x}$ on the boundary of S_n. □

Exercise 14.5.2 Check that each of the conditions (2–5) of the theorem is satisfied for the hypercycle equation (12.1).

Exercise 14.5.3 Consider a replicator equation of the following type:

$$\dot{x}_i = x_i(a_i x_i + b_{i-1} x_{i-1} - \mathbf{x} \cdot A\mathbf{x}) \tag{14.20}$$

with $a_i, b_i > 0$. Show that $a_i < b_i$ for all i is a necessary condition for permanence. Show that (14.20) is permanent if and only if the matrix

$$
C = \begin{bmatrix}
\alpha_1 & -1 & -1 & \cdots & -1 & 0 \\
0 & \alpha_2 & -1 & \cdots & -1 & -1 \\
-1 & 0 & \alpha_3 & \cdots & -1 & -1 \\
\cdot & \cdot & \cdot & & \cdot & \cdot \\
\cdot & \cdot & \cdot & & \cdot & \cdot \\
-1 & -1 & -1 & \cdots & 0 & \alpha_n
\end{bmatrix} \tag{14.21}
$$

with

$$\alpha_k = \frac{b_k}{a_k} - 1 \tag{14.22}$$

is an *M*-matrix. Show that for $n = 3$, this is equivalent to $\alpha_1 \alpha_2 \alpha_3 > 1$, i.e. $\prod_{i=1}^{3}(b_i - a_i) > \prod_{i=1}^{3} a_i$; for $n = 4$ to $\alpha_1 \alpha_2 \alpha_3 \alpha_4 > \alpha_1 + \alpha_2 + \alpha_3 + \alpha_4 + \alpha_1 \alpha_3 + \alpha_2 \alpha_4$; and to $b > a(n-1)$ if $a_i = a$ and $b_i = b$ for all i (for n arbitrary).

Exercise 14.5.4 Consider two counterrotating hypercycles:

$$\dot{x}_i = x_i(a_{i-1} x_{i-1} + b_{i+1} x_{i+1} - \mathbf{x} \cdot A\mathbf{x}) \tag{14.23}$$

with $a_i, b_i > 0$. Show that permanence implies either $a_i < b_i$ for all i or $b_i < a_i$ for all i. If $b_i < a_i$ for all i, then (14.23) is permanent if

$$
A = \begin{bmatrix}
\beta_1 & -1 & -1 & \cdots & \alpha_n & 0 & 0 \\
0 & \beta_2 & -1 & \cdots & -1 & \alpha_1 & 0 \\
0 & 0 & \beta_3 & \cdots & -1 & -1 & \alpha_2 \\
\vdots & \vdots & \vdots & & \vdots & \vdots & \vdots
\end{bmatrix}
$$

is an *M*-matrix (with α_k as in (14.22) and $\beta_k = a_k / b_k - 1$). This is always the case if $n \leq 4$. Show that in the symmetric case $a_i = a, b_i = b$ this reduces to

$$\frac{b}{a} + 2 + \frac{a}{b} > n \quad . \tag{14.24}$$

14.6 Notes

The Hopf bifurcation in section 14.1 and the analysis of cyclic symmetry in section 14.2 are from Hofbauer *et al.* (1980). Limit cycles for Lotka–Volterra equations were displayed in Coste *et al.* (1979). In Hofbauer *et al.* (1980), Schuster *et al.* (1980), and Stadler and Schuster (1996), graphs were used to classify small catalytic networks. Theorem 14.3.1 is from Schuster *et al.* (1980). Section 14.4 follows Amann (1989) (see also Hofbauer and Sigmund (1988, pp. 184ff.) for a proof of theorem 14.4.4). Essentially hypercyclic networks and in particular the examples of exercises 14.5.3 and 14.5.4 were studied in Amann and Hofbauer (1985).

15

Stability of n-species communities

This chapter deals with the connections between the dynamical properties of a Lotka–Volterra equation (having bounded orbits, or a globally stable rest point, or a unique saturated fixed point etc.) and the algebraic properties of the interaction matrix. It also deals with communities having some ecologically relevant special structures.

15.1 Mutualism and M-matrices

We turn to the relationship between stability properties of the Lotka–Volterra equation

$$\dot{x}_i = x_i \left(r_i + \sum_{j=1}^{n} a_{ij}x_j \right) \tag{15.1}$$

and algebraic properties of the interaction matrix $A = (a_{ij})$. For illustration, we shall start with a section on *mutualism* where this relationship is particularly clear. As we saw in exercise 3.3.6, a mutualistic two-species system has unbounded solutions if $a_{21}a_{12} > a_{11}a_{22}$, and a globally stable rest point if $a_{12}a_{21} < a_{11}a_{22}$. The following theorem generalizes this simple dichotomy to higher dimensions.

Theorem 15.1.1 *Consider* (15.1) *with* $a_{ij} \geq 0$ *for all* $i \neq j$, *and assume that it admits an interior rest point* $\hat{x}$. *Then the following statements are equivalent:*

(M1) *All orbits in* $\mathbb{R}^n_+$ *are uniformly bounded as* $t \to +\infty$;
(M2) *The matrix A is stable (i.e. all eigenvalues of A have negative real part);*

(M3) *The leading principal minors of A alternate in sign:*

$$(-1)^k \det(a_{ij})_{1 \le i, j \le k} > 0 \ ;$$

(M4) *For all* $\mathbf{c} > \mathbf{0}$ *there exists an* $\mathbf{x} > \mathbf{0}$ *such that*

$$A\mathbf{x} + \mathbf{c} = \mathbf{0} \ ; \tag{15.2}$$

(M5) *There exists an* $\mathbf{x} > \mathbf{0}$ *with* $A\mathbf{x} < \mathbf{0}$;

(M6) *The rest point* $\hat{\mathbf{x}}$ *is globally asymptotically stable and all (boundary) orbits are uniformly bounded as* $t \to +\infty$.

For the proof, we shall need a result valid for general (not necessarily mutualistic) interactions.

Lemma 15.1.2 *If the matrix A has a left eigenvector* $\mathbf{v} \ge \mathbf{0}$ *with eigenvalue* $\lambda > 0$, *then* (15.1) *has interior solutions which are unbounded as* $t \to +\infty$.

Proof Indeed, let us assume that $\mathbf{v}A = \lambda\mathbf{v}$ with $\mathbf{v} \ge \mathbf{0}, \sum v_i = 1$ and $\lambda > 0$. Let us consider the function

$$P(\mathbf{x}) = \prod_{i=1}^{n} x_i^{v_i}$$

and the sets $M_\alpha = \{\mathbf{x} \in \mathbb{R}^n_+ : P(\mathbf{x}) > \alpha\}$. Since $\prod x_i^{v_i} \le \sum x_i v_i$, we obtain on M_α

$$\frac{\dot{P}}{P} = \mathbf{v} \cdot (\mathbf{r} + A\mathbf{x}) = \mathbf{v} \cdot \mathbf{r} + \lambda\mathbf{v} \cdot \mathbf{x} \ge \mathbf{v} \cdot \mathbf{r} + \lambda\alpha \quad . \tag{15.3}$$

For α sufficiently large, we have $\dot{P}(\mathbf{x}) > 0$. The sets M_α are therefore forward invariant. All orbits in M_α escape to infinity, which proves the lemma. $\square$

We shall also need the Perron–Frobenius theorem: *if M is an* $n \times n$ *matrix with nonnegative elements, there exists a unique nonnegative eigenvalue* λ *which is dominant in the sense that* $|\mu| \le \lambda$ *for all other eigenvalues* μ *of M. There exist right and left eigenvectors* $\mathbf{u} \ge \mathbf{0}$ *(i.e.* $u_i \ge 0$ *for all i) and* $\mathbf{v} \ge \mathbf{0}$ *such that* $M\mathbf{u} = \lambda\mathbf{u}$ *and* $\mathbf{v}M = \lambda\mathbf{v}$. *If M is* irreducible, *i.e. if for every pair of indices* (i, j) *there exists an integer* $k > 0$ *(which may depend on i, j) such that the* (i, j) *entry of* M^k *is positive, then* λ *is simple and positive, and the eigenvectors* $\mathbf{u}$ *and* $\mathbf{v}$ *are unique and positive. If M is* primitive, *i.e. if there exists a* $k > 0$ *such that all entries of* M^k *are positive, then* $|\mu| < \lambda$ *for all other eigenvalues* μ, *and if we normalize* $\mathbf{u}$ *and* $\mathbf{v}$ *such that* $\mathbf{u} \cdot \mathbf{v} = 1$ *and define T as the matrix with elements* $t_{ij} = u_i v_j$ *then* $\lambda^{-k} M^k \to T$ *for* $k \to +\infty$.

Let us now prove theorem 15.1.1.

Proof (M1) $\Rightarrow$ (M2). Since A is mutualistic, we may write it in the form

$$A = B - cI \qquad (15.4)$$

where $c > 0$ and B is a nonnegative matrix. The Perron–Frobenius theorem shows that there exists a dominant eigenvalue $\rho > 0$ of B with nonnegative left and right eigenvectors $\mathbf{v} \geq \mathbf{0}$ and $\mathbf{u} \geq \mathbf{0}$. Clearly $\mathbf{v}$ and $\mathbf{u}$ are also eigenvectors of A, corresponding to the eigenvalue $\lambda = \rho - c$. The previous lemma, together with (M1), shows that $\lambda \leq 0$.

If we suppose that $\lambda = 0$, we obtain $A\mathbf{u} = \mathbf{0}$. The line $\hat{\mathbf{x}} + t\mathbf{u}, t \in \mathbb{R}$, corresponds to fixed points of (15.1). This is a contradiction to (M1). Hence $\lambda < 0$. But no eigenvalue of A has real part larger than λ, and hence A is stable.

(M2) $\Rightarrow$ (M3). Since A is stable, $\det A$ has sign $(-1)^n$. The same is valid for every principal submatrix of A. Indeed, for every $J \subseteq \{1,\dots,n\}$, the principal submatrix $B_J = (b_{ij})_{i,j \in J}$ of B has a dominant eigenvalue $\rho(J)$ which is not larger than the dominant eigenvalue ρ of B. Hence the submatrices $A_J = B_J - cI$ are also stable, and their determinant has sign $(-1)^{\text{card}J}$.

(M3) $\Rightarrow$ (M4). This can be proved by induction on n. We eliminate x_1 in (15.2) by multiplying the first equation

$$\sum_{k=1}^{n} a_{1k} x_k + c_1 = 0 \qquad (15.5)$$

by a_{i1}/a_{11} (note that $a_{11} < 0$), and subtracting it from the i-th equation. This yields a linear system in $x_2,\dots,x_n$:

$$\sum_{k=2}^{n} \bar{a}_{ik} x_k + \bar{c}_i = 0 \qquad (15.6)$$

with

$$\bar{a}_{ik} = a_{ik} - a_{1k}\frac{a_{i1}}{a_{11}} \quad \text{and} \quad \bar{c}_i = c_i - c_1\frac{a_{i1}}{a_{11}} \geq c_i > 0 \ .$$

If we apply the corresponding operations to the leading principal minor, we obtain

$$\det(a_{ij})_{1 \leq i,j \leq k} = \begin{bmatrix} a_{11} & a_{12} & \cdots & a_{1k} \\ 0 & \bar{a}_{22} & \cdots & \bar{a}_{2k} \\ \vdots & \vdots & & \vdots \\ 0 & \bar{a}_{k2} & \cdots & \bar{a}_{kk} \end{bmatrix} = a_{11}\det(\bar{a}_{ij})_{2 \leq i,j \leq k} \ .$$

Hence the $(n-1) \times (n-1)$ matrix $\bar{A}$ satisfies (M3). By inductive hypothesis, (15.6) has a positive solution $\bar{x}_2,\dots,\bar{x}_n$. This yields a positive solution

$x_1, \bar{x}_2, \ldots, \bar{x}_n$ of (15.2), since $x_1 > 0$ is an obvious consequence of (15.5) and (15.6).

(M4) $\Rightarrow$ (M5) is trivial.

(M5) $\Rightarrow$ (M6). For mutualistic systems, (M5) means that A has a *negative dominant diagonal*, i.e.

$$\exists d_i > 0 : a_{ii}d_i + \sum_{j \neq i} |a_{ij}|d_j < 0 \quad . \tag{15.7}$$

We show that this condition ensures global stability of the interior rest point $\hat{\mathbf{x}}$ (which exists by assumption) for general, not necessarily mutualistic, Lotka–Volterra systems. Indeed, let

$$V(\mathbf{x}) = \max_{i=1,\ldots,n} \frac{|x_i - \hat{x}_i|}{d_i} \quad . \tag{15.8}$$

Then $V(\mathbf{x}) \geq 0$, with equality if and only if $\mathbf{x} = \hat{\mathbf{x}}$. The constant level sets of V are boxes of side lengths $2d_i$ centred in $\hat{\mathbf{x}}$. We claim that all these boxes are forward invariant.

Indeed, let i be any index for which $|\frac{x_i - \hat{x}_i}{d_i}|$ is maximal. Then

$$
\begin{aligned}
|x_i - \hat{x}_i|^{\cdot} &= \dot{x}_i \mathrm{sgn}(x_i - \hat{x}_i) \\
&= x_i \Big[a_{ii}(x_i - \hat{x}_i) + \sum_{j \neq i} a_{ij}(x_j - \hat{x}_j) \Big] \mathrm{sgn}(x_i - \hat{x}_i) \\
&\leq x_i \Big[a_{ii}|x_i - \hat{x}_i| + \sum_{j \neq i} |a_{ij}||x_j - \hat{x}_j| \Big] \\
&\leq x_i V(\mathbf{x}) \Big[a_{ii}d_i + \sum_{j \neq i} |a_{ij}|d_j \Big] < 0
\end{aligned}
$$

for all $\mathbf{x} \neq \hat{\mathbf{x}}$ in int $\mathbb{R}_+^n$. Thus $V(\mathbf{x})$ is a strictly decreasing Lyapunov function. Hence all interior orbits converge to $\hat{\mathbf{x}}$. By the same token all boundary orbits are uniformly bounded.

(M6) $\Rightarrow$ (M1) is obvious. $\square$

Exercise 15.1.3 Show that under the assumptions of the theorem, the following properties are also equivalent to (M1)–(M6):
(M7) $\hat{\mathbf{x}}$ is globally stable;
(M8) (15.1) is permanent;
(M9) All interior orbits are uniformly bounded for $t \to +\infty$;
(M10) A^{-1} exists and has nonpositive elements;
(M11) For every $\mathbf{x} \geq \mathbf{0}$ (but $\mathbf{x} \neq \mathbf{0}$) there is an i such that $(A\mathbf{x})_i < 0$.

(Hint: (M6) $\Rightarrow$ (M7) $\Rightarrow$ (M8) $\Rightarrow$ (M9) $\Rightarrow$ (M2); (M10) $\Leftrightarrow$ (M4); (M11) $\Leftrightarrow$ (M5).)

Exercise 15.1.4 For $n = 3$, these conditions are equivalent to $a_{ii} < 0$ and $\det A < 0$.

Exercise 15.1.5 Study mutualistic Lotka–Volterra systems without an interior fixed point (Hint: consider a saturated fixed point $\hat{\mathbf{x}}$ on the boundary.)

If the matrix A satisfies the conditions of the theorem, then $-A$ is said to be a (nonsingular) *M-matrix*. We have met this class in section 14.5. In section 15.3, we shall see that even if the mutualistic equation (15.1) has no interior rest point, the ecologically trivial assumption of uniform boundedness implies the existence of a globally stable rest point.

15.2 Boundedness and B-matrices

We now turn to the characterization of *uniform boundedness*.

Theorem 15.2.1 *The following conditions are equivalent for a matrix A.*

(B1) *For every* $\mathbf{r} \in \mathbb{R}^n$, *the solutions of the Lotka–Volterra equation (15.1) are uniformly bounded for* $t \to +\infty$.

(B2) *The origin* $\mathbf{0}$ *is globally asymptotically stable for the solutions of*

$$\dot{x}_i = x_i(A\mathbf{x})_i \qquad (15.9)$$

in $\mathbb{R}^n_+$.

(B3) *Whenever*

$$x_i(A\mathbf{x})_i = \lambda x_i \qquad i = 1,\ldots,n \qquad (15.10)$$

holds for some $\mathbf{x} \geq \mathbf{0}$ *(with* $\mathbf{x} \neq \mathbf{0}$*), then* $\lambda < 0$.

The matrix A is said to be a *B-matrix* if it satisfies one of the equivalent conditions (B1) to (B3).

Proof We first recall the projective change in coordinates from section 7.5 which brings infinity into view. With

$$z_k = \frac{x_k}{1 + \sum x_i} \quad (k = 1,\ldots,n) \qquad z_{n+1} = \frac{1}{1 + \sum x_i} \ , \qquad (15.11)$$

equation (15.1) transforms into the replicator equation

$$\dot{z}_k = z_k \left(\sum_{j=1}^{n} a_{kj} z_j + r_k z_{n+1} - \bar{a}(\mathbf{z}) \right) \qquad k = 1, \dots, n \qquad (15.12)$$

$$\dot{z}_{n+1} = z_{n+1}(-\bar{a}(\mathbf{z}))$$

on S_{n+1}, with

$$\bar{a}(\mathbf{z}) = \sum_{i,j=1}^{n} a_{ij} z_i z_j + z_{n+1} \sum_{k=1}^{n} r_k z_k \quad .$$

Obviously, (15.1) has uniformly bounded solutions for $t \to +\infty$, if and only if the closed invariant face

$$F_\infty = \{ \mathbf{z} \in S_{n+1} : z_{n+1} = 0 \}$$

corresponding to the points at infinity is a *repellor* for (15.12) in the sense that there exists a $c > 0$ such that

$$\liminf_{t \to +\infty} z_{n+1}(t) > c$$

for all $\mathbf{z} \in S_{n+1}$ with $z_{n+1} > 0$.

(B3) $\Rightarrow$ (B1). It is enough to show that $P(\mathbf{z}) = z_{n+1}$ is an *average Lyapunov function* in the vicinity of F_∞. By the same argument as in theorem 13.6.1 this holds if and only if

$$\frac{\dot{P}}{P}(\bar{\mathbf{z}}) = \left. \frac{\dot{z}_{n+1}}{z_{n+1}} \right|_{\bar{\mathbf{z}}} = -\bar{a}(\bar{\mathbf{z}}) > 0 \qquad (15.13)$$

for all fixed points $\bar{\mathbf{z}}$ of (15.12) with $\bar{z}_{n+1} = 0$. These fixed points $\bar{\mathbf{z}}$ at infinity are characterized by

$$\bar{z}_i (A\bar{\mathbf{z}})_i = \lambda \bar{z}_i \qquad (15.14)$$

where λ is an arbitrary constant. Using $\bar{\mathbf{z}} \in S_{n+1}$, we see that λ is then just $\bar{a}(\bar{\mathbf{z}})$. Hence $P(\mathbf{z}) = z_{n+1}$ is an average Lyapunov function for (15.12) if $\lambda = \bar{a}(\bar{\mathbf{z}}) < 0$ for any such point $\bar{\mathbf{z}} \in F_\infty$.

(B1) $\Rightarrow$ (B2). By assumption there exists a constant $k > 0$ such that

$$B_k := \{ \mathbf{x} \in \mathbb{R}^n_+ : x_i \le k \text{ for all } i \}$$

contains all ω-limits of solutions of (15.1) with $\mathbf{r} = \mathbf{0}$, i.e. of (15.9). But (15.9) is homogeneous and hence invariant under scalings $\mathbf{x} \to \alpha \mathbf{x}$ for $\alpha > 0$. So each set $B_{\alpha k}$ with $\alpha > 0$ has the same property. For $\alpha \to 0$ this proves that all solutions converge to $\mathbf{0}$. The stability of $\mathbf{0}$ follows from the compactness of the sets $B_{\alpha k}$.

(B2) ⇒ (B3). Let $\bar{\mathbf{x}} \geq \mathbf{0}$ satisfy (15.10). The line $\{t\bar{\mathbf{x}} : t > 0\}$ is obviously invariant for (15.9), since $(x_i/x_j)' = 0$ holds there. On this line, the flow reduces to $\dot{x}_i = \lambda x_i$. If $\lambda > 0$ the orbit grows to infinity. For $\lambda = 0$ the line consists of rest points. Since both possibilities contradict (B2), $\lambda < 0$ follows.
□

We note that the vectors $\bar{\mathbf{x}}$ satisfying (15.10) correspond to asymptotic eigendirections of (15.1), and the points $\bar{\mathbf{z}}$ in (15.14) to rest points of (15.12). The corresponding 'transversal' eigenvalue is $-\lambda = -\bar{a}(\bar{\mathbf{z}})$. The condition $\lambda < 0$ means that $\bar{\mathbf{z}}$ is not saturated.

Exercise 15.2.2 Show that the condition obtained from (B1) by dropping 'uniformly' still implies (B1). (Hint: consider the case $r_i = r > 0$.)

Exercise 15.2.3 Show that for symmetric matrices A, (B3) is equivalent to

$$\mathbf{x} \cdot A\mathbf{x} < 0 \text{ for all } \mathbf{x} \geq 0 \text{ with } \mathbf{x} \neq 0 \quad . \tag{15.15}$$

In this case, $\sum x_i$ is decreasing for large $\mathbf{x}$ for (15.1). The matrix $-A$ is said to be *strictly copositive*. (Hint: Show that the maxima of $\mathbf{x} \cdot A\mathbf{x}$ on S_n are given by (15.10).)

Theorem 15.2.4 *The matrix A is a B-matrix if and only if*
(B4) *For all $\mathbf{x} \geq \mathbf{0}$ with $\mathbf{x} \neq \mathbf{0}$ there is an i such that $x_i > 0$ and $(A\mathbf{x})_i < 0$.*

Biologically, this means that for every state $\mathbf{x} \neq \mathbf{0}$, at least one species i has its growth rate reduced by the interaction of the species.

Proof (B4) ⇒ (B3) is obvious. For the converse we show that (B1) to (B3) imply that the transpose A^t of A satisfies (B4). (The class of B-matrices is therefore closed under transposition.)
So let us assume that (B1) or (B2) holds for A, but that A^t does not satisfy (B4). Then we can find a $\mathbf{p} \geq 0$, $\mathbf{p} \neq \mathbf{0}$, such that $p_i(A^t\mathbf{p})_i \geq 0$ for all i. Since $(A^t\mathbf{p})_i = (\mathbf{p}A)_i \geq 0$ for all $i \in I = \mathrm{supp}(\mathbf{p})$, we obtain $\mathbf{p} \cdot A\mathbf{x} \geq 0$ for all $\mathbf{x}$ with $\mathrm{supp}(\mathbf{x}) \subseteq I$. Now $P(\mathbf{x}) = \prod x_i^{p_i}$ satisfies

$$\frac{\dot{P}}{P}(\mathbf{x}) = \mathbf{p} \cdot (\mathbf{r} + A\mathbf{x}) = \mathbf{p} \cdot \mathbf{r} + \mathbf{p} \cdot A\mathbf{x} \quad .$$

This expression is nonnegative for $\mathbf{r} = \mathbf{0}$ and positive for $\mathbf{r} > \mathbf{0}$. Thus P increases along the orbit of $\mathbf{x}$ if $\mathrm{supp}(\mathbf{x}) = \mathrm{supp}(\mathbf{p})$. All these solutions converge to infinity, a contradiction to (B1) and (B2).
□

Exercise 15.2.5 If A is a B-matrix and D a positive diagonal matrix, then DA, AD and every principal submatrix of A are also B-matrices.

Exercise 15.2.6 For $n = 2$, A is a B-matrix if and only if $a_{ii} < 0$ and for mutualistic interactions additionally $\det A > 0$.

Exercise 15.2.7 The following conditions are equivalent to (B1)–(B4):
(B5) If $Ax = \lambda Dx$ for a positive diagonal matrix D and $x \geq 0$ (for $x \neq 0$), then $\lambda < 0$; and every principal submatrix of A has the same property.
(B6) For all nonnegative diagonal matrices $D \geq 0, A - D$ has no zero eigenvalue in a direction $x \geq 0$, and every principal submatrix of A has the same property.
(B7) There exists an $x > 0$ such that $Ax < 0$, and every principal submatrix has the same property.
(B8) For some $r \in \mathbb{R}^n$, the Lotka–Volterra equation (15.1) has all orbits bounded, as $t \to +\infty$, and this property is robust against small perturbations of the a_{ij}.

Exercise 15.2.8 An ecosystem described by (15.1) is said to be *hierarchically ordered* if the graph $G^+(A)$, obtained by drawing an arrow from j to i whenever $a_{ij} > 0$, has no directed cycle and additionally $a_{ii} < 0$ holds for all i. (In this case we can order the species hierarchically such that positive effects occur only from a lower to a higher level). Show that the ecosystem is hierarchically ordered if and only if A is a qualitative B-matrix, i.e. if every matrix $\tilde{A}$ with $\text{sgn}\tilde{a}_{ij} = \text{sgn}a_{ij}$ is a B-matrix.

Exercise 15.2.9 Show that the orbits of a general predator–prey system

$$\dot{x}_i = x_i\left(r_i - \sum a_{ij}x_j - \sum b_{ik}y_k\right)$$
$$\dot{y}_k = y_k\left(-s_k + \sum c_{ki}x_i - \sum d_{kl}y_l\right)$$

$(i = 1, \ldots, n; k = 1, \ldots, m;$ all $a_{ij}, b_{ij}, c_{ij}, d_{ij}, s_j$ and r_i nonnegative and $a_{ii} > 0$, $d_{ii} > 0$) are uniformly bounded for $t \to +\infty$.

When A is a B-matrix, then the Lotka–Volterra equation (15.1) has uniformly bounded orbits for $t \to +\infty$ and thus the index theorem applies (see exercise 13.4.4). One could also use the index theorem for the replicator equation (15.12) and argue that (B3) prevents the existence of saturated fixed points with $z_{n+1} = 0$. Hence all saturated fixed points of (15.12) correspond to 'finite' saturated fixed points of (15.1). Their indices have to add up to $(-1)^n$. In fact, even some sort of converse holds.

Exercise 15.2.10 Show that the following conditions are equivalent to (B1)–(B8).

(B9) For all $\mathbf{r} \in \mathbb{R}^n$, (15.1) has a saturated rest point, and the same holds for all subsystems. (Hint: (B9) $\Rightarrow$ (B7).)

(B10) For every $\mathbf{r} \le \mathbf{0}$, (15.1) has only one saturated rest point, namely $\mathbf{0}$. (Hint: (B2) $\Rightarrow$ (B10) $\Rightarrow$ (B4).)

Theorem 15.2.11 *If all proper principal submatrices of A are B-matrices and* $\det(-A) > 0$ *then A itself is a B-matrix.*

Proof Let us suppose that A is not a B-matrix. By (B3) there exists a $\bar{z} \in \operatorname{int} S_n$ such that

$$(A\bar{z})_i = \lambda > 0 \quad . \tag{15.16}$$

Indeed, $\bar{z} \in \operatorname{bd} S_n$ would contradict the B-property of the submatrices of A, and $\lambda = 0$ would imply $A\bar{z} = \mathbf{0}$ and hence $\det A = 0$. The point $\bar{z}$ corresponds to a regular saturated rest point of (15.12) with $z_{n+1} = 0$. Its index is the negative (for the transversal eigenvalue $-\lambda$) of the determinant of the Jacobian of (15.12) at $\bar{z}$ restricted to the face $z_{n+1} = 0$, independently of $\mathbf{r}$. By exercise 13.5.5 this is

$$i(\bar{x}) = -\operatorname{sgn} \det A \quad . \tag{15.17}$$

We shall show that for $\mathbf{r} = \mathbf{0}$, the point $\bar{z}$ is the only saturated rest point of (15.12). Then by theorem 13.4.1 its index equals $(-1)^n$. By (15.17), therefore, $\operatorname{sgn} \det A = (-1)^{n-1}$ and this yields $\operatorname{sgn} \det(-A) = -1$, which is a contradiction.

For $\mathbf{r} = \mathbf{0}$, (15.9) has no interior fixed point since $\det A \ne 0$. Furthermore, (B2) and the assumption imply that all boundary orbits converge to $\mathbf{0}$. So $\mathbf{0}$ is the only (finite) fixed point of (15.9), and since all of its transversal eigenvalues are 0, it is a (completely degenerate) saturated fixed point. Its index is zero, however, and hence it does not count in theorem 13.4.1, since there are no nearby points $\mathbf{x} > \mathbf{0}$ with $A\mathbf{x} < \mathbf{0}$ (cf. exercise 13.4.5). This follows either directly from (B6) or from the following argument: As A is not a B-matrix, neither is its transpose A^t. Hence (B4) yields a $\mathbf{p} > \mathbf{0}$ with $A^t\mathbf{p} \ge \mathbf{0}$. But then $0 \le \mathbf{p} \cdot A\mathbf{x} < 0$, a contradiction. $\square$

Exercise 15.2.12 Give an alternative proof of (15.17). Consider (15.1) resp. (15.12) with $r_i = -r < 0$. This system has three saturated rest points: $\mathbf{0}$, an interior fixed point $r\mathbf{z}$, and a point at infinity in direction $\mathbf{z}$. Compute the indices of $\mathbf{0}$ and $r\mathbf{z}$ and apply theorem 13.4.1.

Exercise 15.2.13 Show that theorem 15.2.11 can be strengthened as follows:
If all $(n-1) \times (n-1)$ principal submatrices of A are B-matrices, then the following statements are equivalent:

(a) A itself is not a B-matrix
(b) $\det(-A) \le 0$ *and* $\mathrm{adj}(-A) > 0$.

Here $\mathrm{adj}(A)$ is the *adjoint* matrix of A. Its (i,j)-entry is the determinant of the $(n-1) \times (n-1)$ submatrix of A obtained by deleting the j-th row and the i-th column, multiplied by $(-1)^{i+j}$. Since for a nonsingular matrix

$$A^{-1} = \frac{\mathrm{adj} A}{\det A} \ , \tag{15.18}$$

condition (b) means in this case that all entries of A^{-1} are positive. (Hint: (i) Applying (B3) to the matrix $(b_i^{-1} a_{ij})$, we see that for each $\mathbf{b} > \mathbf{0}$ there is an $\mathbf{x} > \mathbf{0}$ with $A\mathbf{x} = \lambda \mathbf{b}$ and $\lambda \ge 0$. (ii) If $\det A \ne 0$ then (i) implies that A^{-1} is nonnegative. It is even positive since a zero entry would contradict the assumption on submatrices. The converse is easy. (iii) If $\det A = 0$, the right and left eigenvectors $\mathbf{u}$ and $\mathbf{v}$ with $A\mathbf{u} = \mathbf{v}A = \mathbf{0}$ are positive. There are, then, no other eigenvectors for 0, and hence A has rank $n-1$. Then $(\mathrm{adj} A)_{ij} = \rho u_i v_j$ with $\mathrm{sgn}\rho = (-1)^{n-1}$. For the converse, observe that $\mathrm{adj} A \ne 0$ implies that the rank of A is $n-1$.)

Exercise 15.2.14 Show that a matrix A with the sign pattern

$$\begin{bmatrix} - & + & - \\ - & - & + \\ + & - & - \end{bmatrix}$$

corresponding to a cyclic predator–prey system is a B-matrix if and only if $\det A < 0$.

Exercise 15.2.15 Prove uniform boundedness of the solutions of the discrete time version of Lotka–Volterra equations (exercise 5.2.5), if the interaction matrix is hierarchically ordered as in exercise 15.2.8.

Exercise 15.2.16 Show that every two-species mutualistic system given by exercise 5.2.5 has unbounded solutions. Hence B-matrices do not imply boundedness for this discrete dynamics.

15.3 VL-stability and global stability

In this section we turn to the strongest stability concept: global asymptotic stability.

The matrix A will be called *Volterra–Lyapunov stable* (VL-stable) if there exists a positive diagonal matrix $D > 0$ such that the symmetric matrix $DA + A^t D$ is negative definite, i.e. if there exist positive numbers d_i such that

$$\sum_{i,j} d_i a_{ij} x_i x_j < 0 \text{ for all } \mathbf{x} \neq \mathbf{0} \quad . \tag{15.19}$$

This means that for suitable $d_i > 0$, the function $V(\mathbf{x}) = \sum d_i x_i^2$ is a strict Lyapunov function for the linear ODE $\dot{\mathbf{x}} = A\mathbf{x}$ and hence that the ellipsoids $V(\mathbf{x}) \leq c$ are strictly forward invariant. This is related to Lyapunov's characterization of stable matrices, see exercise 3.1.1. VL-stability is much stronger, as it requires the positive definite matrix Q to be diagonal, or the ellipsoid to be symmetric with respect to the coordinate planes.

Theorem 15.3.1 *If A is VL-stable, then for every $\mathbf{r} \in \mathbb{R}^n$ the Lotka–Volterra equation (15.1) has one globally stable fixed point.*

Proof (15.19) implies (B4) and hence by (B9) the existence of a saturated rest point $\bar{\mathbf{x}}$. Let $V(\mathbf{x})$ be the standard Lyapunov function proposed by Volterra,

$$V(\mathbf{x}) = \sum_{i=1}^{n} d_i(\bar{x}_i \log x_i - x_i) \quad . \tag{15.20}$$

It has a unique global maximum at the point $\bar{\mathbf{x}}$ and is defined for all $\mathbf{x}$ with $\text{supp}(\mathbf{x}) \supseteq \text{supp}(\bar{\mathbf{x}})$. Now

$$\begin{aligned}\dot{V}(\mathbf{x}) &= \sum_i d_i(\bar{x}_i - x_i)\tfrac{\dot{x}_i}{x_i} = \sum_i d_i(\bar{x}_i - x_i)(r_i + \sum a_{ij}x_j) \\ &= -\sum_{i,j} d_i a_{ij}(x_i - \bar{x}_i)(x_j - \bar{x}_j) + \sum_i d_i(\bar{x}_i - x_i)(r_i + \sum_j a_{ij}\bar{x}_j) \ .\end{aligned} \tag{15.21}$$

By (15.19) the first sum is positive for $\mathbf{x} \neq \bar{\mathbf{x}}$. In the second sum all terms with $\bar{x}_i > 0$ vanish since $\bar{\mathbf{x}}$ is a rest point, and the remaining terms are nonnegative since $\bar{\mathbf{x}}$ is saturated. Thus $\bar{\mathbf{x}}$ is globally stable. $\qquad\square$

Exercise 15.3.2 Show that A is VL-stable if and only if for every positive definite Q the matrix QA has a negative diagonal entry.

Exercise 15.3.3 Show that if A is VL-stable and D, D' are positive diagonal matrices, then A^t, A^{-1}, DAD' and all its principal submatrices are VL-stable.

Exercise 15.3.4 Show that the 2×2 matrix A is VL-stable if and only if $a_{ii} < 0$ $(i = 1, 2)$ and $\det A > 0$.

Exercise 15.3.5 Show that the Lotka–Volterra equation with immigration terms $\varepsilon_i > 0$,

$$\dot{x}_i = x_i (r_i + (A\mathbf{x})_i) + \varepsilon_i \qquad (15.22)$$

(which corresponds to (13.17)), has a unique, globally stable interior fixed point, whenever A is a VL-stable matrix.

Exercise 15.3.6 Consider a two-patch ecosystem with migration,

$$\begin{aligned} \dot{x}_i &= x_i(r_i + (A\mathbf{x})_i) + d_i(y_i - x_i) \\ \dot{y}_i &= y_i(r_i + (A\mathbf{y})_i) + d_i(x_i - y_i). \end{aligned} \qquad (15.23)$$

Show that if A is VL-stable, then for arbitrary $d_i > 0$ and r_i, (15.23) has a unique globally stable fixed point.

Exercise 15.3.7 Show that the function $V(\mathbf{x})$ from (15.20) is a constant of motion for the Lotka–Volterra equation (15.1) if and only if $d_i a_{ij} = -d_j a_{ji}$ holds for all i, j. (The matrix DA is then antisymmetric.)

We have seen in section 15.1 that matrices with negative dominant diagonal also lead to global stability (using a rectangular Lyapunov function). The next exercises compare this with VL-stability.

Exercise 15.3.8 Show that the following conditions are equivalent:
(DD1) The matrix A has a negative dominant diagonal, i.e. satisfies (15.7).
(DD2) The rectangles $\{\mathbf{x} \in \mathbb{R}^n : |x_i| \le d_i \text{const.}\}$ are strictly forward invariant for the linear ODE $\dot{\mathbf{x}} = A\mathbf{x}$.
(DD3) The matrix $\tilde{A}$ defined by $\tilde{a}_{ii} = -a_{ii}$ and $\tilde{a}_{ij} = -|a_{ij}|$ for $i \ne j$ is an M-matrix.
(DD4) A^t has a negative dominant diagonal, i.e. there exist $c_i > 0$ with

$$a_{ii}c_i + \sum_{j \ne i} |a_{ji}|c_j < 0 \quad.$$

(DD5) The function $\sum c_i |x_i|$ is a strict Lyapunov function for $\dot{\mathbf{x}} = A\mathbf{x}$.
(DD6) $\sum c_i |\log(x_i/\hat{x}_i)|$ is a strict Lyapunov function for (15.1), if it has an interior fixed point $\hat{\mathbf{x}}$.
(DD7) There exists an $\mathbf{x} > 0$ such that $SAS\mathbf{x} < 0$ holds for every signature matrix S (a diagonal matrix with entries ± 1).

Exercise 15.3.9 Show that diagonal dominance implies VL-stability. (Hint: use c_i/d_i, with c_i from (DD4) and d_i from (DD1), as multipliers in (15.19).)

15.4 P-matrices

The matrix A is said to be a *P-matrix* if all its principal minors are positive.

Theorem 15.4.1 *The following properties are equivalent:*

(P1) *A is a P-matrix.*

(P2) *For every diagonal matrix $D \geq 0$, $A + D$ is a P-matrix.*

(P3) *For all $\mathbf{x} \neq \mathbf{0}$, there exists an i such that $x_i(A\mathbf{x})_i > 0$.*

(P4) *For all $\mathbf{x} \neq \mathbf{0}$, there exists a diagonal matrix $D > 0$ such that $\mathbf{x} \cdot DA\mathbf{x} = D\mathbf{x} \cdot A\mathbf{x} > 0$.*

(P5) *Every real eigenvalue of a principal submatrix of A is positive.*

Proof (P1) $\Rightarrow$ (P2) is a consequence of the well known determinant formula

$$\det(A + D) = \sum_I \left(\prod_{i \notin I} d_i \right) \det(A_I)$$

where the sum runs over all 2^n subsets $I \subseteq \{1, \ldots, n\}$ and A_I is the corresponding principal submatrix.

(P2) $\Rightarrow$ (P3). Let us suppose that there exists an $\mathbf{x} \neq \mathbf{0}$ such that $x_i(A\mathbf{x})_i \leq 0$ for all i. Let $I = \text{supp}(\mathbf{x})$ and let $\mathbf{x}_I$ be the restricted vector. The components of $A_I\mathbf{x}_I$ and the corresponding components of $\mathbf{x}_I$ have opposite signs and hence $A_I\mathbf{x}_I$ may be written as $-D\mathbf{x}_I$ for a suitable nonnegative diagonal matrix D (over the index set I). Thus $(A_I + D)\mathbf{x}_I = \mathbf{0}$, and hence $A_I + D$ is a singular matrix, which is a contradiction.

(P3) $\Rightarrow$ (P4). If we choose a diagonal matrix D with a 1 in the i-th position given by (P3) and small positive entries elsewhere, we obtain $\mathbf{x} \cdot DA\mathbf{x} > 0$.

(P4) $\Rightarrow$ (P5). Let λ be a real eigenvalue of some submatrix A_I, i.e. $A_I\mathbf{x}_I = \lambda\mathbf{x}_I$. Let $\mathbf{x} \in \mathbb{R}^n$ have the same coordinates x_i as $\mathbf{x}_I$ for $i \in I$, and $x_i = 0$ for $i \notin I$. If we choose $D > 0$ according to (P4), we obtain

$$0 < A\mathbf{x} \cdot D\mathbf{x} = A_I\mathbf{x}_I \cdot D_I\mathbf{x}_I = \lambda\mathbf{x}_I \cdot D_I\mathbf{x}_I \quad .$$

Since this last inner product is positive, we obtain $\lambda > 0$.

(P5) $\Rightarrow$ (P1) follows from the fact that the determinant of a matrix is the product of its eigenvalues. $\qquad\square$

Exercise 15.4.2 Show that (P3) is equivalent to (P6) For every signature matrix S there is an $\mathbf{x} > \mathbf{0}$ with $SAS\mathbf{x} > \mathbf{0}$.

Exercise 15.4.3 Show that the following implications hold for an $n \times n$ matrix A: (i) if A is VL-stable then $-A$ is a P-matrix; (ii) if $-A$ is a P-matrix then A is a B-matrix.

Exercise 15.4.4 Compare the properties (P2), (P3), (P5), (P6) with (B6), (B4), (B5), (B7) and (DD7), respectively.

   The relevance of P-matrices for ecological systems relies upon the fact that they guarantee the uniqueness of a saturated fixed point. In the generic situation, where all fixed points of (15.1) are regular, this is an immediate consequence of the index theorem 13.4.1: If $-A$ is a P-matrix, then all saturated fixed points have the same index $(-1)^n$, and therefore there exists only one such point.

Theorem 15.4.5 *The Lotka–Volterra equation (15.1) has a unique saturated rest point for every* $\mathbf{r} \in \mathbb{R}^n$ *if and only if* $-A$ *is a* P-*matrix.*

Proof Let $-A$ be a P-matrix and $\mathbf{r} \in \mathbb{R}^n$. The existence of saturated rest points follows from (B9). Let us assume that there are two of them, $\mathbf{x}'$ and $\mathbf{x}'$. If $x_i' < x_i''$ then $x_i'' > 0$ and (since $\mathbf{x}''$ is a rest point) $(A\mathbf{x}'')_i + r_i = 0$. Since $\mathbf{x}'$ is saturated we have $(A\mathbf{x}')_i + r_i \leq 0$. Thus we obtain $(A(\mathbf{x}' - \mathbf{x}''))_i \leq 0$ and $(\mathbf{x}' - \mathbf{x}'')_i < 0$. This shows that $-A$ reverses the sign of all components of $\mathbf{x}' - \mathbf{x}'' \neq \mathbf{0}$, a contradiction to (P3).

Now let us assume that $-A$ is not a P-matrix. By (P3) we obtain an $\mathbf{x} \neq \mathbf{0}$ such that $x_i y_i \leq 0$ for all i (where $\mathbf{y} = -A\mathbf{x}$). Separating $\mathbf{x}$ and $\mathbf{y}$ into positive and negative parts, $\mathbf{x} = \mathbf{x}^+ - \mathbf{x}^-$, $\mathbf{y} = \mathbf{y}^+ - \mathbf{y}^-$ (with $\mathbf{x}^+, \mathbf{y}^-, \mathbf{y}^+, \mathbf{y}^- \geq \mathbf{0}$) and defining $\mathbf{r} = -\mathbf{y}^+ - A\mathbf{x}^+ = -\mathbf{y}^- A\mathbf{x}^-$ we obtain $\mathbf{r} + A\mathbf{x}^+ = -\mathbf{y}^+ \leq \mathbf{0}$, $\mathbf{r} + A\mathbf{x}^- = -\mathbf{y}^- \leq \mathbf{0}$ and $x_i^+ y_i^+ = x_i^- y_i^- = 0$, which shows that both $\mathbf{x}^+$ and $\mathbf{x}^-$ are saturated rest points of (15.1). $\qquad\square$

Exercise 15.4.6 Illustrate theorem 15.4.5 for $n = 2$ (bistability versus global stability).

Exercise 15.4.7 Show that the matrix

$$A = \begin{pmatrix} -1 & -\alpha & -\beta \\ -\beta & -1 & -\alpha \\ -\alpha & -\beta & -1 \end{pmatrix}$$

is VL-stable if and only if $-1 < \alpha + \beta < 2$, and $-A$ is a P-matrix if and only if $-1 < \alpha + \beta$ and $\alpha\beta < 1$ (cf. section 5.5).

Exercise 15.4.8 Let $-A$ be a P-matrix and $\bar{\mathbf{x}} = \bar{\mathbf{x}}(\mathbf{r})$ the unique saturated rest point of (15.1). If r_i increases and all other r_j remain constant, then show that $\bar{x}_i(\mathbf{r})$ increases. This seems biologically obvious, but does not hold if $-A$ is not a P-matrix.

Exercise 15.4.9 Show that the equation $\dot{x}_i = x_i f_i(\mathbf{x})$ admits only one saturated rest point if its orbits are uniformly bounded and $-D_{\mathbf{x}}\mathbf{f}$ is a P-matrix for every $\mathbf{x} \in \mathbb{R}^n_+$.

If (15.1) has an interior fixed point $\hat{\mathbf{x}}$ and $-A$ is a P-matrix then theorem 15.4.5 implies that there is no saturated rest point on the boundary. As the following exercises show, this implies some weak form of persistence, but it cannot prevent most orbits from spiralling away from $\hat{\mathbf{x}}$ and towards the boundary.

Exercise 15.4.10 If (15.1) has an interior fixed point and $-A$ is a P-matrix, then no interior orbit $\mathbf{x}(t)$ can converge to a point on the boundary. Show that it is also impossible that $\omega(\mathbf{x})$ lies in the interior of one boundary face. Does it follow that (15.1) is persistent? (Hint: show that the time-average of $\mathbf{x}(t)$ would converge to a saturated fixed point on the boundary.)

Exercise 15.4.11 Find a P-matrix $-A$ such that (15.1) is not permanent although it has an interior rest point. (Hint: see exercise 15.4.7.)

Exercise 15.4.12 Construct a system (15.1), with interior fixed point $\hat{\mathbf{x}}$ and $-A$ a P-matrix, which is not persistent. (Hint: add a fourth competitor to (5.18) with $\alpha\beta < 1$ and $\alpha + \beta > 2$ which obeys

$$\dot{y} = y(r - x_1 - x_2 - x_3 - y) .$$

If $\frac{3}{1+\alpha+\beta} < r < 1$, then the system has an interior fixed point $\hat{\mathbf{x}}$ which attracts all orbits on the plane $x_1 = x_2 = x_3 > 0$ and $y > 0$. Show that for all other interior orbits, however, $y(t) \to 0$ and $\omega(\mathbf{x})$ is the heteroclinic cycle of (5.18) again.)

Exercise 15.4.13 If $-A$ is a P-matrix, then the Lotka–Volterra system 'with immigration' (15.22) has exactly one fixed point for every choice of $\varepsilon_i > 0$. Is the converse also valid?

15.5 Communities with a special structure

Let A be the interaction matrix of an ecosystem modelled by (15.1). We obtain the *undirected graph* $G(A)$ by joining i and j with an edge whenever $a_{ij} \neq 0$ or $a_{ji} \neq 0$, and the *directed graph* $\vec{G}(A)$ by drawing an arrow $j \to i$ whenever $a_{ij} \neq 0$. A *cycle* of A is a *nonvanishing* product of the form $a_{i_1 i_2} a_{i_2 i_3} \dots a_{i_k i_1}$, for a sequence of *pairwise distinct* indices $i_1, i_2, \dots, i_k$. The *length* of this cycle is k.

Theorem 15.5.1 *Suppose that A has no cycles of length ≥ 3. Then A is VL-stable if and only if $-A$ is a P-matrix.*

Proof We shall use induction on the number of species n. The case $n = 2$ was settled in exercise 15.3.4. Let us assume that $-A$ is a P-matrix. We have to find $d_i > 0$ such that (15.19) holds.

(a) Let us assume first that A is *qualitatively symmetric*, i.e. that $a_{ij} \neq 0$ implies $a_{ji} \neq 0$. In this case we may choose $d_i > 0$ such that

$$d_i |a_{ij}| = d_j |a_{ji}| \quad . \tag{15.24}$$

This is possible in a consistent way since the graph $G(A)$ has no cycles. Let D be the diagonal matrix with entries d_i. If $a_{ij} a_{ji} \geq 0$ holds for all i, j then $-DA$ is a symmetric P-matrix. Then the quadratic form $\sum d_i a_{ij} x_i x_j$ is negative definite. Otherwise, there are indices $k \neq l$ with $a_{kl} a_{lk} < 0$. Then the term $x_k x_l$ in the quadratic form vanishes and the community splits into two blocks if we drop the connection between k and l. Applying the induction hypothesis to both subcommunities, both blocks of the quadratic form are positive definite and so is the whole form itself. Hence A is VL-stable.

(b) Let us assume now that there exist k, l with $a_{kl} \neq 0$ but $a_{lk} = 0$. Note that the undirected graph $G(A)$ may have cycles in this case. The matrix A turns out to be reducible now, if we split the community into two blocks as above. Using induction, we are left to show the following lemma. $\qquad\square$

Lemma 15.5.2 *Let A be a reducible matrix of the form*

$$A = \begin{bmatrix} A_1 & A_2 \\ 0 & A_3 \end{bmatrix} \quad .$$

Then A is VL-stable if and only if A_1 and A_3 are VL-stable.

Proof Replacing each of the principal VL-stable submatrices $A_i (i = 1, 3)$ by $D_i A_i + A_i^t D_i$ with suitable $D_i > 0$, we may assume that A_1 and A_3 are symmetric and negative definite. Thus we have only to find an $\varepsilon > 0$ such that the quadratic form

$$(\mathbf{x}, \mathbf{y}) \begin{bmatrix} \varepsilon A_1 & \varepsilon A_2 \\ 0 & A_3 \end{bmatrix} \begin{pmatrix} \mathbf{x} \\ \mathbf{y} \end{pmatrix}$$

is negative definite. Rescaling $\mathbf{x}$ to $\mathbf{x}\varepsilon^{-1/2}$, we are left with

$$\mathbf{x} \cdot A_1 \mathbf{x} + \varepsilon^{1/2} \mathbf{x} \cdot A_2 \mathbf{y} + \mathbf{y} \cdot A_3 \mathbf{y} \quad . \tag{15.25}$$

Since both A_1 and A_3 are negative definite, so is (15.25) for small $\varepsilon > 0$. $\quad\square$

If all pairwise interactions are of predator–prey type, the theorem simplifies even more:

Theorem 15.5.3 *Suppose that $\overset{\circ}{G}(A)$ has no cycles of length ≥ 3, $a_{ii} < 0$ and $a_{ij}a_{ji} \leq 0$ for all $i \neq j$. Then A is VL-stable.*

Proof By the construction in the proof of the previous theorem, for each irreducible block of A, $DA + A^t D$ yields a negative diagonal matrix. Together with lemma 15.5.2 this implies VL-stability. $\quad\square$

A particular case of theorem 15.5.3 is *food chains* which are always VL-stable, as was seen in section 5.3. The systems satisfying the assumptions of theorem 15.5.3 are of interest from another viewpoint. They represent those communities which guarantee stability when only the signs $(+, 0, -)$ of the interaction coefficients a_{ij} are known.

Exercise 15.5.4 Show that A is *qualitatively semi-stable* (i.e. every matrix with the same sign pattern has only eigenvalues with real part ≤ 0) if and only if the following three conditions are satisfied:

(i) $a_{ii} \leq 0$;
(ii) $a_{ij}a_{ji} \leq 0$ for $i \neq j$;
(iii) there are no cycles of length ≥ 3.

Exercise 15.5.5 Show that a matrix A is qualitatively VL-stable if and only if

(i) $a_{ii} < 0$;
(ii) $a_{ij}a_{ji} \leq 0$ for $i \neq j$;
(iii) there are no cycles of length ≥ 3.

Exercise 15.5.6 Show that the matrix $-A$ is a qualitative P-matrix if and only if all cycles are negative: $a_{ii} < 0$, $a_{ij}a_{ji} \leq 0$ ($i \neq j$), $a_{ij}a_{jk}a_{ki} \leq 0$ (i, j, k pairwise distinct), etc.

Another type of community defined by sign conditions is the following:

$$
\begin{array}{ll}
\text{(i)} & a_{ii} \leq 0; \\
\text{(ii)} & \text{all cycles of } A \text{ of length } \geq 2 \text{ are positive}.
\end{array} \qquad (15.26)
$$

Exercise 15.5.7 Consider a matrix A describing the competition of two symbiotic systems (i.e. there is a partition of the species into two disjoint groups I and J, one of which could be empty, such that $a_{ij} \geq 0$ for $i \neq j$, $i, j \in I$ or $i, j \in J$ and $a_{ij} \leq 0, a_{ji} \leq 0$ for $i = j$ or $i \in I$ and $j \in J$). Show that such a matrix A satisfies (15.26), but that the converse is not true. Show that (15.1) generates a monotone flow in the variables x_i ($i \in I$) and $-x_j$ ($j \in J$).

The following theorem generalizes the case of mutualism treated in section 15.1.

Theorem 15.5.8 *If A satisfies condition* (15.26) *then A is VL-stable if and only if $-A$ is a P-matrix.*

Proof Let $\tilde{A}$ be the matrix defined by $\tilde{a}_{ii} = a_{ii} \leq 0$ and $\tilde{a}_{ij} = |a_{ij}|$ for $i \neq j$. Then every principal minor of $\tilde{A}$ equals the corresponding minor of A, since in every cycle the number of negative entries of A (which change their sign in $\tilde{A}$) is even by (15.26). Thus when $-A$ is a P-matrix then so is $-\tilde{A}$. By (M3) and (15.7), $\tilde{A}$ and hence also A has a negative dominant diagonal. By exercise 15.3.9 A is VL-stable. $\qquad\qquad\qquad\qquad\qquad\qquad\qquad\square$

Exercise 15.5.9 Show that a matrix A is qualitatively diagonally dominant if and only if $a_{ii} < 0$ and there are no cycles of length ≥ 2.

Condition (15.26) is the most general condition on signs under which P-matrices can be characterized by diagonal dominance. Both theorem 15.5.1 and theorem 15.5.8 give conditions on the sign patterns of A which guarantee that being a P-matrix already implies VL-stability. It would be interesting to find a common generalization of these two results.

We conclude this section with a class of systems which is not defined by qualitative conditions on the signs only. (15.1) is said to be a *symmetric system* if $\operatorname{sgn} a_{ij} = \operatorname{sgn} a_{ji}$ for all i, j and if every cycle is equal to its reversed cycle, i.e.

$$
a_{i_1 i_2} a_{i_2 i_3} \cdots a_{i_k i_1} = a_{i_1 i_k} a_{i_k i_{k-1}} \cdots a_{i_2 i_1} . \qquad (15.27)
$$

Exercise 15.5.10 Show that it suffices that condition (15.27) holds for cycles of length 3.

Exercise 15.5.11 Show that (15.1) is a symmetric system if and only if there exist $d_i > 0$ such that $(d_i a_{ij})$ is a symmetric matrix.

Exercise 15.5.12 Show that a symmetric system is VL-stable if and only if $-A$ is a P-matrix.

Exercise 15.5.13 Show that a symmetric system admits a global Lyapunov function of the form

$$\sum_{i,j} d_i a_{ij}(x_i - \bar{x}_i)(x_j - \bar{x}_j) \,. \tag{15.28}$$

Show that in this case (15.1) is actually a gradient system with (15.28) as its potential function and a suitable Riemannian metric analogous to (7.48). (The case $n = 2$ was treated in exercise 3.3.4. In particular every two-species Lotka–Volterra system that is competitive or mutualistic is a gradient system.)

Exercise 15.5.14 Analyse the symmetric system with $a_{ij} = -f(|i - j|)$, where f is a positive function, e.g. $f(x) = a^{x^2}$. Under what conditions on f is the system stable?

Exercise 15.5.15 Analyse competition systems with $-a_{ij} = \varepsilon_i \delta_{ij} + c_i d_j$ (for $\varepsilon_i, c_i, d_i > 0$).

15.6 D-stability and total stability

Let $\bar{x}$ be an interior fixed point of the Lotka–Volterra equation (15.1). The Jacobian at $\bar{x}$ is given by $(\bar{x}_i a_{ij})$, and hence depends on the rates r_i. In order to ensure that $\bar{x}$ is always asymptotically stable, the interaction matrix A should be *D-stable*, which means that DA is stable for every diagonal matrix $D > 0$.

It turns out that this concept is rather strong, and — surprisingly — has some *global* meaning for Lotka–Volterra equations.

Exercise 15.6.1 If A is D-stable, then show that every principal submatrix of A is D-semi-stable (all eigenvalues have real part ≤ 0). (Hint: choose some d_i very small.)

Exercise 15.6.2 If A is a D-stable matrix, then show that $-A$ has nonnegative principal minors.

Exercise 15.6.3 If A is VL-stable, then show that it is D-stable.

Exercise 15.6.4 Show that a 2×2 matrix is D-stable if and only if $a_{11} \leq 0$ and $a_{22} < 0$ (or vice versa) and $\det A > 0$.

The last example shows that D-stable matrices need not be B-matrices (in some marginal cases). Thus we consider a slightly stronger stability concept: A is said to be *totally stable* if every principal submatrix of A is D-stable.

This concept has again some global meaning for the Lotka–Volterra equations.

Theorem 15.6.5 *A is totally stable if and only if for every* $\mathbf{r}$, *(15.1) has exactly one saturated fixed point and this point is asymptotically stable within its face.*

Exercise 15.6.6 Prove this. (Hint: use theorem 15.4.5.)

Conjecture 15.6.7 *If the Lotka–Volterra equation (15.1) has an interior fixed point* $\bar{\mathbf{x}}$ *and the interaction matrix is D-stable, then* $\bar{\mathbf{x}}$ *is globally stable.*

Exercise 15.6.8 Show that qualitative D-stability is the same as qualitative stability.

Exercise 15.6.9 Show that qualitative total stability is equivalent to qualitative VL-stability (compare exercise 15.5.5).

Exercise 15.6.10 Show that a 3×3 matrix A is stable if and only if

$$D < 0, \quad T < 0 \quad \text{and} \quad D > MT, \qquad (15.29)$$

where $D = \det A$, $T = \text{trace } A$, and M is the sum of the three principal 2×2 minors of A. This is the *Routh–Hurwitz* criterion. Show that for $D = MT$, A has a pair of imaginary eigenvalues.

Exercise 15.6.11 Show that a 3×3 matrix A is totally stable if and only if $a_{ii} < 0$, $\det A < 0$, the 2×2 minors $A_i = a_{jj}a_{kk} - a_{kj}a_{jk}$ are positive and

$$\sqrt{-a_{11}A_1} + \sqrt{-a_{22}A_2} + \sqrt{-a_{33}A_3} > \sqrt{-\det A} \quad .$$

Exercise 15.6.12 Show that a two-prey one-predator system (or any other three-species system with both 3-cycles nonnegative) whose interaction matrix is a P-matrix is totally stable.

Exercise 15.6.13 Construct matrices A which are totally stable but not VL-stable: Consider two-prey one-predator systems with

$$-A_\varepsilon = \begin{bmatrix} a & b & 1 \\ c & d & 1 \\ -1 & -1 & \varepsilon \end{bmatrix},$$

$(a, b, c, d, \varepsilon > 0)$. Show that A_ε is VL-stable for every $\varepsilon > 0$ if and only if $(b - c)^2 < 4ad$; $-A_\varepsilon$ is a P-matrix for every $\varepsilon > 0$ if and only if A_ε is totally stable for every $\varepsilon > 0$ if and only if $ad > bc$ and $a + d > b + c$.

Exercise 15.6.14 Show that for $n = 3$, D-stability together with the existence of an interior fixed point implies permanence. (Hint: wait for the next chapter.)

15.7 Notes

Many of the matrix stability concepts occurring in this chapter were originally motivated by mathematical economics, see Nikaido (1968) and Arrow and Hahn (1971). Their importance for ecological systems was stressed by Svirezhev and Logofet (1983), Logofet (1993) and Šiljak (1978). Excellent books on matrix theory are Berman and Plemmons (1979), and Horn and Johnson (1991). They contain surveys on M-matrices. The relevance of the 'linear complementarity problem' to the study of Lotka–Volterra equations (it is equivalent to the problem of finding saturated fixed points) was realized by Takeuchi and Adachi (1980). Theorems 15.4.5 and the equivalence of (B4) and (B9) are two basic results of this theory, due to Gale, see Berman and Plemmons (1979) and Cottle (1980). Mutualistic systems were studied by Goh (1979) and Smith (1986a). The characterization of boundedness in terms of B-matrices goes back to Moltchanov (1961), see also Khazin and Shnol (1991). The explicit characterization of B-matrices (exercise 15.2.13) generalizes a criterion for copositive matrices of Hadeler (1983). For a generalization of exercise 15.2.14 to n species see Oshime (1988). Exercise 15.2.15 is from Hofbauer *et al.* (1987) and 15.2.16 is from Lu (1996). The concept of VL-stability goes back to Volterra, see Scudo and Ziegler (1978), who called these matrices 'dissipative'. For further results see Redheffer (1985, 1989) and Lu (1996). A different approach to global stability is proposed by Gouzé (1993). The global stability results for diagonally dominant and VL-stable matrices extend to more general models allowing time-dependent interaction terms, diffusion or time delays, see e.g. Gopalsamy (1992).

Theorem 15.4.1 on *P*-matrices is due to Fiedler and Ptak (1962) and Gale and Nikaido (1965). Section 15.5 follows Takeuchi *et al.* (1978). Theorem 15.5.1 was also shown by Berman and Hershkowitz (1983). Competition between two symbiotic subcommunities (exercise 15.5.7) was studied by Smith (1986b) using monotone flows. The Lyapunov function for symmetric systems was found by MacArthur (1970). Exercise 15.5.14 arises in a theory of species packing and niche overlap, see MacArthur (1970) and May (1973). Exercise 15.5.15 is from Shigesada *et al.* (1984, 1989). *D*-stable matrices were studied by Johnson (1974), Cross (1978) and Clark and Hallam (1982). Sign stable matrices were emphasized by May (1973); for a survey of qualitative (or sign) matrices see Quirk (1981).

16

Some low-dimensional ecological systems

We concentrate here on the effects of heteroclinic cycles, and the interplay of permanence and heteroclinic repellors on the boundary of low-dimensional systems. For three-dimensional Lotka–Volterra equations, we characterize robust permanence. We also discuss permanence for more general ecological equations, analyse the two-prey two-predator case and show how an epidemiologic model describing the evolution of virulence can be analysed in the Lotka–Volterra framework.

16.1 Heteroclinic cycles

We shall now investigate permanence and persistence for *three-species systems*, starting with three-dimensional Lotka–Volterra equations

$$\dot{x}_i = x_i \left(r_i + \sum_{j=1}^{3} a_{ij} x_j \right) \qquad i = 1, 2, 3 \quad . \tag{16.1}$$

In this context, the most remarkable phenomenon is the occurrence of *heteroclinic cycles*, i.e. cyclic arrangements of saddle rest points and saddle connections (orbits having one saddle as α-limit and the next one as ω-limit). We have met with them in section 5.5 already. The saddles are the one-species rest points $\mathbf{F}_i$, which can occur (in a biologically reasonable way) only if $r_i > 0$ and $a_{ii} < 0$. Under this assumption, we may write (16.1) in the form

$$\dot{x}_i = r_i x_i \left(1 - \sum_{j=1}^{3} c_{ij} x_j \right) \qquad i = 1, 2, 3 \tag{16.2}$$

203

(with $c_{ii} > 0$). The transversal eigenvalue at $\mathbf{F}_i$ in direction $\mathbf{e}_j$ is given by

$$\left.\frac{\dot{x}_j}{x_j}\right|_{\mathbf{F}_i} = r_j \left(1 - \frac{c_{ji}}{c_{ii}}\right) \tag{16.3}$$

In order to obtain a heteroclinic cycle $\mathbf{F}_1 \to \mathbf{F}_2 \to \mathbf{F}_3 \to \mathbf{F}_1$, each $\mathbf{F}_i$ must have one positive and one negative transversal eigenvalue, and they must be arranged in a cyclic pattern. This means that

$$\begin{aligned}
c_{31} &> c_{11} > c_{21} \\
c_{12} &> c_{22} > c_{32} \\
c_{23} &> c_{33} > c_{13} .
\end{aligned} \tag{16.4}$$

To stress the analogy with the special case treated in section 5.5, we introduce the abbreviations

$$\alpha_i = \frac{c_{i-1,i}}{c_{ii}} \quad \beta_i = \frac{c_{i+1,i}}{c_{ii}} \tag{16.5}$$

(where the indices are counted modulo 3). Then (16.4) reads as

$$\alpha_i > 1 > \beta_i \quad . \tag{16.6}$$

We note that the β_i may be negative.

From the discussion in section 3.3 we know that on the face $x_3 = 0$, the isoclines do not intersect and hence that an orbit starting in $(x_1, x_2, 0)$ (with $x_1, x_2 > 0$) converges to $\mathbf{F}_2$. In particular the face contains an orbit with α-limit $\mathbf{F}_1$ and ω-limit $\mathbf{F}_2$. Similarly, there is an orbit from $\mathbf{F}_2$ to $\mathbf{F}_3$ and one from $\mathbf{F}_3$ to $\mathbf{F}_1$. These orbits together with the $\mathbf{F}_i$ form the heteroclinic cycle γ.

Theorem 16.1.1

(a) *If* $\det C > 0$ *and*

$$\prod_{i=1}^{3}(\alpha_i - 1) < \prod_{i=1}^{3}(1 - \beta_i) \tag{16.7}$$

then (16.2) *is permanent.*

(b) *If*

$$\prod_{i=1}^{3}(\alpha_i - 1) > \prod_{i=1}^{3}(1 - \beta_i) \tag{16.8}$$

then the heteroclinic cycle is an attractor.

Proof

(a) We shall prove first that $\det C > 0$ implies that all orbits are uniformly bounded. Indeed, the two-species subsystems have uniformly bounded orbits and the corresponding interaction matrices are B-matrices. Together with $\det(-A) > 0$, this implies by theorem 15.2.11 that A is a B-matrix.

In order to prove permanence, it is enough to show that there exists a positive solution (p_1, p_2, p_3) of the following system of inequalities, which is just (13.45) at the boundary rest points $\mathbf{0}, \mathbf{F}_1, \mathbf{F}_2, \mathbf{F}_3$:

$$
\begin{aligned}
r_1 p_1 + r_2 p_2 + r_3 p_3 &> 0 \\
r_2(1 - \beta_1)p_2 - r_3(\alpha_1 - 1)p_3 &> 0 \\
-r_1(\alpha_2 - 1)p_1 + r_3(1 - \beta_2)p_3 &> 0 \\
r_1(1 - \beta_3)p_1 - r_2(\alpha_3 - 1)p_2 &> 0 .
\end{aligned}
\tag{16.9}
$$

The first inequality is satisfied for every positive $\mathbf{p}$. The other inequalities can be written as

$$
\frac{p_{i+1}}{p_{i-1}} > \frac{r_{i-1}}{r_{i+1}} \frac{\alpha_i - 1}{1 - \beta_i}
\tag{16.10}
$$

for $i = 1, 2, 3 \pmod 3$. By multiplying those inequalities, it is easy to see that (16.10) has a positive solution $\mathbf{p}$ if and only if (16.7) is satisfied. This proves (a).

(b) If (16.8) is valid, then one can find $p_i > 0$ such that the inequalities in (16.10) are reversed. The corresponding function $P(x) = \prod x_i^{p_i}$ is then exponentially decreasing near $\mathbf{F}_1$, $\mathbf{F}_2$ and $\mathbf{F}_3$. The heteroclinic cycle γ is an attractor in bd $\mathbb{R}_+^3$. Indeed, the function

$$
\Psi(\mathbf{x}) = \sum_{i=1}^{3} p_i r_i \left(1 - \sum_{j=1}^{3} c_{ij} x_j \right)
$$

satisfies $\dot{P} = P\Psi$ in int $\mathbb{R}_+^3$ and

$$
\int_0^T \Psi(\mathbf{x}(t)) dt < 0
\tag{16.11}
$$

for every $\mathbf{x} \in \gamma$. Thus P is an average Lyapunov function (but a decreasing one) near γ. The same argument as in the proof of theorem 12.2.1 shows that all orbits in $\mathbb{R}_+^3$ near γ converge to γ, i.e. that the heteroclinic cycle is an attractor. $\square$

Exercise 16.1.2 Show that (16.2) is always persistent under assumptions (16.4).

Exercise 16.1.3 Show that the following conditions are equivalent:

(a) (16.2) has uniformly bounded orbits;
(b) (16.2) has a unique interior rest point;
(c) $\det A < 0$;
(d) $\det C > 0$.

(Hint: (d) $\Rightarrow$ (a) was shown in the previous proof; (a) $\Rightarrow$ (b) follows from the index theorem 13.4.1 and (b) $\Rightarrow$ (c) from Cramer's rule and (16.6).)

Exercise 16.1.4 Assume $\det C > 0$ and show that (16.7) is equivalent to the unique interior rest point lying above the plane through F_1, F_2 and F_3. Check that the geometric condition from theorem 13.6.6 is satisfied in this case.

Exercise 16.1.5 For $r_i = 1$, (16.2) is equivalent (in which sense?) to the general rock–scissors–paper game (see section 7.7). In this case, (16.7) resp. (16.8) imply global stability of the interior equilibrium, resp. of the heteroclinic cycle.

Exercise 16.1.6 Show that if the r_i are not equal, stable limit cycles can occur. (Hint: construct examples that are permanent but the interior equilibrium is unstable.)

Exercise 16.1.7 Determine the asymptotic behaviour of the time average of (16.2) under the assumption (16.8). (Hint: recall section 5.5.)

16.2 Permanence for three-dimensional Lotka–Volterra systems

In section 13.5 the following conditions were shown to be necessary both for permanence and robust persistence of the n-dimensional Lotka–Volterra equation (16.1):

(i) there exists an interior rest point $\hat{x}$;
(ii) $\det(-A) > 0$;
(iii) if the truncated system without species k admits a unique interior rest point, then the corresponding principal minor of $-A$ is positive.

Indeed, (i) is a consequence of theorem 13.5.1, (ii) is just (13.21) and (iii) is theorem 13.5.7.

In dimension three, (iii) excludes those two-species subsystems where the interior rest point is a saddle. By the index theorem such a system would have two saturated and hence stable boundary rest points. Thus (iii) states that *the bistable case cannot occur as a subsystem of a persistent three-species system.*

We will now show that for a uniformly bounded three-dimensional system, (i) to (iii) are sufficient conditions for persistence and — up to the occurrence of heteroclinic cycles — even for permanence. We shall strengthen (iii) to

(iii') the two-species subsystems are uniformly bounded and not bistable.

More explicitly: the determinant of a two-species subsystem (say $x_3 = 0$) must be positive ($a_{11}a_{22} > a_{12}a_{21}$) if the system is mutualistic (a_{12} and $a_{21} \geq 0$) or if it (is competitive and) has a rest point $\bar{x}$ with $\bar{x}_1 \geq 0$ and $\bar{x}_2 \geq 0$.

Note that (iii') excludes also the degenerate cases of a line of rest points or of the two isoclines intersecting at an unstable rest point on a coordinate axis. Neither case can occur in robustly persistent systems.

We shall also need the following condition:

(iv) If the system admits a heteroclinic cycle, then (16.7) holds.

With those conditions we obtain

Theorem 16.2.1 *Consider the three-dimensional Lotka–Volterra equation with intraspecific competition ($a_{ii} < 0$)*

$$\dot{x}_i = x_i \left(r_i + \sum_{j=1}^{3} a_{ij}x_j \right) \qquad i = 1, 2, 3 . \qquad (16.12)$$

(a) *This system is persistent and uniformly bounded, and both properties are robust, if and only if* (i), (ii) *and* (iii') *hold.*
(b) *This system is robustly permanent and uniformly bounded if and only if* (i), (ii), (iii') *and* (iv) *hold.*

Proof It remains to show that (i), (ii) and (iii') guarantee permanence if there is no heteroclinic cycle. First we note that theorem 15.2.4 together with (ii) and (iii') implies that A is a B-matrix. Hence the orbits of (16.12) are uniformly bounded for $t \to +\infty$.

Next we show that none of the boundary rest points of (16.12) is saturated. For the two-species rest points this follows from (iii') and (13.36). So we are left with the origin $\mathbf{0}$ and the one-species rest points F_1, F_2 and F_3 (if they exist). If any of these were regular and saturated, it would be a sink and hence have index -1. By the index theorem 13.4.1, however, only one saturated rest point can exist: this must be the interior rest point. If one of the rest points on the boundary had a zero eigenvalue, a suitable small perturbation of the r_i would make all its transversal eigenvalues negative. Since (i), (ii) and (iii') are open conditions, the above argument applies to the perturbed system.

Now we apply exercise 13.6.3. We have to find $p_1, p_2, p_3 > 0$ such that

$$\sum_{i:x_i=0} p_i \left(r_i + \sum_{j=1}^{3} a_{ij} x_j \right) > 0 \tag{16.13}$$

at every boundary rest point $\mathbf{x}$. Again we need not worry about the two-species rest points, since their unique transversal eigenvalue is positive by the above considerations, so that (16.13) is trivially satisfied. At $\mathbf{0}, \mathbf{F}_1, \mathbf{F}_2$ and $\mathbf{F}_3$, (16.13) yields:

$$r_1 p_1 + r_2 p_2 + r_3 p_3 \quad > \quad 0 \tag{16.14}$$

$$\left(r_2 - r_1 \frac{a_{21}}{a_{11}} \right) p_2 + \left(r_3 - r_1 \frac{a_{31}}{a_{11}} \right) p_3 \quad > \quad 0 \tag{16.15}$$

$$\left(r_1 - r_2 \frac{a_{12}}{a_{22}} \right) p_1 + \left(r_3 - r_2 \frac{a_{32}}{a_{22}} \right) p_3 \quad > \quad 0 \tag{16.16}$$

$$\left(r_1 - r_3 \frac{a_{13}}{a_{33}} \right) p_1 + \left(r_2 - r_3 \frac{a_{23}}{a_{33}} \right) p_2 \quad > \quad 0 \quad . \tag{16.17}$$

Since none of these rest points is saturated, at least one of the coefficients of the p_i is positive in each inequality. Note that $\mathbf{F}_i$ exists if and only if $r_i > 0$, since $a_{ii} < 0$ by assumption. We distinguish three cases concerning the number of one-species rest points in the system. Since $\mathbf{0}$ is not saturated, at least one of the r_i is positive.

Case A: $r_1 > 0$ and $r_2, r_3 \leq 0$. Then only $\mathbf{F}_1$ exists. We choose first p_2 and $p_3 > 0$ such that (16.15) holds. Then for large p_1, (16.14) holds too.

Case B: $r_1, r_2 > 0$ and $r_3 \leq 0$. Then $\mathbf{F}_1$ and $\mathbf{F}_2$ exist. We note that either $r_2 - r_1 a_{21}/a_{11} > 0$ or $r_1 - r_2 a_{12}/a_{22} > 0$. Indeed we would otherwise have $a_{12}, a_{21} < 0$ and $a_{21}/a_{11} \geq r_2/r_1 \geq a_{22}/a_{12}$ and so by (3.12) a bistable two-dimensional competition system, in contradiction to (iii'). If $r_1 - r_2 a_{12}/a_{22} > 0$, say, then we choose $p_2, p_3 > 0$ to fulfil (16.15), and then some p_1 which is suitably large.

Case C: $r_1, r_2, r_3 > 0$. Then all $\mathbf{F}_i$ exist, but (16.14) is trivially satisfied. If there exists one species i which invades both other species, i.e. if the transversal eigenvalues at both $\mathbf{F}_j$ $(j \neq i)$ in direction $\mathbf{F}_i$ is positive, then choosing p_i very large will give a solution of (16.15–17). If at one of the $\mathbf{F}_j$ both transversal eigenvalues are nonnegative, then it is again straightforward to find a solution $p_1, p_2, p_3 > 0$. The only remaining case is that of a heteroclinic cycle, which was treated in the previous section. □

Exercise 16.2.2 Suppose that (16.12) has uniformly bounded orbits. Then it is robustly permanent if and only if the two sets C and D from theorem 13.6.6 are disjoint.

Exercise 16.2.3 Try to give a purely algebraic proof (without referring to the index theorem) that (i), (ii) and (iii) exclude the possibility of saturated rest points on the boundary.

Exercise 16.2.4 Show that (16.12) with

$$A = \begin{bmatrix} -1 & 2 & 2 \\ 1 & -1 & 0 \\ 0 & -1 & -1 \end{bmatrix}$$

and $\hat{\mathbf{x}} = (1, 1, 1)$ satisfies (i), (ii), (iii) but not (iii') since there is a stable rest point at infinity in the subsystem $x_3 = 0$. Thus (i), (ii) and (iii) do not imply permanence or persistence.

Exercise 16.2.5 Show that the competition model with

$$-A = \begin{bmatrix} 8 & 5 & 7 \\ 10 & 2 & 3 \\ 2 & 10 & 9 \end{bmatrix}$$

and $\hat{\mathbf{x}} = (1, 1, 1)$ is not permanent, although $\hat{\mathbf{x}}$ is asymptotically stable. Thus permanence and local stability share the conditions (ii) on the determinant and $T < 0$ on the trace T, but the conditions (iii') and (15.29) concerning the 2×2 minors are different.

Exercise 16.2.6 If the growth rates r_i are nonzero, (16.12) can be written in the form

$$\dot{x}_i = r_i x_i \left(1 - \sum_{j=1}^{3} c_{ij} x_j\right) . \tag{16.18}$$

Show that the criteria for permanence and persistence given by theorem 16.2.1 depend only on the interaction matrix c_{ij}, and not on the growth rates r_i (as long as they keep the same signs), whereas local stability of the interior rest point depends both on the r_i and the c_{ij}. Construct an example where a change in the r_i results in the loss of the stability of the interior rest point $\hat{\mathbf{x}}$.

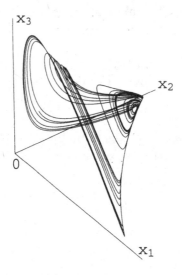

Fig. 16.1.

Exercise 16.2.7 Simplify the permanence criterion from theorem 16.2.1 as much as possible for (a) a food chain; (b) one predator and two prey; (c) two predators and one prey.

Exercise 16.2.8 Classify the possible phase portraits of permanent three-species Lotka–Volterra systems according to their boundary behaviour (ignoring the behaviour in $\text{int } \mathbb{R}^3_+$).

Exercise 16.2.9 Show that the two-prey one-predator system with $r_1 = r_2 = -r_3 = 1$ and

$$A = \begin{bmatrix} -1 & -1 & -10 \\ -1.5 & -1 & -1 \\ 5 & 0.5 & -0.01 \end{bmatrix}$$

is permanent. Prove that the interior rest point is not stable. Run the system on a computer. You will find chaotic motion (see fig. 16.1).

Exercise 16.2.10 Show that every two-species competition system which is not bistable may turn into a permanent system by the introduction of a suitable (a) common predator, or (b) third competitor.

Exercise 16.2.11 Generalize theorem 16.2.1 to the discrete dynamics given by exercise 5.2.5.

16.3 General three-species systems

For general three-species ecological equations

$$\dot{x}_i = x_i f_i(x_1, x_2, x_3) \qquad i = 1, 2, 3 \tag{16.19}$$

we cannot expect criteria as precise as in the Lotka–Volterra case. A biologically intuitive, coarse statement which can be expected to hold under rather general circumstances, however, is that the system is permanent if and only if there are no saturated rest points on the boundary. It is clear that this condition is necessary for robust permanence. For the converse, we need a few biologically reasonable restrictions to prevent limit cycles or attracting heteroclinic cycles on the boundary.

(H1) The solutions of (16.19) in $\mathbb{R}^3_+$ are uniformly bounded for $t \to +\infty$.

(H2) The ω-limit of every orbit on $\mathrm{bd}\,\mathbb{R}^3_+$ consists of rest points.

(H3) The origin is hyperbolic, i.e. $f_i(\mathbf{0}) \neq 0$, and there is at most one further rest point on each axis.

Thus there are two possibilities for each species. Take for example species 1:

(a) $f_1(x_1, 0, 0) < 0$ for all $x_1 > 0$. Then there is no rest point on the x_1-axis. Species 1 is *not self-supporting*.

(b) There is a $k_1 > 0$ such that $(x_1 - k_1)f(x_1, 0, 0) < 0$ for all $x_1 \leq 0$ with $x_1 \neq k_1$. Then species 1 is *self-supporting*, and k_1 is its *carrying capacity*. By $\mathbf{F}_1$ we denote the corresponding rest point $(k_1, 0, 0)$ and by λ_{ij} the transversal eigenvalues $f_j(\mathbf{F}_i)$ at $\mathbf{F}_i$ in direction $\mathbf{e}_j (j \neq i)$.

It follows that the three-species system belongs to one of the three following types.

Case A: Only one species is self-supporting: e.g. $f_1(\mathbf{0}) > 0$, $f_2(\mathbf{0}) < 0$ and $f_3(\mathbf{0}) < 0$.

Case B: Two species are self-supporting, e.g. $f_1(\mathbf{0}) > 0, f_2(\mathbf{0}) > 0$ and $f_3(\mathbf{0}) < 0$.

Case C: All three species are self-supporting: $f_i(\mathbf{0}) > 0$ for $i = 1, 2, 3$.

If none of the $\mathbf{F}_i$ is saturated, then one of the following possibilities holds:

(C1) There exists a species i which invades both other species: $\lambda_{ji} > 0$ and $\lambda_{ki} > 0$.

(C2) There exists a species i which can be invaded by both other species:
$\lambda_{ij} > 0$ and $\lambda_{ik} > 0$.

(C3) One has $\lambda_{12}, \lambda_{23}, \lambda_{31} > 0$ and $\lambda_{21}, \lambda_{32}, \lambda_{13} \leq 0$ (or the other way round).

Case (C3) gives rise to a heteroclinic cycle on the boundary (if there are no two-species rest points which break up the connection between saddle points). In this case we need an additional hypothesis:

(H4) If (C3) applies, then

$$\lambda_{12}\lambda_{23}\lambda_{31} + \lambda_{21}\lambda_{32}\lambda_{13} > 0 . \tag{16.20}$$

Theorem 16.3.1 *Let* (16.19) *describe a three-species community satisfying* (H1)–(H4). *If none of the boundary rest points is saturated, then the system is permanent.*

Proof We essentially repeat the proof from the last section, except that our average Lyapunov function will be of the form

$$P(\mathbf{x}) = (x_1 + x_2)^{p_0} x_1^{p_1} x_2^{p_2} x_3^{p_3} . \tag{16.21}$$

In $\operatorname{int} \mathbb{R}_+^3$, one has

$$\frac{\dot{P}}{P} = \Psi(\mathbf{x}) = p_0 \frac{x_1 f_1(\mathbf{x}) + x_2 f_2(\mathbf{x})}{x_1 + x_2} + \sum_{i=1}^{3} p_i f_i(\mathbf{x}) . \tag{16.22}$$

We observe that the first term in $\Psi(\mathbf{x})$ admits no continuous extension to the x_3-axis. We have therefore to take the lower semicontinuous extension $\Psi(0, 0, x_3) = p_0 \min(f_1(0, 0, x_3), f_2(0, 0, x_3))$.

By exercise 12.2.3, condition (12.12) on the extension of Ψ still guarantees permanence. By (H2), this condition reduces to $\Psi(\mathbf{x}) > 0$ for all rest points $\mathbf{x} \in \operatorname{bd} \mathbb{R}_+^3$, for a suitable choice of $p_0 \geq 0$, $p_1, p_2, p_3 > 0$.

The condition is trivially satisfied for all two-species rest points, since these points are not saturated. We are left with up to four conditions corresponding to the rest points $\mathbf{0}, \mathbf{F}_1, \mathbf{F}_2$ and $\mathbf{F}_3$:

$$p_0 \min(f_1(0), f_2(0)) + \sum_{i=1}^{3} p_i f_i(0) > 0 \tag{16.23}$$

$$\lambda_{12} p_2 + \lambda_{13} p_3 > 0 \tag{16.24}$$

$$\lambda_{21} p_1 + \lambda_{23} p_3 > 0 \tag{16.25}$$

$$\lambda_{31} p_1 + \lambda_{32} p_2 > 0 . \tag{16.26}$$

In case A, where only $\mathbf{0}$ and $\mathbf{F}_1$ exist, we proceed as in section 16.2, setting $p_0 = 0$. In case B, since neither $\mathbf{F}_1$ nor $\mathbf{F}_2$ is saturated, one of the $\lambda_{12}, \lambda_{13}$

and one of the $\lambda_{21}, \lambda_{23}$ are positive. Hence there is a choice of $p_1, p_2, p_3 > 0$ such that (16.24) and (16.25) are satisfied. Since $f_1(\mathbf{0})$ and $f_2(\mathbf{0})$ are both positive, (16.23) holds whenever p_0 is sufficiently large. Case C runs as in the Lotka–Volterra case (see section 16.1). □

Exercise 16.3.2 Apply the above results to

(a) a general one-predator two-prey system;
(b) a general two-predator one-prey system;
(c) a food chain;
(d) two mutualists invaded by a species which takes advantage of an association of mutualists without providing any benefits in return.

Show that (C3) does not apply in any of these cases. As long as there are no periodic orbits in the coordinate planes, permanence is therefore essentially equivalent to the absence of saturated boundary rest points.

Exercise 16.3.3 Derive permanence criteria for a two-predator, one-prey system of the form

$$\dot{x} = rx\left(1 - x - \frac{a_1 y_1}{1 + b_1 x} - \frac{a_2 y_2}{1 + b_2 x}\right)$$

$$\dot{y}_1 = s_1 y_1\left(-1 + \frac{c_1 x}{1 + b_1 x} - y_1\right)$$

$$\dot{y}_2 = s_2 y_2\left(-1 + \frac{c_2 x}{1 + b_2 x} - y_2\right)$$

assuming that there are no periodic orbits on the boundary planes.

Exercise 16.3.4 Construct examples of a permanent one-predator two-prey system with nonlinear growth rates where the competition of the two prey is bistable. (Hint: observe the slight difference in argumentation in sections 16.2 and 16.3.)

Exercise 16.3.5 Let $\mathbf{x}(t)$ be a periodic orbit with period T in the (x_1, x_2)-plane for (16.19). Show that this periodic orbit is attracting (resp. repelling) in the x_3-direction if the integral $\int_0^T f_3(\mathbf{x}(t))dt$ is negative (resp. positive). (Hint: use x_3 as an average Lyapunov function.)

16.4 A two-prey two-predator system

The higher the dimension of the ecological system, the more likely are heteroclinic cycles. The following simple Lotka–Volterra model illustrates this; it also shows how bistable systems of two competitors can be made permanent by the introduction of *two* predatory species (from section 16.2 we know that *one* such species cannot do this job).

We shall denote the densities of the prey by x_1 and x_2, and those of the predators by y_1 and y_2. For the sake of simplicity we shall assume that the predators do not interfere with each other, that they specialize in different prey and that they show no crowding effects. This leads to

$$
\begin{aligned}
\dot{x}_1 &= x_1(r_1 - a_{11}x_1 - a_{12}x_2 - b_1y_1) \\
\dot{x}_2 &= x_2(r_2 - a_{21}x_1 - a_{22}x_2 - b_2y_2) \\
\dot{y}_1 &= y_1(-c_1 + d_1x_1) \\
\dot{y}_2 &= y_2(-c_2 + d_2x_2) \, .
\end{aligned}
\tag{16.27}
$$

We denote the rest points of (16.27) by $\mathbf{F}_1$, $\mathbf{F}_2$, $\mathbf{F}_{12}$, $\mathbf{F}_1^1$, $\mathbf{F}_2^2$ etc., the subscripts referring to the prey species and the superscripts to the predator species.

Theorem 16.4.1 *Suppose that the interior rest point* $\mathbf{F}_{12}^{12} = (\bar{x}_1, \bar{x}_2, \bar{y}_1, \bar{y}_2)$ *of* (16.27) *exists and that the* (x_1, x_2)-*subsystem is bistable. Then* (16.27) *is permanent if*

$$
\frac{a_{21}}{r_2}\bar{x}_1 + \frac{a_{12}}{r_1}\bar{x}_2 < 1 \, .
\tag{16.28}
$$

Proof Let us start by assuming only that $\mathbf{F}_{12}^{12}$ exists. (Then $a_{ij}\bar{x}_j < r_i$ for $i, j = 1, 2$, so that both terms on the left hand side of (16.28) are less than 1.) Suppose that initially only prey 1 is present, i.e. that the state is at $\mathbf{F}_1$, and that a small amount of predator 1 is introduced. Its rate of growth is

$$
\frac{\dot{y}_1}{y_1}\bigg|_{\mathbf{F}_1} = -c_1 + d_1\frac{r_1}{a_{11}} = \frac{d_1}{a_{11}}(r_1 - a_{11}\bar{x}_1) = \frac{d_1}{a_{11}}(a_{12}\bar{x}_2 + b_1\bar{y}_1) > 0 \, .
$$

So predator 1 will invade at $\mathbf{F}_1$ and the system will converge to $\mathbf{F}_1^1$. At that point, prey 2 can invade, because $x_1(\mathbf{F}_1^1) = c_1/d_1 = \bar{x}_1$ and hence

$$
\frac{\dot{x}_2}{x_2}\bigg|_{\mathbf{F}_1^1} = r_2 - a_{21}\bar{x}_1 = a_{22}\bar{x}_2 + b_2\bar{y}_2 > 0 \, .
$$

Biologically, this means that prey 1 is weakened by its predator 1 to such an extent that it loses the competition against prey 2. Because of bistability, prey 2 will take over. Obviously, $\mathbf{F}_2$ is a stable rest point in the three-species system corresponding to the face $y_2 = 0$. In particular, if $\mathbf{F}_{12}^1$ does not exist, theorem 4.2.1 implies that the orbit with α-limit $\mathbf{F}_1^1$ converges to $\mathbf{F}_2$. At $\mathbf{F}_2$, by the same argument, predator 2 will invade, so that the system switches to $\mathbf{F}_2^2$. Again, prey 1 will invade and, if $\mathbf{F}_{21}^2$ does not exist, will eventually take over. Thus if neither $\mathbf{F}_{21}^2$ nor $\mathbf{F}_{12}^1$ exists, then a heteroclinic cycle $\mathbf{F}_1 \to \mathbf{F}_1^1 \to \mathbf{F}_2 \to \mathbf{F}_2^2 \to \mathbf{F}_1$ arises. We shall presently see that (16.28) is the condition needed to make this cycle unstable, and the system permanent. Indeed, the hyperplane H spanned by $\mathbf{F}_1, \mathbf{F}_1^1, \mathbf{F}_2, \mathbf{F}_2^2$ is given by

$$\frac{a_{11}}{r_1}x_1 + \frac{b_1}{r_1}y_1 + \frac{a_{22}}{r_2}x_2 + \frac{b_2}{r_2}y_2 = 1 \quad . \tag{16.29}$$

The remaining boundary rest points $\mathbf{0}, \mathbf{F}_{12}$ and (if they exist) $\mathbf{F}_{12}^1$ and $\mathbf{F}_{21}^2$ lie below the plane H. We shall now use theorem 13.6.6. Let (x_1, x_2, y_1, y_2) be in the cone D given by

$$\begin{aligned}
a_{11}x_1 + a_{12}x_2 + b_1y_1 &\geq r_1 \\
a_{21}x_1 + a_{22}x_2 + b_2y_2 &\geq r_2 \\
x_1 &\leq \bar{x}_1 \\
x_2 &\leq \bar{x}_2 .
\end{aligned} \tag{16.30}$$

These inequalities imply

$$\frac{a_{11}x_1 + b_1y_1}{r_1} + \frac{a_{22}x_2 + b_2y_2}{r_2} \geq 1 - \frac{a_{12}\bar{x}_2}{r_1} + 1 - \frac{a_{21}\bar{x}_1}{r_2} \quad .$$

The latter expression is larger than 1 if and only if (16.28) holds. But (16.28) implies that the set D is disjoint from the convex hull C of the boundary rest points. By theorem 13.6.6 the system (16.27) is permanent. $\qquad\square$

Exercise 16.4.2 Show that if (16.27) admits an interior rest point and the (x_1, x_2)-subsystem is not bistable, then (16.27) is permanent.

Exercise 16.4.3 If $a_{11}a_{22} > a_{12}a_{21}$ then show that $\mathbf{F}_{12}^{12}$ is globally stable. (Hint: theorem 15.5.1.)

Exercise 16.4.4 Show that if $\mathbf{F}_{12}^{12}$, but neither $\mathbf{F}_{12}^1$ nor $\mathbf{F}_{21}^2$ exists, and if the condition $\frac{a_{21}}{r_2}\bar{x}_1 + \frac{a_{12}}{r_1}\bar{x}_2 > 1$ holds, then the heteroclinic cycle is an attractor. The system is persistent but not permanent.

Exercise 16.4.5 Find examples where:

(a) $\mathbf{F}_{12}^{12}$ is asymptotically stable, but the heteroclinic cycle is attracting.
(b) $\mathbf{F}_{12}^{12}$ is unstable but (16.27) is permanent.

Exercise 16.4.6 Study the two-prey two-predator system

$$\dot{x}_1 = x_1(1 - \frac{2}{3}x_1 - \frac{1}{3}x_2 - by_1 - y_2)$$

$$\dot{x}_2 = x_2(1 - \frac{1}{3}x_1 - \frac{2}{3}x_2 - y_1 - by_2)$$

$$\dot{y}_1 = y_1(-1 + 2x_1 + x_2)$$

$$\dot{y}_2 = y_2(-1 + x_1 + 2x_2) \quad .$$

Show that for $b > 2$ there is a heteroclinic cycle $\mathbf{F}_1^1 \to \mathbf{F}_2^1 \to \mathbf{F}_2^2 \to \mathbf{F}_1^2 \to \mathbf{F}_1^1$. It is attracting for $b > b_0 \approx 6.4$, whereas for $b < b_0$ it is repelling and the system is permanent. (Use an average Lyapunov function as in section 16.3.) However, the inequalities (13.44) have a positive solution only for $b < 5$. For $b \geq 5$, the interior rest point lies in the convex hull C of the boundary rest points. Thus the sufficient conditions for permanence given by theorems 13.6.1 and 13.6.6 are in general not necessary for $n \geq 4$.

Exercise 16.4.7 Consider a bistable two-species Lotka–Volterra system. Show that if one adds two suitable competing species, the resulting Lotka–Volterra equation can be permanent. Show that the same holds for the introduction of suitable predators and competitors.

Exercise 16.4.8 Consider a Lotka–Volterra equation (16.2) for three competing species with a heteroclinic attractor on the boundary. Show that if one adds a suitable predator (or a suitable competing species) the resulting four-species Lotka–Volterra equation can be permanent. (These are examples of four-species permanent eco-communities which cannot be obtained from a three-species subsystem by the invasion of the 'missing species': the community construction, in such cases, must be more roundabout.)

16.5 An epidemiological model

The evolution of the virulence of parasites has attracted a lot of attention. It depends crucially on the particular mechanism of transmission, on whether a host can carry several strains or not, etc. Many pathogenic parasites can be transmitted both horizontally (through infection) and vertically (i.e. from parent to offspring). In order to study this case, we consider two parasitic strains 1 and 2 and assume that a higher horizontal transmission rate entails a higher virulence (i.e. a lower fitness of their host). We also assume that a host cannot be infected by both strains, and that vertical transmission is perfect. Let x be the number of uninfected hosts and y_1 and y_2 the number of hosts infected by strain 1 resp. 2. The death rate of an uninfected host is u_x, that of a host carrying strain i is u_i. The corresponding birth rates are b_x and b_i. They are reduced according to a logistic term, so that they vanish when the total number of hosts reaches some value K (a kind of carrying capacity). Horizontal transmission occurs at a rate proportional to the densities of infected and uninfected hosts, with horizontal transmission rate β_i. This yields

$$
\begin{aligned}
\dot{x} &= x\left[b_x(1 - \tfrac{x+y_1+y_2}{K}) - u_x - \beta_1 y_1 - \beta_2 y_2\right] \\
\dot{y}_1 &= y_1\left[b_1(1 - \tfrac{x+y_1+y_2}{K}) - u_1 + \beta_1 x\right] \\
\dot{y}_2 &= y_2\left[b_2(1 - \tfrac{x+y_1+y_2}{K}) - u_2 + \beta_2 x\right] \quad .
\end{aligned}
\tag{16.31}
$$

Let us consider first the situation of one strain only, say $y_2 = 0$. The rest point of the uninfected host is $\hat{x} = K(1 - u_x/b_x)$. The parasite (strain 1) can invade if

$$
H_1 + V_1 > 1 \tag{16.32}
$$

where $H_1 = \frac{\beta_1}{u_1}\hat{x}$ and $V_1 = \frac{b_1 u_x}{b_x u_1}$. This condition is easy to interpret: in an uninfected population at equilibrium, an invading parasite of strain 1 — i.e. one infected host — must give rise to at least one further infected individual through a combination of horizontal and vertical transmission. Indeed, H_1 can be viewed as the average number of horizontal transmissions during the lifetime of an infected individual (note that its life span is inversely proportional to u_1), and V_1 is $\frac{b_1}{u_1}(1 - \frac{\hat{x}}{K})$, i.e. the number of offspring produced by a parasitized host. The factor $\frac{b_1}{u_1}$ is just the proportionate reduction in host fitness, and hence corresponds to the *virulence* of strain 1.

Exercise 16.5.1 Show that if the parasite can spread, then either the whole population becomes infected, or infected and uninfected hosts will coexist at equilibrium. This last will hold if and only if

$$
\frac{b_1}{u_1} < \frac{b_x + \beta K}{u_x + \beta K}. \tag{16.33}
$$

Compute the ratio of vertically acquired to horizontally acquired cases.

Let us now consider the full system (16.31). On the face $x = 0$ (infected only) the function $b_1^{-1}\log y_1 - b_2^{-1}\log y_2$ is a Lyapunov function whose derivative has the sign of $\frac{u_2}{b_2} - \frac{u_1}{b_1}$ which, generically, is different from 0. We shall assume that strain 2 is more virulent than strain 1: this implies that on the (y_1, y_2) face, all orbits converge to the rest point with only y_1 present. From the previous exercise, we know that on an (x, y)-face, we have a globally stable rest point, which can consist of infected only, uninfected only or a coexistence of infected and uninfected (no bi-stability).

Exercise 16.5.2 Show that in the absence of horizontal transmission (i.e. for $\beta_i = 0$) the less virulent strain 1 always wins.

Exercise 16.5.3 Show that the three one-species rest points can form a heteroclinic cycle on the boundary, but that this cycle is never an attractor.

It is straightforward to check that an interior rest point in the (x, y_1)-face is saturated if and only if

$$\frac{\beta_2 u_1}{\beta_1 u_2} < 1 + \frac{b_1\beta_2 - b_2\beta_1}{u_2\beta_1}\frac{\beta_1 K - u_x - u_1}{\beta_1 K + b_x - b_1} \tag{16.34}$$

(This is best done by expressing the eigenvalue condition in terms of $1 - \frac{x+y_1}{K}$.) It is possible for strain 1 and strain 2 to invade each other's equilibria: in this case, there exists an interior rest point with all species present.

Exercise 16.5.4 Compute this interior rest point. Show that, generically, the system (16.31) admits exactly one saturated rest point. Is it globally stable? Show that there are no heteroclinic cycles with the exception of the one found in exercise 16.5.3.

The two strains can coexist if one is favoured by horizontal transmission and the other by vertical transmission. This corresponds to two different ecological 'niches'.

Exercise 16.5.5 Show that, generically, no more than two strains of the parasite can coexist.

Exercise 16.5.6 Specialized antibodies can recognise specific loci (so-called epitopes) of pathogenic virus and mount an immune response. If the virus has two epitopes, each with two variants, this leads to the following eight-dimensional predator–prey equation:

$$\begin{aligned}
\dot{v}_{ij} &= v_{ij}(r_{ij} - x_i - y_j) \\
\dot{x}_i &= x_i[c_i(v_{i1} + v_{i2}) - b] \\
\dot{y}_j &= y_j[k_j(v_{1j} + v_{2j}) - b]
\end{aligned} \tag{16.35}$$

where v_{ij} is the concentration of the virus with molecular sequence i at the first and j at the second epitope ($1 \leq i, j \leq 2$), and x_i and y_j are the concentrations of antibodies directed at sequence i of the first resp. sequence *j* of the second epitope. Show that

$$(v_{12}v_{21})^{-1}v_{11}v_{22} \text{ and } \frac{1}{c_1}\log x_1 + \frac{1}{c_2}\log x_2 - \frac{1}{k_1}\log y_1 - \frac{1}{k_2}\log y_2$$

are generically Lyapunov functions. The system admits a unique saturated rest point **P**, which lies on some boundary face (either one or two of the four viral species and the same number of the antibody species have to vanish). Within the boundary face, **P** is neutrally stable. Find a constant of motion. How much of this carries over to the case where there are more than two variants per epitope?

16.6 Notes

Theorems 16.1.1 and 16.2.1 are due to Hofbauer. Theorem 16.2.1 generalizes the characterization of permanence of two-prey one-predator systems given by Hutson and Vickers (1983). Other criteria for persistence of three-dimensional Lotka–Volterra systems were derived by Hallam *et al.* (1979). The example for chaotic motion (exercise 16.2.9) is from Gilpin (1979). Other examples were found by Arneodo *et al.* (1980, 1982) and Takeuchi and Adachi (1983, 1984). Section 16.3 follows Hutson and Law (1985). Similar results were obtained by Freedman and Waltman (1984, 1985) using a completely different approach. Two-prey two-predator systems have been studied by Takeuchi and Adachi (1984); the condition for permanence is due to Kirlinger (1986), see also Kirlinger (1988, 1989) for extensions. The results of exercises 16.4.7 and 16.4.8 are from Hofbauer and Sigmund (1989), for extensions see Schreiber (1997). The material from section 16.5 is from Lipsitch *et al.* (1996). The evolution of virulence is a fast-growing field: for an introduction, we refer to Ewald (1993), Anderson and May (1991), Nowak and May (1994), and Stadler *et al.* (1994). Exercise 16.5.6 is from Nowak *et al.* (1995a). For more on the classification of three-dimensional Lotka–Volterra competition equations, see Zeeman (1993, 1995, 1996) and Zeeman and Zeeman (1994). An open problem is the number of limit cycles. Examples with two limit cycles were constructed by Hofbauer and So (1994). Van den Driessche and Zeeman (1997) have excluded the possiblity of periodic orbits in some cases, using an extension of the Bendixson–Dulac test due to Busenberg and van den Driessche (1993).

17

Heteroclinic cycles: Poincaré maps and characteristic matrices

This chapter deals with heteroclinic cycles, which frequently occur as robust features of replicator equations. The first part uses Poincaré sections, the second characteristic matrices to classify heteroclinic cycles and to investigate their stability.

17.1 Cross-sections and Poincaré maps for periodic orbits

Let $\mathbf{p}$ be a nonstationary point of the ODE $\dot{\mathbf{x}} = \mathbf{f}(\mathbf{x})$ and H a hyperplane through $\mathbf{p}$ which is not parallel to $\mathbf{f}(\mathbf{p})$. An open neighbourhood S of $\mathbf{p}$ in H is said to be a *cross-section* if for every $\mathbf{x} \in S$, $\mathbf{f}(\mathbf{x})$ is transverse to H.

If, in particular, $\mathbf{p}$ lies on some periodic orbit γ with period T, then the orbit of every $\mathbf{x}$ in some sufficiently small neighbourhood U of $\mathbf{p}$ will follow the orbit of $\mathbf{p}$ for some time and hence intersect the cross-section S. Thus there exists for every $\mathbf{x} \in U$ a well defined $\tau_{\mathbf{x}} > 0$ such that $\mathbf{x}(\tau_{\mathbf{x}}) \in S$ and $\mathbf{x}(t) \notin S$ for $0 < t < \tau_{\mathbf{x}}$. This gives rise to a map $g : \mathbf{x} \to \mathbf{x}(\tau_{\mathbf{x}})$ from $S_0 = U \cap S$ to S, which is called the *Poincaré map* of the vector field near γ. Since the construction works for backward time as well, the map g is one-to-one and hence a diffeomorphism. It is easy to see that if we start from a different cross-section S' at $\mathbf{p}' \in \gamma$ we obtain a Poincaré map g' which is equivalent in the sense that there exists a local diffeomorphism h from S to S' with $h(\mathbf{p}) = \mathbf{p}'$ and $g' \circ h = h \circ g$.

The local behaviour of the flow near the periodic orbit γ can be completely understood by analysing the behaviour of the Poincaré map g near its fixed point $\mathbf{p}$.

In particular, if all eigenvalues of $D_{\mathbf{p}}g$ are in the interior of the unit circle, then $\mathbf{p}$ is an asymptotically stable fixed point for g and the periodic orbit γ

is an attractor, i.e. $d(\mathbf{x}(t), \gamma) \to 0$ as $t \to +\infty$ for all $\mathbf{x}$ near γ. One can show even more: every orbit $\mathbf{x}(t)$ near γ is *in phase* with γ in the sense that there exists a unique $\mathbf{z} \in \gamma$ such that $d(\mathbf{x}(t), \mathbf{z}(t)) \to 0$ for $t \to +\infty$.

Exercise 17.1.1 Show that the system

$$\begin{aligned} \dot{x} &= x - y - x\left(x^2 + y^2\right) \\ \dot{y} &= x + y - y\left(x^2 + y^2\right) \end{aligned}$$

has a unique periodic orbit. Compute its Poincaré map. (Hint: use polar coordinates.)

In most cases, it is no easy matter to compute a Poincaré map. But we shall study Poincaré maps for heteroclinic cycles, which can often be handled more easily.

17.2 Poincaré maps for heteroclinic cycles

Let Γ be a heteroclinic cycle in the plane. For the sake of simplicity we shall assume that the connecting orbits are straight lines, as in the rock–scissors–paper game (section 7.7) where Γ is the boundary of S_3, or in the Battle of the Sexes (section 10.2) where Γ is the boundary of the unit square.

Let $\mathbf{F}_i$ be one of the saddle points. We can take local coordinates in such a way that $\dot{x} = \lambda_i x$ and $\dot{y} = -\mu_i y$ hold approximately near $\mathbf{F}_i$ (with $\mu_i, \lambda_i > 0$). We now consider two cross-sections, namely $S_i = \{(x, y) : y = 1\}$ 'before' the saddle point $\mathbf{F}_i$ and $S'_i = \{(x, y) : x = 1\}$ 'after' $\mathbf{F}_i$. The units of length are chosen such that these sections are very close to $\mathbf{F}_i$. The transition map $\varphi_i : S_i \to S'_i$ is given approximately by

$$(x, 1) \to \left(xe^{\lambda_i t}, e^{-\mu_i t}\right) = (1, y)$$

so that

$$\varphi_i(x) \sim x^{\mu_i/\lambda_i} \ . \tag{17.1}$$

The transition map ψ_i from S'_i to S_{i+1} can be approximated (near the connecting line from $\mathbf{F}_i$ to $\mathbf{F}_{i+1}$) by a linear map

$$\psi_i(x) \sim a_i x \ . \tag{17.2}$$

Exercise 17.2.1 Show that by taking coordinates (x, y) with $0 \le x \le 1, y \ge 0$

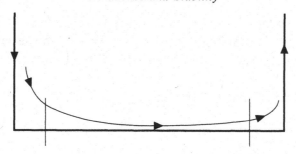

Fig. 17.1.

as in fig. 17.1, one obtains

$$a_i = \exp \int_0^1 c_i(x,0)dx \tag{17.3}$$

where $c_i(x,0) = \lim_{y \to 0} c_i(x,y)$ and

$$c_i(x,y) = \frac{\dot{y}}{y\dot{x}} + \frac{\mu_i}{\lambda_i}\frac{1}{x} - \frac{\lambda_{i+1}}{\mu_{i+1}}\frac{1}{1-x} \quad . \tag{17.4}$$

(Hint: take cross-sections at $x = \varepsilon$ and $x = 1 - \delta$ and let $\varepsilon, \delta \to 0$.)

By composing the $2n$ transition maps along the heteroclinic cycle, we obtain for the Poincaré map $g = \psi_n \circ \varphi_n \circ \cdots \circ \psi_1 \circ \varphi_1$ the approximate expression

$$g(x) \sim ax^\rho \tag{17.5}$$

where

$$\rho = \prod_{i=1}^n \frac{\mu_i}{\lambda_i} \tag{17.6}$$

and

$$a = a_n a_{n-1}^{\mu_n/\lambda_n} \dots a_1^{\mu_n \dots \mu_2/\lambda_n \dots \lambda_2} \quad . \tag{17.7}$$

This Poincaré map g is defined on a small interval $[0, \alpha[$ and the fixed point $x = 0$ corresponds to the heteroclinic cycle Γ.

If $\rho > 1$ then 0 is a local (one-sided) attractor for g and hence Γ is attracting (with a rate of convergence much larger than for the periodic orbit case described in section 17.1). If $\rho < 1$ then 0 and hence Γ are repelling. The constant a in (17.5) is relevant only in the critical case $\rho = 1$.

Exercise 17.2.2 Give a different proof of this result by constructing an average Lyapunov function of the form $\prod_{i=1}^{n} x_i^{p_i}$.

The approximate form (17.5) of the Poincaré map contains enough information to treat an interesting bifurcation problem. Let us consider a one-parameter family of vector fields $\dot{\mathbf{x}} = \mathbf{f}_\mu(\mathbf{x})$ in $\mathbb{R}^2$ which all have the same heteroclinic cycle (as in (13.8) or (7.33)). Suppose that $\rho > 1$ for $\mu < 0$ and $\rho < 1$ for $\mu > 0$. If $\frac{d\rho}{d\mu}(0) < 0$ we can rescale in such a way that

$$\rho(\mu) = 1 - \mu \quad . \tag{17.8}$$

The Poincaré map (17.5) then takes the form

$$g_\mu(x) = a(\mu)x^{1-\mu} \quad . \tag{17.9}$$

The map g_μ from the interval $[0, \alpha[$ to itself has two fixed points, namely $x = 0$ and

$$\bar{x}(\mu) = a(\mu)^{\frac{1}{\mu}} \quad . \tag{17.10}$$

By expanding $a(\mu)$ we obtain

$$a(\mu) = a + ab\mu + O(\mu^2) \tag{17.11}$$

where $a = a(0)$ and $O(\mu^n)$ denotes some function with the property that $\mu^{-n}O(\mu^n)$ is bounded for $\mu \to 0$. Hence

$$\log \bar{x}(\mu) = \frac{1}{\mu} \log a + b + O(\mu) \quad . \tag{17.12}$$

Thus $\bar{x}(\mu)$ is small and positive if either $a < 1$ and $\mu > 0$ or $a > 1$ and $\mu < 0$. The fixed point $\bar{x}(\mu)$ of the Poincaré map corresponds to a periodic orbit bifurcating from the heteroclinic cycle. Generically (i.e. whenever $a \neq 1$) the bifurcation is either supercritical (Γ is attracting for $\mu \leq 0$, the periodic orbit exists for $\mu > 0$ and is stable) or subcritical (the periodic orbit exists and is unstable for $\mu < 0$ and Γ is unstable for $\mu \geq 0$).

This *heteroclinic bifurcation* is analogous to the familiar Hopf bifurcation described in section 4.5.

Exercise 17.2.3 Compute the Poincaré map of the heteroclinic orbit for the Battle of the Sexes (10.25). Show that $a = \rho = 1$ in (17.5).

Exercise 17.2.4 Compute the Poincaré map for the rock–scissors–paper game (7.33). Show that $\rho < 1$ if and only if $\det A > 0$. If $\rho = 1$ then $a = 1$ and the heteroclinic bifurcation is degenerate.

Exercise 17.2.5 Study the system

$$\dot{x} = x(1-x)(a+bx+cy)$$
$$\dot{y} = y(1-y)(d+ex+fy)$$

on the unit square. For which choice of the coefficients is the boundary a heteroclinic cycle? Compute its Poincaré map. Show that both sub- and supercritical heteroclinic bifurcations occur and deduce the existence of limit cycles.

Exercise 17.2.6 Compute in a similar way the Poincaré map for the heteroclinic cycle in the three-species system described in section 16.1. Choose coordinates $x \geq 0$ and $y \in \mathbb{R}$ such that x measures the distance from $\mathrm{bd}\,\mathbb{R}^3_+$. Show that the Poincaré map takes the form

$$(x,y) \rightarrow (ax^\rho, x^\sigma f(x,y)) \tag{17.13}$$

where $a, \rho, \sigma > 0$ and f is affine linear in y. Derive a stability criterion for the heteroclinic cycle and study the heteroclinic bifurcation.

17.3 Heteroclinic cycles on the boundary of S_n

The generalization of the rock–scissors–paper game to higher dimensions is a replicator equation on S_n with a heteroclinic cycle Γ connecting the vertices $e_1 \rightarrow e_2 \rightarrow \cdots \rightarrow e_n \rightarrow e_1$ along the edges. We shall assume that within $\mathrm{bd}\,S_n$, this cycle is locally attracting. This means that at every vertex e_i there is one positive eigenvalue in direction e_{i+1}, while the remaining $n-2$ eigenvalues are negative. These assumptions generate the 'essentially hypercyclic' systems of section 14.5.

For the sake of simplicity we restrict ourselves to the case $n = 4$. We consider first a neighbourhood of a vertex in S_4 and transform it into the origin of $\mathbb{R}^3_+$. The linearized system then reads

$$\dot{x} = \lambda x \quad \dot{y} = -\mu y \quad \dot{z} = -\sigma z \tag{17.14}$$

with $\lambda, \mu, \sigma > 0$. We consider the two cross-sections $S = \{(x,y,z) : y = 1\}$ and $S' = \{(x,y,z) : x = 1\}$ (see fig. 17.2). A point $(x,1,z) \in S$ is mapped by (17.14) to the point

$$\left(e^{\lambda t}x, e^{-\mu t}, e^{-\sigma t}z\right) = \left(1, x^{\mu/\lambda}, x^{\sigma/\lambda}z\right) \tag{17.15}$$

in S'. The map from S' to the cross-section 'before' the next edge is again approximately given by a linear map

$$(y,z) \rightarrow (ay, bz) \quad .$$

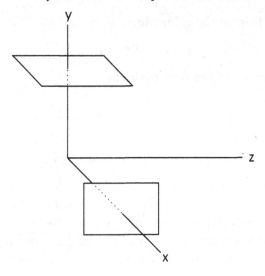

Fig. 17.2.

Let us apply this to a replicator equation with matrix

$$A = \begin{bmatrix} 0 & -b_2 & -c_3 & a_4 \\ a_1 & 0 & -b_3 & -c_4 \\ -c_1 & a_2 & 0 & -b_4 \\ -b_1 & -c_2 & a_3 & 0 \end{bmatrix} \quad . \tag{17.16}$$

This is an essentially hypercyclic network (cf. (14.19)). The eigenvalues at e_i along the cycle Γ are a_i and $-b_i$, while $-c_i$ is the 'transversal' eigenvalue. We take now eight cross-sections as above: S_i before e_i and S_i' after it. As coordinates we take $(x, y) = (x_{i+1}, x_{i+2})$ at S_i and S_{i-1}'. Then $\varphi_i : S_i \to S_i'$ and $\psi_i : S_i' \to S_{i+1}$ are approximately given by

$$\begin{aligned} \varphi_i(x, y) &\sim \left(x^{c_i/a_i} y, x^{b_i/a_i} \right) \\ \psi_i(x, y) &\sim (\alpha_i x, \beta_i y) \quad . \end{aligned} \tag{17.17}$$

The constants α_i and β_i can be determined just as in exercise 17.2.1. The Poincaré map g of the heteroclinic cycle Γ is the composite of these eight maps. Thus

$$g(x, y) \sim \left(A x^\alpha y^\beta, B x^\gamma y^\delta \right) \quad . \tag{17.18}$$

This is easier to analyse after a change of coordinates

$$z_1 = -\log x \quad z_2 = -\log y \quad . \tag{17.19}$$

The heteroclinic cycle Γ corresponds to the limit $z_1, z_2 \to +\infty$. With $\mathbf{z} =$

(z_1, z_2) we obtain from g the return map

$$L : \mathbf{z} \rightarrow P\mathbf{z} + \mathbf{q} \quad . \tag{17.20}$$

Here P is the product of the four matrices

$$\begin{bmatrix} c_i/a_i & 1 \\ b_i/a_i & 0 \end{bmatrix} \qquad i = 4, 3, 2, 1 \quad .$$

Exercise 17.3.1 Compute $\mathbf{q}$ in terms of $\alpha_i, \beta_i, a_i, b_i, c_i$.

On a higher-dimensional simplex S_n, the same procedure works again: in (17.20), P is then an $(n-2) \times (n-2)$ matrix. The map (17.20) carries all the essential information about the flow near the heteroclinic cycle Γ.

Since P is a positive matrix, it has (by the Perron–Frobenius theorem, see section 15.1) a dominant eigenvalue $\rho > 0$ and positive left and right eigenvectors $\mathbf{v}, \mathbf{u}$. If $\rho \neq 1$, we can forget about the translation term $\mathbf{q}$ in (17.20) by replacing $\mathbf{z}$ by $\mathbf{z} + \hat{\mathbf{z}}$, where $\hat{\mathbf{z}} = (1 - P)^{-1}\mathbf{q}$ is the fixed point of (17.20). The Perron–Frobenius theorem implies that

$$\rho^{-n} P^n \mathbf{z} \rightarrow \mathbf{u}$$

for every $\mathbf{z} > \mathbf{0}$. If $\rho > 1$, then $L^n\mathbf{z}$ grows to infinity in the direction of the right eigenvector $\mathbf{u} > \mathbf{0}$. This implies that the heteroclinic cycle is attracting.

If $\rho < 1$ then all orbits of $\mathbf{z} \rightarrow P\mathbf{z}$ converge to $\mathbf{0}$. Thus 'infinity' is repelling for (17.20) and so is the heteroclinic cycle. Thus the stability criterion for heteroclinic cycles given in section 17.2 generalizes to higher dimensions.

In order to decide which of these two cases holds, it is not necessary to compute ρ explicitly (which is impossible in general). Instead one considers the matrix

$$Q = P - I \tag{17.21}$$

which has nonnegative off-diagonal terms. Then section 14.5 and theorem 15.1.1 tell us that $\rho < 1$ if and only if Q is a stable matrix, which is the case if and only if $-Q$ is an M-matrix.

Exercise 17.3.2 Using this criterion, derive an explicit permanence criterion for (17.16) and compare it with the result of section 14.5.

Let us now study the heteroclinic bifurcation which occurs when Γ loses stability. We shall again suppose that the parameter μ of the vector field on S_n is related to the dominant eigenvalue $\rho = \rho(\mu)$ by

$$\rho(0) = 1 \qquad \left. \frac{d\rho}{d\mu} \right|_{\mu=0} < 0 \quad . \tag{17.22}$$

The fixed point of the map (17.20) is given (for $\mu \neq 0$) by

$$\hat{\mathbf{z}}(\mu) = (I - P(\mu))^{-1} \mathbf{q}(\mu) . \tag{17.23}$$

Whenever $\hat{\mathbf{z}}(\mu)$ is positive and very large, it will correspond to a periodic orbit near the heteroclinic cycle. To see the asymptotic behaviour of $\hat{\mathbf{z}}(\mu)$ for $\mu \to 0$ we write $(I - P(\mu))^{-1}$ as the quotient of the adjoint matrix and its determinant. Since $\rho < 1$ for $\mu > 0$, the matrix $I - P(\mu)$ is positively stable for $\mu > 0$ and has exactly one negative eigenvalue for small $\mu < 0$. Thus $\det(I - P(\mu))$ has the sign of μ for μ near 0. The adjoint matrix of $I - P(\mu)$ is a small perturbation of $\mathrm{adj}(I - P(0))$. Since $I - P(0)$ has a zero eigenvalue of multiplicity 1, the adjoint matrix has rank one and is of the form $c u_i v_j$, where $\mathbf{u} > \mathbf{0}$ and $\mathbf{v} > \mathbf{0}$ are the right and left eigenvectors of $P(0)$ corresponding to the dominant eigenvalue $\rho(0) = 1$ (cf. exercise 15.2.13). The constant c is positive since $(I - P(\mu))^{-1}$ is a nonnegative matrix for $\mu > 0$ (by condition (M10) in section 15.1). Hence

$$\hat{\mathbf{z}}(\mu) \sim \frac{1}{\mu} \, \mathrm{adj} \, (I - P(0)) \mathbf{q} = \frac{1}{\mu} (\mathbf{q} \cdot \mathbf{v}) \mathbf{u} . \tag{17.24}$$

Thus the fixed point $\hat{\mathbf{z}}(\mu)$ tends to infinity in the direction of $\mathbf{u} > \mathbf{0}$ if $\mu \to 0$. If $\mathbf{q} \cdot \mathbf{v} > 0$ then the bifurcation is supercritical: the periodic orbits corresponding to $\hat{\mathbf{z}}(\mu)$ occur for $\mu > 0$ and are stable. If $\mathbf{q} \cdot \mathbf{v} < 0$ then the bifurcation is subcritical: the periodic orbits occur for $\mu < 0$ and are unstable.

Exercise 17.3.3 Compute the explicit conditions on a_i, b_i and c_i which make the heteroclinic bifurcation of (17.16) subcritical resp. supercritical.

Exercise 17.3.4 Compute the Poincaré map (17.20) for the hypercycle (12.1). Show that $\rho = 0$. (Hence Γ is 'very repelling'.)

17.4 The characteristic matrix of a heteroclinic cycle

We now present a common framework for all heteroclinic cycles studied so far. Let us assume that the state space of our dynamical system is a polyhedron X in some $\mathbb{R}^N$, defined as the intersection of finitely many half-spaces $\{\mathbf{x} \in \mathbb{R}^N : x_j \geq 0\}$, $j = 1, \ldots, n$. Here x_j denotes a linear functional on $\mathbb{R}^N$ which vanishes on one of the supporting hyperplanes of X. Let $\mathrm{bd}\, X = \bigcup_j \{x_j = 0\} \cap X$ denote the boundary of X, and $\mathrm{int}\, X = X \setminus \mathrm{bd}\, X$ its interior.

If $\mathrm{bd}\, X$ is invariant under the dynamics then we can write our differential equation in the form

$$\dot{x}_j = x_j f_j(\mathbf{x}), \quad j = 1, \dots, n, \tag{17.25}$$

(there will be some relations between the f_j if $n > N$). We can always assume that $n \geq N$ and that a point $\mathbf{x} \in X$ is uniquely determined by its 'coordinates' x_j, $(j = 1, \dots, n)$. If not, we simply add some additional variables to achieve that.

Consider a fixed point $\hat{\mathbf{x}}$ of (17.25), and rearrange indices such that the zero coordinates come first. Then the Jacobian at $\hat{\mathbf{x}}$ takes the form

$$\begin{bmatrix} E & O \\ * & * \end{bmatrix} \tag{17.26}$$

where E is a diagonal matrix whose entries are the external eigenvalues $\frac{\dot{x}_j}{x_j}|_{\hat{\mathbf{x}}} = f_j(\hat{\mathbf{x}})$.

Now let $\Gamma \subset \mathrm{bd}\, X$ be a heteroclinic cycle which consists of m fixed points $\mathbf{F}_k$ connected by heteroclinic orbits. Then we associate with Γ a rectangular scheme of external eigenvalues: the entry in row k and column j is the external eigenvalue $f_j(\mathbf{F}_k)$. (It is 0 if $x_j > 0$ at $\mathbf{F}_k$.) We call this scheme the *characteristic matrix* C of Γ. Since the numbering of the hyperplanes (or columns) and of the fixed points $\mathbf{F}_k$ (or rows of C) is arbitrary, the characteristic matrix is determined only up to permutations of rows and columns.

Exercise 17.4.1 Show that the (sign structure of the) characteristic matrix of the rock–scissors–paper examples in sections 5.5 and 7.7 and the two-prey two-predator system (16.27) takes the form

	x_1	x_2	x_3
F_1	0	+	−
F_2	−	0	+
F_3	+	−	0

	x_1	x_2	y_1	y_2
F_1	0	−	+	−
F_1^1	0	+	0	−
F_2	−	0	−	+
F_2^2	+	0	−	0

Exercise 17.4.2 In exercise 17.2.2, X is essentially an n-gon. Show that the characteristic matrix can be arranged such that it contains the positive eigenvalues λ_k in the main diagonal and the negative eigenvalues $-\mu_k$ next to it.

Exercise 17.4.3 Show that the characteristic matrix of the heteroclinic cycle in an essentially hypercyclic system (14.19) is the matrix A. Compute the characteristic matrix of the heteroclinic cycles in (14.21) and (14.23).

The characteristic matrix C contains the essential information about Γ. In particular, the stability properties of Γ can be read off from C in many cases, as will be shown in section 17.5.

A priori C is an $m \times n$ matrix (where m is the number of fixed points, and n the number of boundary hyperplanes of the polyhedron X), but we can omit columns of zeros, thus ignoring those hyperplanes which do not touch Γ. Each row of C contains at least one positive entry (under the generic assumption that the fixed points $\mathbf{F}_k$ are hyperbolic). If each row and each column of C contains only one positive entry then we call Γ a *simple heteroclinic cycle*. All examples in exercises 17.4.1–3 are simple cycles, with C a square matrix.

Exercise 17.4.4 Construct a heteroclinic cycle whose characteristic matrix is not square, where each row contains only one positive entry. (Hint: take X as a cube, choose $m = 8$ and $n = 6$.)

If for a simple heteroclinic cycle, C has only one negative entry in each row and column (so that there are just two external directions at each $\mathbf{F}_k$, one stable and one unstable) then we speak of a *'planar'* heteroclinic cycle.

Exercise 17.4.5 Which of the examples in exercises 17.4.1–3 are planar?

If some row of C contains at least two positive entries (so that at least one $\mathbf{F}_k$ has an unstable manifold of dimension ≥ 2) then Γ is a *multiple heteroclinic cycle* or a *heteroclinic network*.

Exercise 17.4.6 Give examples of nonplanar heteroclinic cycles and networks in the simplex S_4.

Exercise 17.4.7 Construct a two-prey two-predator system (not in Lotka–Volterra form) which has a heteroclinic cycle of the following form: $F_1 \rightarrow F_1^1 \rightarrow F_{12} \rightarrow F_1$. Note that the last connection is not in the x_1–x_2 plane (the boundary face spanned by F_{12} and F_1), but in the x_1–x_2–y_2 subsystem. Show that the characteristic matrix has the form

	x_1	x_2	y_1	y_2
F_1	0	−	+	−
F_1^1	0	+	0	−
F_{12}	0	0	−	+

Note also that the first column of C is zero (and hence can be ignored), since $x_1 > 0$ along the cycle. Hence the cycle is simple.

17.5 Stability conditions for heteroclinic cycles

Let Γ be a heteroclinic cycle on $\mathrm{bd}\,X$ with vertices $\mathbf{F}_k$ and characteristic matrix C. As before, we omit those columns which consist of zero entries only, thus ignoring those hyperplanes which do not touch Γ.

Theorem 17.5.1

(a) *If there is a vector* $\mathbf{p} \in \mathbb{R}^n$ *such that* $\mathbf{p} > 0$ *and* $C\mathbf{p} > 0$ *then* $\mathrm{bd}\,X$ *is repelling near* Γ.

(b) *If* Γ *is asymptotically stable within* $\mathrm{bd}\,X$ *and there is a* $\mathbf{p} < 0$ *such that* $C\mathbf{p} > 0$ *then* Γ *is asymptotically stable in* X.

(c) *If* Γ *is asymptotically stable within* $\mathrm{bd}\,X$ *and there is a* $\mathbf{p} \in \mathbb{R}^n$ *such that* $p_i < 0$ *for at least one* i *and* $C\mathbf{p} > 0$ *then* Γ *attracts at least one (actually an open set of) interior orbit(s) from* X.

Proof Part (a) is left as an exercise. (Hint: show that $\prod x_k^{p_k}$ is an average Lyapunov function.)

For the proof of (b) and (c) let $P(\mathbf{x}) = \prod x_i^{-p_i}$. Let $I = \{i : p_i < 0\}$ and denote by $X_0 = \{\mathbf{x} \in X : x_i = 0 \text{ for some } i \in I\}$ the part of $\mathrm{bd}\,X$ where P vanishes. Now $\dot{P}/P = -\sum_j p_j \dot{x}_j / x_j$ extends continuously to $\mathrm{bd}\,X$ and takes the value $-\sum_j p_j \lambda_j(\mathbf{F}_k) = -(C\mathbf{p})_k < 0$ at the fixed point $\mathbf{F}_k$. Intuitively this means that P decreases exponentially near Γ and we expect that orbits close to Γ will converge to X_0. More precisely, with $\Gamma_0 = \Gamma \cap X_0$ we have:

Lemma 17.5.2 *Suppose there is a* $\mathbf{p} \in \mathbb{R}^n$ *such that* $p_i < 0$ *for at least one* i *and* $C\mathbf{p} > 0$ *and that* Γ_0 *is asymptotically stable within* X_0. *Then all orbits starting in a set* $\{\mathbf{x} \in \mathrm{int}\,X : P(\mathbf{x}) < \delta\}$ *and close enough to* Γ_0 *will converge to* Γ_0.

Proof Since Γ_0 is asymptotically stable within the invariant subset X_0 one can find a neighbourhood $U \subset X_0$ of Γ_0 in X_0 which is forward invariant and whose smooth boundary is transverse to the vector field in X_0. (If $\Gamma_0 = X_0$, then we set $U = X_0$.) For small enough ε, the cylinder

$$U_\varepsilon = \bigcup_{i \in I} \{\mathbf{x} \in X : 0 \leq x_i < \varepsilon \text{ and } \pi_i(\mathbf{x}) \in U\}$$

(where π_i is a suitable projection onto the face $\{x_i = 0\}$), with base set U and height ε, is then still forward invariant along its side surface (but not necessarily at its top). Let $U_\varepsilon^\delta = U_\varepsilon \cap \{\mathbf{x} \in X : P(\mathbf{x}) < \delta\}$, with δ so small that $\min\{x_i : i \in I\} < \varepsilon$ holds for all $\mathbf{x}$ with $P(\mathbf{x}) \leq \delta$. As in theorem 12.2.1 one can find constants $\bar{T} > 1$, $k < 1$ and $K > 0$ such that for all $\mathbf{x} \in X$

close to Γ there is a time T with $1 < T < \bar{T}$ such that $P(\mathbf{x}(T)) < kP(\mathbf{x})$ and $P(\mathbf{x}(t)) < KP(\mathbf{x})$ for all $0 < t < \bar{T}$. Hence for $\mathbf{x} \in U_\varepsilon^{\delta/K}$ and $0 < t < \bar{T}$, $\mathbf{x}(t) \in U_\varepsilon^\delta$ while $\mathbf{x}(T) \in U_\varepsilon^{k\delta/K}$. Iterating this argument we see that the forward orbit of $\mathbf{x}$ cannot leave U_ε^δ and $P(\mathbf{x}(t))$ will converge exponentially to 0, as $t \to \infty$. Hence $\omega(\mathbf{x}) \subset \Gamma_0$ (the maximal compact invariant subset of U_ε^δ) and Γ_0 is stable for the semiflow restricted to $U_\varepsilon^{\delta/K}$. $\qquad\square$

Now we conclude the proof of theorem 17.5.1. If all $p_i < 0$ then $X_0 = \text{bd } X$ and the lemma implies that Γ is asymptotically stable in X. This shows part (b) of the theorem. If $\mathbf{p}$ has positive and negative components then lemma 17.5.2 isolates a partially attracting part Γ_0 of Λ. (It is easy to see that $\Gamma_0 \neq \emptyset$.) $\qquad\square$

Theorem 17.5.3 *Let Γ be asymptotically stable within* $\text{bd } X$*; and $m = n'$ (the number of nonzero columns) so that the (reduced) characteristic matrix C is a square matrix. Furthermore let $\det C \neq 0$. Then $\text{bd } X$ is repelling near Γ if and only if $C^{-1} \geq 0$, i.e. the inverse of C has only nonnegative entries.*

Proof If Γ is repelling then case (c) of theorem 17.5.1 cannot apply and hence for any positive vector $\mathbf{q} > \mathbf{0}$ in $\mathbb{R}^n$, we must have $\mathbf{p} = C^{-1}\mathbf{q} \geq \mathbf{0}$. This implies $C^{-1} \geq \mathbf{0}$. The converse follows immediately from case (a). $\qquad\square$

The case where C is not square ($n' \neq m$) is largely unresolved. Theorem 17.5.1 gives only a partial result in this case. For simple heteroclinic cycles the situation is considerably simpler because case (c) can be ignored: we can apply the theory of M-matrices to solve the linear inequalities which arise.

Exercise 17.5.4 Assume that Γ is a simple heteroclinic cycle, which is asymptotically stable within $\text{bd } X$. Then C is a square matrix (after elimination of superfluous columns) with positive entries occurring only in the main diagonal (after a suitable rearrangement of the rows or columns). Let $\det C \neq 0$. Show that if C is an M-matrix then Γ is repelling, and if C is not an M-matrix then Γ is asymptotically stable.

Exercise 17.5.5 Let Λ be a planar heteroclinic cycle, which is asymptotically stable within $\text{bd } X$. Let $\lambda_k > 0$ and $-\mu_k < 0$ be the two external eigenvalues at F_k. If $\prod \lambda_k > \prod \mu_k$ then Γ is repelling. If $\prod \lambda_k < \prod \mu_k$ then Γ is asymptotically stable. (For heteroclinic cycles in the plane, see exercise 17.2.2.)

17.6 Notes

For more background on cross-sections and Poincaré maps see Robinson (1995). Sections 17.2 and 17.3 are from Hofbauer (1987), and sections 17.4 and 17.5 are from Hofbauer (1994). For the behaviour of time averages near (planar and simple) heteroclinic cycles see Gaunersdorfer (1992), Akin (1993) and Takens (1994). For multiple heteroclinic cycles see Brannath (1994) and Kirk and Silber (1994). Chawanya (1995, 1996) has found chaotic motion near heteroclinic networks. The example from exercise 17.4.7 is from Sikder and Roy (1994). Heteroclinic cycles occur in a robust way also in dynamical systems with symmetry; see Field and Swift (1991), and Krupa (1996) for a survey.

Part four

Population Genetics and Game Dynamics

18

Discrete dynamical systems in population genetics

We discuss the Hardy–Weinberg law of population genetics and derive the
discrete selection model which describes the evolution of gene frequencies if
genotypes have different survival probabilities and generations do not overlap.
The fundamental result, here, is that the average fitness never decreases. This
result is no longer valid if mutation or recombination terms are included.

18.1 Genotypes

The cells of higher organisms may present very different aspects, depending
on the type of tissue. Apart from a few exceptions, however, they all contain
a nucleus where the genetic program is stored in the form of *chromosomes*.
The number of chromosomes varies with the species (there are 46 of them
in our cells), but they always occur in homologous pairs. If such a *diploid*
cell divides, each chromosome duplicates, and the two daughter cells each
receive a complete set of chromosomes.

Our organism also contains *haploid* cells having only half the number
of chromosomes, one from each pair. These are the germ cells or *gametes*
(sperms and eggs). Such cells issue from diploid cells by the process of *meiosis*,
which splits the chromosomal pairs. Different pairs of chromosomes split
up independently of each other. At mating, pairs of gametes fuse, thereby
restoring the full number of chromosomes. The newly formed diploid cell,
called a *zygote*, is the starting point of a new organism, which inherits half
of its genome from each parent.

Let us consider now some hereditary trait expressed in the *phenotype* (i.e.
the set of manifested attributes) of an organism, for instance eye colour
or blood group. In the simplest case, it is determined by the joint action

of two *genes* sitting at corresponding positions on a pair of homologous chromosomes: this site is called the chromosomal *locus* of the trait. In many cases, there are several types of genes which may occupy a locus, the so-called *alleles* $A_1, \ldots, A_n$.

The *genotype* is determined by that pair which actually occurs. If it is of type $A_i A_i$, that is if the same allele appears twice, the genotype is *homozygous*; if it is of type $A_i A_j (i \neq j)$ it is *heterozygous*. In the latter case, one allele may suppress the effect of the other, in the sense that $A_i A_j$ manifests itself as $A_i A_i$. The allele A_i, then, is called *dominant* and A_j *recessive*. It may also happen that the heterozygous genotype leads to an expression which is different from the homozygous ones. The genotypes $A_j A_i$ and $A_i A_j$ cannot be distinguished (there are exceptions, but we shall ignore them). Hence, it does not matter which of the two genes stems from the father and which from the mother. Nevertheless, we shall use the convention of describing the *gene pair* of an individual by writing first the gene from the father, and then the gene from the mother. We shall distinguish therefore the gene pairs (A_1, A_2) and (A_2, A_1), although both correspond to the same genotype $A_1 A_2$.

18.2 The Hardy–Weinberg law

Let us assume that the probabilities of the alleles $A_1, \ldots, A_n$ are given by $x_1, \ldots, x_n$ (with $x_1 + \cdots + x_n = 1$), and those of the gene pairs (A_i, A_j) by x_{ij} $(1 \leq i, j \leq n)$. A randomly chosen gene sits with probability $\frac{1}{2}$ at the 'first place' (according to our convention, it has been transmitted by the father) and with probability $\frac{1}{2}$ at the 'second place'. In the former case, it is of type A_i if the gene pair is of the form (A_i, A_j) (for arbitrary j). In the latter case, it is of type A_i if the gene pair is of the form (A_j, A_i) (for arbitrary j). Thus

$$x_i = \frac{1}{2} \sum_j x_{ij} + \frac{1}{2} \sum_j x_{ji}. \tag{18.1}$$

Let us denote by x_i' resp. x_{ij}' the corresponding probabilities in the next generation. Assuming random mating of gametes we obtain

$$x_{ij}' = x_i x_j \tag{18.2}$$

(the paternal gene is with probability x_i of type A_i, the maternal one with probability x_j of type A_j). It follows that

$$x_i' = \frac{1}{2} \left(\sum_j x_{ij}' + \sum_j x_{ji}' \right) = \frac{1}{2} \cdot 2 \cdot \sum_j x_i x_j = x_i. \tag{18.3}$$

This implies the *Hardy–Weinberg law*:

(a) the gene probabilities remain unchanged from generation to generation;

(b) from the first daughter generation onward, the probabilities of the homozygous genotypes A_jA_j are given by x_j^2, those of the heterozygous genotypes A_iA_j (with $i \neq j$) by $2x_ix_j$.

Exercise 18.2.1 We have derived the Hardy–Weinberg law under the assumption of random union of gametes. Show that it is also valid under the assumption of random mating of the parental individuals. (Hint: write down all possible mating types $(A_i, A_j) \times (A_k, A_l)$, multiply by the probability $x_{ij}x_{kl}$ of such a mating, write down the (equally probable) gene pairs for the offspring, i.e. (A_i, A_k), (A_i, A_l) etc., and do the sums.)

Exercise 18.2.2 Sex chromosomes are not homologous. Thus for mammals, the X-chromosome is longer than the Y-chromosome. Females have two X-chromosomes, males one X- and one Y-chromosome. Half of the male sperm cells contain the X, half the Y-chromosome, while all female egg cells carry X. Fusion leads with equal probability to XX (a daughter) or XY (a son). For genes on the X-chromosome, the Hardy–Weinberg law is not valid: but the frequencies $x_i^m(n)$ and $x_i^f(n)$ of allele A_i of the male resp. female population in the n-th generation converge (for $n \to \infty$) to some common limit x_i, and the relations $x_{ij} = x_ix_j$ are asymptotically valid.

18.3 The selection model

So far, we have assumed that all genotypes are equally good at surviving to maturity and producing offspring. Let us assume now that the gene pairs (A_i, A_j) differ in their probability w_{ij} of surviving to an adult age. The *selective values* w_{ij} satisfy $w_{ij} \geq 0$ and $w_{ij} = w_{ji}$, since the gene pairs (A_i, A_j) and (A_j, A_i) belong to the same genotype. The selection matrix $W = (w_{ij})$ is therefore symmetric.

If N is the number of zygotes in the new generation, then x_ix_jN of them carry the gene pair (A_i, A_j), of which $w_{ij}x_ix_jN$ survive to maturity. The total number of individuals reaching the mating stage is $\sum_{r,s=1}^n w_{rs}x_rx_sN$. We shall always assume that this number is distinct from 0. Denoting by x'_{ij} the frequency of the gene pair (A_i, A_j) in the adult stage of the new generation, and by x'_i the frequency of the allele A_i, we obtain

$$x'_{ij} = \frac{w_{ij}x_ix_jN}{\sum_{r,s=1}^n w_{rs}x_rx_sN}.$$

For x'_{ji}, we obtain the same expression, since $w_{ij} = w_{ji}$. Clearly

$$x'_i = \frac{1}{2}\sum_j x'_{ij} + \frac{1}{2}\sum_j x'_{ji}.$$

This leads to the relation

$$x'_i = x_i \frac{\sum_j w_{ij} x_j}{\sum_{r,s} w_{rs} x_r x_s} \qquad i = 1, \ldots, n \qquad (18.4)$$

which describes the evolution of the gene frequencies from one generation to the next.

18.4 The increase in average fitness

Theorem 18.4.1 *For the dynamical system* $\mathbf{x} \to \mathbf{x}'$ *given by* (18.4), *i.e. by*

$$x'_i = x_i \frac{(W\mathbf{x})_i}{\mathbf{x} \cdot W\mathbf{x}}, \qquad (18.5)$$

the average fitness $\bar{w}(\mathbf{x}) = \mathbf{x} \cdot W\mathbf{x}$ *increases along every orbit in the sense that*

$$\bar{w}(\mathbf{x}') \geq \bar{w}(\mathbf{x}) \qquad (18.6)$$

with equality if and only if $\mathbf{x}$ *is a rest point.*

Proof Since we assume $\mathbf{x} \cdot W\mathbf{x} > 0$, we have to show that

$$(\mathbf{x} \cdot W\mathbf{x})^2 (\mathbf{x}' \cdot W\mathbf{x}') \geq (\mathbf{x} \cdot W\mathbf{x})^3. \qquad (18.7)$$

Clearly

$$(\mathbf{x} \cdot W\mathbf{x})^2 (\mathbf{x}' \cdot W\mathbf{x}') = (\mathbf{x} \cdot W\mathbf{x})^2 \sum_{i,k} x'_i w_{ik} x'_k.$$

On replacing x'_i and x'_k by the expression in (18.4) we obtain

$$\sum_{i,k} x_i(W\mathbf{x})_i w_{ik} x_k (W\mathbf{x})_k = \sum_{i,j,k} x_i w_{ij} x_j w_{ik} x_k (W\mathbf{x})_k =: s(1).$$

Exchange of the indices j and k yields

$$\sum_{i,j,k} x_i w_{ik} x_k w_{ij} x_j (W\mathbf{x})_j =: s(2).$$

The expressions $s(1)$ and $s(2)$ are equal, and hence coincide with their arithmetic mean

$$\frac{1}{2} \sum_{i,j,k} x_i w_{ij} x_j w_{ik} x_k \left[(W\mathbf{x})_j + (W\mathbf{x})_k \right].$$

By the inequality of the arithmetic and the geometric mean, this is not smaller than

$$\sum_{i,j,k} x_i w_{ij} x_j w_{ik} x_k (W\mathbf{x})_j^{1/2} (W\mathbf{x})_k^{1/2} =$$

$$= \sum_i x_i \sum_j w_{ij} x_j (W\mathbf{x})_j^{1/2} \sum_k w_{ik} x_k (W\mathbf{x})_k^{1/2}$$

$$= \sum_i x_i \left(\sum_j w_{ij} x_j (W\mathbf{x})_j^{1/2} \right)^2 .$$

By Jensen's inequality (7.11) applied to the function x^2, this last expression is not smaller than

$$\left(\sum_i x_i \sum_j w_{ij} x_j (W\mathbf{x})_j^{1/2} \right)^2 = \left(\sum_j x_j (W\mathbf{x})_j^{1/2} \sum_i x_i w_{ij} \right)^2 .$$

Since $w_{ij} = w_{ji}$, we obtain therefore

$$\left(\sum_j x_j (W\mathbf{x})_j^{1/2} (W\mathbf{x})_j \right)^2 = \left(\sum_j x_j (W\mathbf{x})_j^{3/2} \right)^2 .$$

Using Jensen's inequality again, this time for the function $x^{3/2}$, we see that this expression is not smaller than

$$\left(\sum_j x_j (W\mathbf{x})_j \right)^{(3/2)\cdot 2} = (\mathbf{x} \cdot W\mathbf{x})^3$$

which proves (18.7).

If $\bar{w}(\mathbf{x}) = \bar{w}(\mathbf{x}')$, there must exist a value c such that $(W\mathbf{x})_j = c$ for all j with $x_j > 0$. This means that $\mathbf{x}$ is a rest point. □

The theorem implies that every orbit $T^k\mathbf{x}$ of the map $\mathbf{x} \to T\mathbf{x} = \mathbf{x}'$ converges to the set of rest points, or more precisely that every accumulation point of the orbit is a rest point. Indeed, the sequence $\bar{w}(T^k\mathbf{x})$ increases monotonically by (18.6) and hence converges to some limit L. Let $\mathbf{y}$ be an accumulation point of $T^k\mathbf{x}$: there exists, then, a subsequence $k_j \to +\infty$ such that $T^{k_j}\mathbf{x} \to \mathbf{y}$. The continuity of T implies that $T^{k_j+1}\mathbf{x} = T(T^{k_j}\mathbf{x})$ converges to $T\mathbf{y}$. But

$$\bar{w}(\mathbf{y}) = \bar{w}\left(\lim_{j \to +\infty} T^{k_j}\mathbf{x} \right) = L$$

and

$$\bar{w}(T\mathbf{y}) = \bar{w}\left(\lim T^{k_j+1}\mathbf{x} \right) = L$$

so that $\bar{w}(\mathbf{y}) = \bar{w}(T\mathbf{y})$. Thus $\mathbf{y}$ is a rest point.

Exercise 18.4.2 Give an alternative proof for theorem 18.4.1:

(a) Show that $\left(\dfrac{\mathbf{x}\cdot A\mathbf{x}}{\mathbf{x}\cdot\mathbf{x}}\right)^m \leq \dfrac{\mathbf{x}\cdot A^m\mathbf{x}}{\mathbf{x}\cdot\mathbf{x}}$ for symmetric $n \times n$ matrices A with nonnegative elements a_{ij} and $x_i \geq 0$. (Use induction on n.)

(b) Set $m = 3$ and choose $a_{ij} = x_i^{1/2}w_{ij}x_j^{1/2}$ to obtain the result.

Exercise 18.4.3 Consider the map on S_n

$$x_i' = x_i\frac{\partial P}{\partial x_i}\left(\sum_r x_r\frac{\partial P}{\partial x_r}\right)^{-1}\qquad i = 1,\ldots,n$$

with P a polynomial in $x_1,\ldots,x_n$ with nonnegative coefficients. Then $P(\mathbf{x}') \geq P(\mathbf{x})$ with equality if and only if $\mathbf{x}' = \mathbf{x}$. (Hint: it is enough to consider homogeneous polynomials.)

18.5 The case of two alleles

We shall write p and $1 - p$ instead of x_1 and x_2. Equation (18.4) defines a dynamical system on the interval $[0, 1]$, namely

$$p' = F(p) = \frac{a_1}{a_1 + a_2}\tag{18.8}$$

with $a_1 = p(w_{11}p + w_{12}(1 - p))$ and $a_2 = (1 - p)(w_{12}p + w_{22}(1 - p))$. The average fitness is

$$\bar{w}(p) = a_1 + a_2 = p^2\left[(w_{11} - w_{12}) + (w_{22} - w_{12})\right] - 2p(w_{22} - w_{12}) + w_{22}.$$

As we can check directly, the increment within one generation is given by

$$p' - p = \frac{p(1 - p)}{2\bar{w}(p)}\frac{d}{dp}\bar{w}(p).\tag{18.9}$$

The fixed points are therefore the two endpoints of $[0, 1]$ and the critical points of $\bar{w}(p)$ in the interior $]0, 1[$. There are several possibilities.

(1)   In the (degenerate) case $w_{12} = \frac{1}{2}(w_{11} + w_{22})$, the function $\bar{w}(p)$ is linear. If $w_{11} = w_{12} = w_{22}$, selection does not operate, $\bar{w}(p)$ is constant, and all points are rest points. Otherwise, the slope of $\bar{w}(p)$ is nonzero, and p converges to 0 or 1. Thus, the homozygote with the highest fitness becomes established.

(2)   In the generic case $w_{12} \neq \frac{1}{2}(w_{11} + w_{22})$, $\bar{w}(p)$ is a parabola whose extremum is at the point

$$\bar{p} = \frac{w_{22} - w_{12}}{(w_{11} - w_{12}) + (w_{22} - w_{12})}\;.\tag{18.10}$$

If

$$(w_{11} - w_{12})(w_{22} - w_{12}) \leq 0 \qquad (18.11)$$

i.e. w_{12} lies between w_{11} and w_{22}, then $\bar{p} \notin]0,1[$. All orbits in the interior of the state space converge to the endpoint where $\bar{w}$ is highest. In this case, which includes dominance of one allele ($w_{12} = w_{11}$ or $w_{12} = w_{22}$), the homozygote with highest fitness becomes established in the population.

If (18.11) does not hold, however, then $\bar{p}$ is a rest point in $]0,1[$. Two cases are possible.

(a) In the case of *heterozygote advantage*, $w_{12} > w_{11}$ and $w_{12} > w_{22}$, the average fitness $\bar{w}(p)$ has its maximum at $\bar{p}$;
(b) otherwise, $\bar{w}(p)$ has its minimum at $\bar{p}$.

As we shall presently see, an orbit cannot jump from one side of $\bar{p}$ to the other side. This implies that an orbit starting in $]0,\bar{p}[$ or $]\bar{p},1[$ remains in that interval and converges to the endpoint with highest fitness. In case (a), this leads to the *polymorphic* state $\bar{p}$ (both alleles are present). In case (b), one or the other allele vanishes from the gene pool, depending on the initial state.

That $p < \bar{p}$ implies $p' < \bar{p}$ follows from the monotonicity of F, which in turn is a consequence of

$$\frac{d}{dp}F(p) = (a_1 + a_2)^{-2}\left[a_2\frac{da_1}{dp} - a_1\frac{da_2}{dp}\right] > 0.$$

Indeed, a straightforward computation shows that

$$a_2\frac{da_1}{dp} - a_1\frac{da_2}{dp} = w_{11}w_{12}p^2 + 2w_{11}w_{22}p(1-p) + w_{22}w_{12}(1-p)^2$$

is positive in $[0,1]$ both in case (a) and case (b).

The heterozygote genotype survives only in the case of heterozygote advantage.

Exercise 18.5.1 Prove these results without referring to theorem 18.4.1 (Hint: calculations are simplified if $x = p/(1-p)$ is used as variable.)

18.6 The mutation–selection equation

The copying of genes does not occur with absolute precision. An allele may mutate to another one. Recurrent mutations help to maintain a supply of genetic variation; even if selection pressure at a given time acts against an allele, it will be maintained by mutation and thus kept available for changed conditions.

Let us consider a one-locus two-allele model and assume first that the two alleles A_1 and A_2 are selectively neutral. We shall denote by p and $1 - p$ the frequencies of A_1 and A_2, by μ the rate of mutation from A_1 to A_2 and by v the rate of mutation from A_2 to A_1. The frequency of A_1 in the next generation will then be

$$p' = G(p) = p - \mu p + v(1 - p) = (1 - \mu - v)p + v. \qquad (18.12)$$

This is a dynamical system in the interval $[0, 1]$. Its only rest point is

$$\bar{p} = \frac{v}{\mu + v} . \qquad (18.13)$$

Since (18.12) may be written as

$$p' - \bar{p} = (1 - \mu - v)(p - \bar{p}) \qquad (18.14)$$

we see that all orbits converge at the rate $(1 - \mu - v)$ towards the rest point $\bar{p}$:

$$T^n p - \bar{p} = (1 - \mu - v)^n (p - \bar{p}) \to 0 .$$

Let us now assume that selection acts between the zygote and the adult stage and that during the subsequent meiosis, mutations may occur. In the model with two alleles A_1 and A_2, the frequency p of A_1 is changed first into $F(p)$ and then into $G(F(p))$, with F given by (18.8) and G by (18.12). Thus the frequency p' of A_1 after one generation is given by

$$p' = H(p) = G(F(p)) = (1 - \mu - v)F(p) + v . \qquad (18.15)$$

Since the mutation rates μ and v are small, we can certainly assume $1 - \mu - v > 0$. The function F is monotonic, as we have seen in section 18.5. Therefore, so is H. The map $p \to p'$ of $[0, 1]$ into itself is invertible. Setting

$$\Phi(p) = p^{2v}(1 - p)^{2\mu}\bar{w}(p)^{1 - \mu - v} \qquad (18.16)$$

where $\bar{w}$ is the average fitness as in (18.6), we can check by a straightforward computation that for $p \in \,]0, 1[$,

$$p' - p = \frac{p(1 - p)}{2\Phi(p)} \frac{d}{dp}\Phi(p) \qquad (18.17)$$

which is similar to (18.9). The rest points of the dynamical system $p \to H(p)$ in the interior of the state space $[0, 1]$ are therefore the critical points of Φ. The orbits are monotonic: the function Φ is increasing along them.

Exercise 18.6.1 Prove this. (Hint: consider each monotonicity interval of Φ separately.)

In the absence of mutation (i.e. for $\mu = v = 0$), this is nothing but the increase in average fitness. In the absence of selection (i.e. if the fitness is constant), Φ reduces to the function $p^{2v}(1 - p)^{2\mu}$, whose maximum is at $\frac{v}{\mu + v}$. The interplay between selection and mutation leads to a compromise between two maximum principles.

In particular, recurrent mutations from A_2 to A_1 will sustain the allele A_1 even if it is threatened by elimination through selection. Let us, for simplicity, assume that the mutation rate μ from A_1 to A_2 is zero. If A_1 is dominant, then the fitness of the genotypes $A_1 A_1$ and $A_1 A_2$ will be some fraction $1 - s$ of the fitness of $A_2 A_2$ (with $s \in]0, 1[$). The unique stable rest point $\bar{p}$ in $]0, 1[$ is the critical point given (approximately) by v/s. If A_1 is recessive, however, then $A_1 A_2$ has the same fitness as $A_2 A_2$, and $\bar{p}$ is approximately $\sqrt{v/s}$.

Exercise 18.6.2 Check this. Can there exist a second rest point in $]0, 1[$? What happens for $\mu > 0$?

Typical values for s and v would be 10^{-2} and 10^{-6}, respectively. Then $\bar{p}$ is in the dominant case 10^{-4} and in the recessive case 10^{-2}. The unfavourable allele A_1 can subsist in the gene pool at a much higher frequency if it is recessive, because selection cannot discern it in heterozygous individuals.

18.7 The selection–recombination equation

During meiosis, a *recombination* — or '*crossing over*' — of genetic material between corresponding parts of the homologous chromosomes can occur (see fig. 18.1). Recombination is, like mutation, a source of genetic variability. It does not produce new genes, but new gene combinations. It works, in general, at a much higher speed than mutation. The recombination rate between different loci can be as high as $1/2$.

In order to study recombination, we have to leave one-locus genetics. Even the simplest case of two loci with two alleles each conveys a dismaying glimpse of the intricacies of gene shuffling.

Let us assume that the alleles A_1 and A_2 belong to the first genetic locus, and the alleles B_1 and B_2 to the second one. The possible gametes, then, are $A_1 B_1$, $A_2 B_1$, $A_1 B_2$ and $A_2 B_2$. We shall denote them by G_1 to G_4, and their frequencies at the zygote stage (that is, immediately after fusion) by x_1 to x_4. The state of the *gamete pool*, then, is described by the point $\mathbf{x} = (x_1, x_2, x_3, x_4)$ in the simplex S_4.

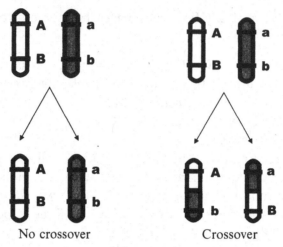

No crossover   Crossover

Fig. 18.1.

Each individual is obtained by fusion of a G_i-sperm with G_j-egg, and will be denoted by (G_i, G_j). The probability of its survival from zygote to adult stage will be denoted by w_{ij}. We have, of course, $w_{ij} = w_{ji}$, and we shall assume furthermore that $w_{23} = w_{14}$, since both (G_2, G_3) and (G_1, G_4) are on the first locus of genotype $A_1 A_2$ and on the second locus of genotype $B_1 B_2$. The fitness, then, is given by the following table:

	A_1A_1	A_1A_2	A_2A_2	
B_1B_1	w_{11}	$w_{12} = w_{21}$	w_{22}	(18.18)
B_1B_2	$w_{13} = w_{31}$	$w_{14} = w_{41} = w_{23} = w_{32}$	$w_{24} = w_{42}$	
B_2B_2	w_{33}	$w_{34} = w_{43}$	w_{44}	

We shall assume random union again. A gamete G_i will fuse with probability x_j with a gamete G_j, and then belongs to an individual with fitness w_{ij}. We may define

$$w_i = \sum_j w_{ij}x_j = (W\mathbf{x})_i \qquad (18.19)$$

as the fitness of the gamete G_i. The average fitness in the gamete pool of the population is

$$\bar{w} = \sum_i w_i x_i = \sum_{ij} w_{ij}x_i x_j . \qquad (18.20)$$

At meiosis, a crossover between the two loci may occur with probability r (where $r = \frac{1}{2}$ if the loci are independent, for example on nonhomologous chromosomes). If the individual is homozygous on at least one locus, a crossover does not change anything. It is only for the 'double heterozygotes' (G_2, G_3), (G_3, G_2), (G_1, G_4) and (G_4, G_1) that recombination makes a difference. Let us consider for instance type (G_2, G_3), whose frequency in the adult stage is $w_{23} x_2 x_3 \bar{w}^{-1}$. A fraction $\frac{r}{2}$ of its gametes are of types G_1 and G_4 each, and a fraction $\frac{1-r}{2}$ of type G_2 and G_3. Similar considerations hold for the other gamete pairs. The gametic frequencies in the zygote stage of the next generation are given by

$$
\begin{aligned}
x_1' &= \frac{1}{\bar{w}}(x_1 w_1 - rbD) \\[2mm]
x_2' &= \frac{1}{\bar{w}}(x_2 w_2 + rbD) \\[2mm]
x_3' &= \frac{1}{\bar{w}}(x_3 w_3 + rbD) \\[2mm]
x_4' &= \frac{1}{\bar{w}}(x_4 w_4 - rbD)
\end{aligned}
\tag{18.21}
$$

where b is the selection coefficient of the double heterozygotes ($b = w_{23} = w_{32} = w_{14} = w_{41}$) and $D = x_1 x_4 - x_2 x_3$. The case $r = 0$ leads to the selection equation (18.5) for the four 'alleles' G_1 to G_4.

18.8 Linkage

The *linkage disequilibrium coefficient* $D = x_1 x_4 - x_2 x_3$ is a measure of the statistical dependence between the two gene loci. Indeed, since the frequencies of the alleles A_1, A_2, B_1 and B_2 are given by $x_1 + x_3$, $x_2 + x_4$, $x_1 + x_2$ and $x_3 + x_4$, and since

$$
x_1 x_4 - x_2 x_3 = x_1 - (x_1 + x_3)(x_1 + x_2)
\tag{18.22}
$$

it follows that

$$
D = \Pr(A_1 B_1) - (\Pr A_1)(\Pr B_1) .
\tag{18.23}
$$

Hence, $D = 0$ if and only if

$$
\Pr(A_i B_j) = (\Pr A_i)(\Pr B_j)
\tag{18.24}
$$

for $i, j = 1, 2$. In this case, the population is said to be in *linkage equilibrium*. The coefficient D vanishes for the states belonging to the *Wright manifold*

$$
\mathbf{W} = \{\mathbf{x} \in S_4 : x_1 x_4 = x_2 x_3\}
\tag{18.25}
$$

(see fig. 18.2).

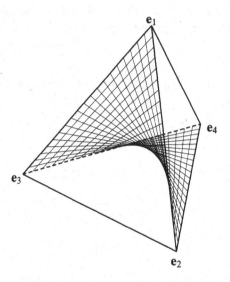

Fig. 18.2. The Wright manifold

If all w_{ij} are equal, that is, if no selection operates, then

$$x_i' = x_i - rD \qquad (i = 1, 4)$$
$$x_i' = x_i + rD \qquad (i = 2, 3)$$

and the linkage disequilibrium coefficient in the next generation is given by

$$D' = x_1'x_4' - x_2'x_3' = (1 - r)D . \qquad (18.26)$$

Since $D = 0$ implies $D' = 0$, the Wright manifold $\mathbf{W}$ is invariant: A population in linkage equilibrium remains in linkage equilibrium for ever. If $r > 0$, the linkage disequilibrium coefficient converges to zero. Recombination thus drives the population into linkage equilibrium. The situation resembles that of section 18.2, where random mating produces genotypes by independent fusion of the genes. The Hardy–Weinberg equilibrium is reached in one generation. Recombination leads to gametes constituted of independently assorted genes, but the manifold of linkage equilibria is reached only in the long run. The states on $\mathbf{W}$ are uniquely determined by the allelic frequencies of A_1 and B_1.

If the population is subject to selection, however, then $\mathbf{W}$ is not invariant in general, and the state need not converge to linkage equilibrium.

Exercise 18.8.1 Show this by an example.

18.9 Fitness under recombination

If the fitness depends on more than one locus, it need no longer increase from generation to generation, as can be shown by simple examples. It is sufficient to choose parameter values w_{ij} such that $\bar{w}$ attains an uniquely determined maximum at a point $\mathbf{x}$ in int S_n which does not lie on the Wright manifold.

Exercise 18.9.1 Find such a fitness matrix.

If the recombination rate were zero, $\mathbf{x}$ would have to be a fixed point of the resulting selection equation, since the average fitness can no longer increase. By (18.21) this would imply $w_1 = w_2 = w_3 = w_4 = \bar{w}$. As it is, $w_i = \bar{w}$ must still be valid for $r > 0$, since neither $\bar{w}$ nor the w_i's depend on r. From (18.21) it follows that

$$x_1' = x_1 - \frac{rbD}{\bar{w}} , \tag{18.27}$$

which implies $\mathbf{x} \neq \mathbf{x}'$ since $D \neq 0$. The function $\bar{w}$ attains its maximum at $\mathbf{x}$, and hence

$$\bar{w}(\mathbf{x}') \leq \bar{w}(\mathbf{x}) . \tag{18.28}$$

This means that the average fitness can drop from one generation to the next.

In some special cases, the average fitness does increase. This happens, for instance, if the fitness depends additively upon the contributions of the loci:

	A_1A_1	A_1A_2	A_2A_2	
B_1B_1	$a_1 + b_1$	$a_2 + b_1$	$a_3 + b_1$	(18.29)
B_1B_2	$a_1 + b_2$	$a_2 + b_2$	$a_3 + b_2$	
B_2B_2	$a_1 + b_3$	$a_2 + b_3$	$a_3 + b_3$	

In this case the average fitness is given by

$$
\begin{aligned}
\bar{w}(\mathbf{x}) &= \sum_{i,j} w_{ij} x_i x_j \\
&= a_1(x_1 + x_3)^2 + 2a_2(x_1 + x_3)(x_2 + x_4) + a_3(x_2 + x_4)^2 \\
&\quad + b_1(x_1 + x_2)^2 + 2b_2(x_1 + x_2)(x_3 + x_4) + b_3(x_3 + x_4)^2 .
\end{aligned} \tag{18.30}
$$

$\bar{w}(\mathbf{x}')$ depends only on the gene frequencies $x_1' + x_3'$ etc., which are obtained via (18.21): the value of r does not enter the computation. For $r = 0$ the function $\bar{w}$ increases, as we know from theorem 18.4.1 applied to the 4×4 matrix W: hence $\bar{w}$ increases for $r > 0$ too.

Exercise 18.9.2 Assume $a_2 > a_1, a_3$ and $b_2 > b_1, b_3$ in the additive fitness scheme (18.29) (overdominance at both loci). Show that there is a unique rest point $\hat{x}$ in the interior of S_4. Prove that all orbits in S_4 converge to $\hat{x}$.

Exercise 18.9.3 Is the Wright manifold $\mathbf{W}$ invariant for additive fitness?

Exercise 18.9.4 Consider a multiplicative fitness scheme analogous to (18.29). Show that D does not change sign along orbits. In particular, $\mathbf{W}$ is invariant. Prove that on $\mathbf{W}$, the mean fitness $\bar{w}$ increases.

18.10 Notes

The increase in fitness is a consequence of the Fundamental Theorem of Natural Selection which R.A. Fisher, the founding father of modern statistics and population genetics, presented in his book *The Genetical Theory of Natural Selection* (1930). For a clear account, which involves a discussion of the genetic variance, we refer to Ewens (1989, 1992), Edwards (1994), and Lessard (1996). Our proof of theorem 18.4.1 follows Kingman (1961). The proof in exercise 18.4.2 is due to Scheuer and Mandel (1959). The generalization in exercise 18.4.3 is due to Baum and Eagon (1967). More results on the selection equation can be found in the books of Edwards (1977), Nagylaki (1992) and Lyubich (1992). An example for decrease of mean fitness under recombination is in Moran (1964). Ewens (1969) showed that mean fitness is increasing if fitness is additive between loci.

19

Continuous selection dynamics

The continuous analogue of the discrete selection model is a replicator equation with a symmetric matrix. After showing that every orbit converges to some rest point, and discussing the location of stable rest points, we turn to the Shahshahani inner product, which allows the selection equation to be viewed as a gradient. We also consider the case where the selection coefficients of the fitness matrix depend on the population density.

19.1 The selection equation

As in the previous chapter, we shall consider one gene locus with n alleles $A_1, \ldots, A_n$, but this time assume that generations blend into each other. If the population is very large and births and deaths occur more or less continuously, we may treat the number $N_i(t)$ of alleles A_i at time t as a differentiable function of t. If the total number of individuals in the population is given by N, then $2N$ is the total number of genes and $x_i = N_i/2N$ the relative frequency of A_i.

Let us assume that the population is at any time in Hardy–Weinberg equilibrium. Then the frequency of the gene pair (A_i, A_j) is given by $x_i x_j = (4N^2)^{-1} N_i N_j$. Let d_{ij} be the death rate and b_{ij} the birth rate of (A_i, A_j)-individuals. The difference $m_{ij} = b_{ij} - d_{ij}$ is said to be the *Malthusian fitness parameter*. For the present, we shall assume that it is a constant.

The allele A_i occurs in the gene pairs (A_i, A_j) and (A_j, A_i) (with $j = 1, \ldots, n$). Its increase in the gene pool is given by

$$\dot{N}_i = \sum_j m_{ij} \frac{N_i N_j}{4N^2} N + \sum_j m_{ji} \frac{N_j N_i}{4N^2} N$$

249

i.e., since $m_{ij} = m_{ji}$, by

$$\dot{N}_i = \frac{N_i}{2N} \sum_j m_{ij} N_j \ .$$

Hence the growth rate of the total population is given by

$$\dot{N} = \frac{1}{2} \sum_i \dot{N}_i = \frac{1}{4N} \sum_{i,j} m_{ij} N_i N_j = N \sum_{i,j} m_{ij} x_i x_j \ . \qquad (19.1)$$

For the relative frequencies we obtain

$$\dot{x}_i = \left(\frac{N_i}{2N} \right)^{\cdot} = \frac{\dot{N}_i N - \dot{N} N_i}{2N^2} = x_i \left(\sum_j m_{ij} x_j - \sum_{r,s} m_{rs} x_r x_s \right) \ .$$

In matrix notation this reads

$$\dot{x}_i = x_i \left((M\mathbf{x})_i - \mathbf{x} \cdot M\mathbf{x} \right) \ . \qquad (19.2)$$

Equation (19.2) is the *continuous-time selection equation of population genetics*, defined on the probability simplex S_n. Obviously this is a replicator equation, with the additional condition $m_{ij} = m_{ji}$. This equation has been studied in section 7.8. In particular, the asymptotically stable rest points of (19.2) are exactly the evolutionarily stable strategies of the game given by the matrix M, and the average fitness

$$\bar{m}(t) = \mathbf{x}(t) \cdot M\mathbf{x}(t) = \sum_{i,j} m_{ij} x_i(t) x_j(t) \qquad (19.3)$$

is a strict Lyapunov function for (19.2). Furthermore, $\bar{m}/2$ is a potential for (19.2) with respect to the Shahshahani metric.

We shall stress right away that while (19.2) is a classic object of population genetics, its derivation is rather shaky. First of all, (19.1) would imply exponential growth for the whole population. But this is a minor point, which is repaired in section 19.4. More seriously, the assumption of Hardy–Weinberg equilibrium has to be challenged. Moreover, the notion of a birth rate for $A_i A_j$-genotypes depends crucially on this assumption. We shall discuss this in detail in chapter 21. Another justification of (19.2) is given in the exercise below.

Exercise 19.1.1 Show that (19.2) arises as a limiting case of the discrete selection model (18.5). (Hint: set $w_{ij} = 1 + h m_{ij}$, interpret h as generation length, and let $h \to 0$.)

Exercise 19.1.2 Discuss the case of 2 alleles and compare the results with those of section 18.5.

Exercise 19.1.3 Derive a similar selection model for asexually reproducing populations. Identify them as a special case of (19.2) by choosing special fitness matrices M.

Exercise 19.1.4 Show that a rest point $\mathbf{p} \in \text{int } S_n$ is asymptotically stable if and only if

$$\boldsymbol{\xi} \cdot M \boldsymbol{\xi} < 0 \qquad (19.4)$$

for all $\boldsymbol{\xi} \neq \mathbf{0}$ with $\sum \xi_i = 0$. Show that (19.4) is equivalent to the concavity of $\bar{m}$ on S_n.

Exercise 19.1.5 Show that (19.4) in turn implies the existence of a unique globally stable rest point (which may lie on the boundary). (This is a special case of exercise 7.2.7, but there exists a simpler direct proof here.)

Exercise 19.1.6 Suppose $\mathbf{p}$ is an interior fixed point of equation (19.2) or (18.5) with $m_{ij} \geq 0$. Show that each of the following conditions is equivalent to the asymptotic stability of $\mathbf{p}$.

(a) M has one positive and $n - 1$ negative eigenvalues.
(b) The leading principal minors of M alternate in sign.

Exercise 19.1.7 Show that for $n = 3$ and $m_{ij} \geq 0$, the existence of an asymptotically stable interior rest point implies that the heterozygote fitnesses m_{12}, m_{23}, m_{31} satisfy the triangle inequalities $m_{12} < m_{13} + m_{23}$ etc. Show that the converse holds if $m_{11} = m_{22} = m_{33} = 0$.

19.2 Convergence to a rest point

We know that the orbits of (19.2) converge to the set of rest points. But does each orbit converge to one rest point? Under the generic assumption that the rest points are isolated, this is a simple consequence of the fact that each ω-limit is connected. In degenerate cases, it may happen that linear manifolds of rest points occur. It is conceivable that an orbit approaches such a manifold without homing in on one of its points. That this cannot happen for (19.2) is shown in the following theorem.

Theorem 19.2.1 *Each orbit of the selection equation (19.2) converges to some rest point.*

Proof Let $\mathbf{p}$ be an arbitrary point in $\omega(\mathbf{x})$ (with $\mathbf{p} \neq \mathbf{x}$). Suppose first that $\mathbf{p}$ lies in int S_n. Then with $P = \prod x_i^{p_i}$ we have for all t

$$\frac{\dot{P}}{P} = \mathbf{p} \cdot M\mathbf{x} - \mathbf{x} \cdot M\mathbf{x} = \mathbf{x} \cdot M\mathbf{p} - \mathbf{x} \cdot M\mathbf{x}. \tag{19.5}$$

The same argument as in the proof of theorem 7.8.1 shows that this expression is strictly positive. Since $\mathbf{p} \in \omega(\mathbf{x})$ implies that $\mathbf{x}(t)$ comes arbitrarily close to $\mathbf{p}$, which is the unique maximum of P, the orbit $\mathbf{x}(t)$ must converge to $\mathbf{p}$.

The general case is slightly more technical. For convenience we replace P by its log, or more precisely by

$$L(\mathbf{x}) = - \sum_{i:p_i>0} p_i \log \frac{x_i}{p_i} . \tag{19.6}$$

Obviously L is a continuous extended real valued function on S_n which is differentiable on the set

$$\{\mathbf{x} : L(\mathbf{x}) < \infty\} = \{\mathbf{x} : \mathrm{supp}(\mathbf{x}) \supseteq \mathrm{supp}(\mathbf{p})\} .$$

As we have seen in the proof of theorem 7.2.4, L attains its minimal value 0 only at $\mathbf{p}$. Now let $J = \{i : (M\mathbf{p})_i \neq \mathbf{p} \cdot M\mathbf{p}\}$. Since $\mathbf{p}$ is a rest point, $p_i = 0$ for $i \in J$. If $J = \emptyset$, the same proof as for $\mathbf{p} \in$ int S_n is valid. Thus we assume $J \neq \emptyset$. We define

$$S(\mathbf{x}) = \sum_{i \in J} x_i \qquad Z(\mathbf{x}) = \min_{i \in J} [(M\mathbf{x})_i - \mathbf{x} \cdot M\mathbf{x}]^2 . \tag{19.7}$$

Then

$$\frac{1}{2}(\bar{m})^{\cdot} = \sum x_i [(M\mathbf{x})_i - \mathbf{x} \cdot M\mathbf{x}]^2 \geq S(\mathbf{x})Z(\mathbf{x}) . \tag{19.8}$$

By the definition of J, $Z(\mathbf{p}) =: z > 0$ and hence $\{\mathbf{x} : Z(\mathbf{x}) > \frac{z}{2}\}$ is an open neighbourhood of $\mathbf{p}$. Now

$$\{\mathbf{p}\} = \bigcap_{\varepsilon>0} \{\mathbf{x} : L(\mathbf{x}) \leq \varepsilon\}$$

is a decreasing intersection of compact sets. Thus there exists an $\bar{\varepsilon} > 0$ such that

$$L(\mathbf{x}) \leq \bar{\varepsilon} \Rightarrow Z(\mathbf{x}) > \frac{z}{2} . \tag{19.9}$$

From $\mathbf{p} \in \omega(\mathbf{x})$ follows $\mathrm{supp}(\mathbf{x}(t)) \supseteq \mathrm{supp}(\mathbf{p})$, and hence we can differentiate

L to obtain

$$
\begin{aligned}
\dot{L}(\mathbf{x}) &= -\sum p_i \frac{\dot{x}_i}{x_i} = -\sum p_i [(M\mathbf{x})_i - \mathbf{x}\cdot M\mathbf{x}] = \mathbf{x}\cdot M\mathbf{x} - \mathbf{p}\cdot M\mathbf{x} \\
&= \mathbf{x}\cdot M\mathbf{x} - \mathbf{p}\cdot M\mathbf{p} + \mathbf{p}\cdot M\mathbf{p} - \mathbf{x}\cdot M\mathbf{p} \\
&= \mathbf{x}\cdot M\mathbf{x} - \mathbf{p}\cdot M\mathbf{p} + \sum_i [\mathbf{p}\cdot M\mathbf{p} - (M\mathbf{p})_i] x_i \,.
\end{aligned}
$$

Because $\mathbf{x}\cdot M\mathbf{x} < \mathbf{p}\cdot M\mathbf{p}$ on the orbit, we have

$$
\dot{L}(\mathbf{x}) < KS(\mathbf{x}) \tag{19.10}
$$

where

$$
K = \max_{i\in J} |\mathbf{p}\cdot M\mathbf{p} - (M\mathbf{p})_i| \,.
$$

We choose a sequence $t_n \to +\infty$ such that $\mathbf{x}(t_n) \to \mathbf{p}$. Then $L(\mathbf{x}(t_n)) \to L(\mathbf{p}) = 0$. For $\varepsilon < \bar{\varepsilon}$, we choose $N = N(\varepsilon)$ such that

$$
L(\mathbf{x}(t_N)) < \frac{\varepsilon}{2} \quad \text{and} \quad \frac{K}{z}[\bar{m}(\mathbf{p}) - \bar{m}(\mathbf{x}(t_N))] < \frac{\varepsilon}{2} \,. \tag{19.11}
$$

As long as $t > t_N$ and $L(\mathbf{x}(s)) < \varepsilon$ for $s \in [t_N, t]$, the estimates (19.8–10) at $\mathbf{x}(s)$ imply

$$
\dot{L}(\mathbf{x}(s)) \le \frac{K}{2}\frac{[\bar{m}(\mathbf{x}(s))]^{\cdot}}{Z(\mathbf{x}(s))} < \frac{K}{z}[\bar{m}(\mathbf{x}(s))]^{\cdot} \,.
$$

Integrating and applying (19.11) we have

$$
0 \le L(\mathbf{x}(t)) \le L(\mathbf{x}(t_N)) + \frac{K}{z}[\bar{m}(\mathbf{x}(t)) - \bar{m}(\mathbf{x}(t_N))] < \varepsilon \,. \tag{19.12}
$$

Because $\varepsilon < \bar{\varepsilon}$, the point $\mathbf{x}(t)$ remains in the neighbourhood described by (19.9). Hence (19.12) holds for any $t > t_N$.

Since ε may be chosen arbitrarily small, $L(\mathbf{x}(t))$ approaches 0 as $t \to +\infty$, i.e. $\mathbf{x}(t)$ converges to $\mathbf{p}$. □

Exercise 19.2.2 Give an alternative proof by showing that instead of (19.6) the function

$$
L(\mathbf{x}) + 2\sum_{i\in J_+} x_i \,,
$$

where $J_+ = \{i : \mathbf{p}\cdot M\mathbf{p} > (M\mathbf{p})_i\}$, increases along the orbit of $\mathbf{x}$ in a neighbourhood of $\mathbf{p}$.

19.3 The location of stable rest points

The dynamics of the selection equation is essentially determined by the stable rest points. How many can there be, and how are they distributed in the simplex? We know that if one such rest point lies in the interior of S_n, then it is the only one. But several stable rest points may coexist if they all lie on the boundary, each one having some open subset of S_n as its basin of attraction. The simplest example is obtained if M is the identity matrix.

In order to get an idea of the variety of different patterns of stable rest points, we study a special class of fitness matrices which can be described by graphs. Let G be a (nonoriented) graph on the set of vertices $\{1,\dots,n\}$. Then we associate with G a symmetric matrix M, the *incidence matrix* of the graph:

$$m_{ij} = \begin{cases} 1 & \text{if} & i \text{ and } j \text{ are joined } (i \neq j) \\ 0 & \text{if} & i \text{ and } j \text{ are not joined} \\ \frac{1}{2} & \text{if} & i = j \end{cases}$$

A *clique* of G is a maximal subset $I \subseteq \{1,\dots,n\}$ such that any two elements i and j of I are joined by an edge.

We show now that the cliques of G correspond to the asymptotically stable rest points of the selection equation (19.2) with the above fitness matrix M. Without loss of generality, we may assume $I = \{1,\dots,k\}$. Let $\mathbf{p} = (\frac{1}{k},\dots,\frac{1}{k},0,\dots,0)$ be the barycentre of the face $S_n(I)$. The average fitness restricted to this face is

$$\mathbf{x}\cdot M\mathbf{x} = \sum_{i \neq j} x_i x_j + \frac{1}{2}\sum x_i^2 = \left(\sum_{i \in I} x_i\right)^2 - \frac{1}{2}\sum_{i \in I} x_i^2 = 1 - \frac{1}{2}\sum_{i \in I} x_i^2 .$$

It takes its maximal value $1 - \frac{1}{2k}$ at $\mathbf{p}$. This shows that $\mathbf{p}$ is asymptotically stable within the restriction of (19.2) to $S_n(I)$. If $j \notin I$, there is some $i \in I$ with $m_{ij} = 0$. Hence

$$(M\mathbf{p})_j \leq \frac{1}{k}(k-1) = 1 - \frac{1}{k} < 1 - \frac{1}{2k} = \mathbf{p}\cdot M\mathbf{p} .$$

Hence $\mathbf{p}$ is saturated and therefore asymptotically stable in S_n.

Exercise 19.3.1 Show that all stable rest points arise in this way.

Exercise 19.3.2 Draw all possible graphs for $n = 3$ and determine the corresponding pattern of stable rest points.

Exercise 19.3.3 For any partition of $n = k_1 + k_2 + \cdots + k_s$ (with $s, k_i = 1, 2, \dots$), construct an $n \times n$ matrix M where the number of asymptotically stable rest points of (19.2) is the product $k_1 \ldots k_s$.

Exercise 19.3.4 Construct an $n \times n$ matrix M where the number of coexisting stable rest points is $g(n) = r3^k$ with $n = 3k + r$ and $r = 2, 3$ or 4. (It has been shown that $g(n)$ is the maximal number of cliques in a graph with n vertices.)

Theorem 19.3.5 *Suppose that for an incidence matrix M there exists an asymptotically stable rest point $\mathbf{x}$ of* (18.1) *or* (19.2) *involving $n - i$ alleles. Then 2^i is an upper bound for the total number of asymptotically stable rest points.*

Proof Let I (with cardinality i) be the complement of supp($\mathbf{p}$). The map $C \rightarrow C \cap I$ is a one-to-one map of cliques to subsets of I: indeed, if $C_1 \cap I = C_2 \cap I$ for two cliques C_1 and C_2, then $C_1 \cup C_2$ is also a clique and hence $C_1 = C_2$. Therefore the number of cliques is bounded by the number 2^i of subsets of I. The theorem is then implied by the correspondence of asymptotically stable rest points and cliques. $\qquad\square$

It is natural to ask whether general fitness matrices allow for essentially other patterns of stable rest points which cannot be described by cliques. An example for $n = 4$ is the following

$$M = \begin{pmatrix} 1 & 14 & 18 & 6 \\ 14 & 10 & 9 & 10.2 \\ 18 & 9 & 6 & 11 \\ 6 & 10.2 & 11 & 10 \end{pmatrix}.$$

With this matrix, (19.2) has 3 stable rest points $\left(\frac{1}{3}, \frac{1}{3}, \frac{1}{3}, 0\right)$, $\left(0, 0, \frac{1}{6}, \frac{5}{6}\right)$ and $\left(0, \frac{1}{2}, 0, \frac{1}{2}\right)$. But the graph induced by these rest points (by joining any two vertices in the same support) has only two cliques: 123 and 234.

Exercise 19.3.6 Check this.

Similarly the number of coexisting stable rest points for general fitness matrices may exceed the maximal number of cliques given in exercise 19.3.4.

Exercise 19.3.7 Show that the circulant fitness matrix generated by the row $(0, 8, 13, 2, 2, 13, 8)$ has 14 stable rest points given by the cyclic permutations of $(8, 3, 8, 0, 0, 0, 0)$ and $(13, 0, 24, 0, 13, 0, 0)$. But $g(7) = 12$.

19.4 Density dependent fitness

The classical selection equation (19.2) describes the changes in relative frequencies in a population. But our derivation started with absolute numbers. What happened to the size of the whole population? The corresponding differential equation (19.1) for the total number N of individuals, $\dot{N} = N\bar{m}$, says that once an equilibrium $\mathbf{p}$ of the relative frequencies is established, the population will grow (or decay) exponentially, with the mean fitness $\bar{m} = \mathbf{p} \cdot M\mathbf{p}$ as growth rate. This is a rather unrealistic feature of the model, and the reason that (19.1) is usually repressed in population genetics.

The simplest way to repair this defect of the genetic model is to make the fitness values m_{ij} dependent on the population number N. In order to obtain saturation, we assume that owing to overpopulation the fitness values $m_{ij}(N)$ become negative for large N, or more precisely that there exist constants K_{ij} (the *carrying capacities* of the genotypes A_iA_j), such that

$$m_{ij}(N) > 0 \text{ for } N < K_{ij} \text{ and } m_{ij}(N) < 0 \text{ for } N > K_{ij} . \tag{19.13}$$

(The simplest choice would be

$$m_{ij}(N) = r_{ij}\left(1 - \frac{N}{K_{ij}}\right) \tag{19.14}$$

with some positive constants r_{ij}.) Equations (19.1–2) then read

$$\begin{aligned} \dot{N} &= N\bar{m}(N,\mathbf{x}) \\ \dot{x}_i &= x_i\left(\sum_j m_{ij}(N)x_j - \bar{m}(N,\mathbf{x})\right) \end{aligned} \tag{19.15}$$

with

$$\bar{m}(N,\mathbf{x}) = \sum_{r,s} m_{rs}(N)x_r x_s .$$

The state space is $\mathbb{R}_+ \times S_n$, a prism with base S_n. Since (19.13) implies that

$$\bar{m}(N,\mathbf{x}) < 0 \text{ for } N > \max_{i,j} K_{ij} = K_0 \quad ,$$

all orbits are bounded and end up in the region $N \leq K_0$. Let us consider the regions I (resp. II where $\bar{m}(N,\mathbf{x})$ is positive (resp. negative) and the separating manifold $H = \{(N,\mathbf{x}) : \bar{m}(N,\mathbf{x}) = 0\}$. Clearly N increases in I and decreases in II. This suggests that all orbits will end up on H. In order to exclude the possibility of oscillations between I and II we consider the mean fitness

$$(\bar{m})^{\cdot} = \sum_{i=1}^{n} \frac{\partial \bar{m}}{\partial x_i} \dot{x}_i + \frac{\partial \bar{m}}{\partial N} \dot{N}$$

$$= 2 \sum_{i} x_i [(M\mathbf{x})_i - \mathbf{x} \cdot M\mathbf{x}]^2 + \sum_{i,j} x_i x_j \frac{dm_{ij}}{dN} N \bar{m}(N, \mathbf{x}) .$$

On the set H, the second term vanishes and $\dot{m} \geq 0$. Hence H is semipermeable. The region I is forward invariant. At all points of H, the orbits enter horizontally from II into I, with the exception of the points where $\dot{m} = 0$, which are the rest points of (19.15).

Thus there are two possibilities for a nonstationary solution: either (a) it remains all the time in the upper region II, where $\dot{N} < 0$ along the whole orbit, and converges to an invariant subset of H, which is just the set of rest points; or (b) — and this is the generic case — it reaches region I and remains there, again converging to the set of rest points.

If all m_{ij} are decreasing functions of N, then $\bar{m}(N)$ is decreasing too and H is the graph of a function from S_n to $\mathbb{R}_+$ assigning to each frequency distribution $\mathbf{x}$ a carrying capacity $N(\mathbf{x})$. The rest points of (19.15) are exactly the critical points of this function and the stable rest points are its local maxima.

Exercise 19.4.1 Prove this last statement.

Exercise 19.4.2 Give a complete discussion of the case $n = 2$, when m_{ij} is given by (19.14). Show that a stable polymorphism exists if and only if the carrying capacity K_{12} of the heterozygote is larger than K_{11} and K_{22}.

Exercise 19.4.3 Show that when $m_{ij}(N) = r_{ij} - c_{ij}N$, with $c_{ij} = 0$ for $i \neq j$ and $c_{ii} > 0$ (heterozygotes have infinite carrying capacity), then there exists a unique polymorphism. Prove its global stability. (Hint: use the Perron–Frobenius theorem.)

19.5 The Shahshahani gradient

We have seen in section 7.8 that the selection equation (19.2) is a gradient system with respect to the Shahshahani inner product, which is given by

$$\langle \xi, \eta \rangle_\mathbf{x} = \sum_{i=1}^{n} \frac{1}{x_i} \xi_i \eta_i \qquad (19.16)$$

for tangent vectors $\xi, \eta \in T_{\mathbf{x}} S_n$. This is also a consequence of the following, more general theorem.

Theorem 19.5.1 *Let $\dot{x}_i = f_i(\mathbf{x}) = \frac{\partial V}{\partial x_i}$ be a Euclidean gradient vector field on $\mathbb{R}^n$. Then the corresponding replicator equation*

$$\dot{x}_i = \hat{f}_i(\mathbf{x}) = x_i(f_i(\mathbf{x}) - \bar{f}(\mathbf{x})) \tag{19.17}$$

is a Shahshahani gradient on $\mathrm{int}\, S_n$ with the same potential function V.

Proof For $\mathbf{x} \in \mathrm{int}\, S_n$ and $\xi \in \mathbb{R}_0^n$ we compute

$$
\begin{aligned}
\langle \hat{\mathbf{f}}(\mathbf{x}), \xi \rangle_{\mathbf{x}} &= \sum_{i=1}^n \frac{1}{x_i} \hat{f}_i(\mathbf{x}) \xi_i = \sum_{i=1}^n \frac{1}{x_i} x_i (f_i(\mathbf{x}) - \bar{f}(\mathbf{x})) \xi_i \\
&= \sum_{i=1}^n f_i(\mathbf{x}) \xi_i - \bar{f}(\mathbf{x}) \sum_{i=1}^n \xi_i \\
&= \sum_{i=1}^n \frac{\partial V}{\partial x_i} \xi_i = D_{\mathbf{x}} V(\xi) \, .
\end{aligned}
$$

$\square$

In particular, the proof shows that

$$\dot{V}(\mathbf{x}) = D_{\mathbf{x}} V(\dot{\mathbf{x}}) = \langle \hat{\mathbf{f}}(\mathbf{x}), \hat{\mathbf{f}}(\mathbf{x}) \rangle_{\mathbf{x}} = \sum_{i=1}^n x_i (f_i(\mathbf{x}) - \bar{f}(\mathbf{x}))^2 \, . \tag{19.18}$$

Thus the change of V is equal to the variance of the values $f_i(\mathbf{x})$.

Since $m_{ij} = m_{ji}$, the linear differential equation $\dot{\mathbf{x}} = M\mathbf{x}$ is a Euclidean gradient on $\mathbb{R}^n$ with $\frac{1}{2}\mathbf{x} \cdot M\mathbf{x}$ as its potential. This yields another proof of theorem 7.8.3.

Exercise 19.5.2 If the potential $V(\mathbf{x}) : \mathbb{R}_+^n \to \mathbb{R}$ is *homogeneous of degree* s, i.e. if $V(\alpha x_1, \dots, \alpha x_n) = \alpha^s V(x_1, \dots, x_n)$ for all $\alpha > 0$, then show that $\bar{f} = sV$. (Hint: use Euler's theorem on homogeneous functions.) Hence the average fitness $\bar{f}$ is increasing at maximal rate. Show also that every function $V : S_n \to \mathbb{R}$ is the restriction of a homogeneous function (of any prescribed degree) on $\mathbb{R}_+^n$.

Exercise 19.5.3 Show that equations of the form

$$\dot{x}_i = x_i(f_i(x_i) - \bar{f}) \tag{19.19}$$

are Shahshahani gradients. If the f_i are monotonically decreasing functions, then the potential is strictly concave on S_n and hence there is a unique, globally attracting rest point **p** for (19.19). Show that $P(\mathbf{x}) = \prod_{i=1}^{n} x_i^{p_i}$ is also a global Lyapunov function in this case.

Euclidean gradients $\mathbf{f}(\mathbf{x}) = \operatorname{grad} V(\mathbf{x})$ can be easily recognized by the symmetry of the Jacobian matrix $D_{\mathbf{x}}\mathbf{f}$, i.e. by the *integrability conditions*

$$\frac{\partial f_i}{\partial x_j} = \frac{\partial f_j}{\partial x_i}$$

(at least if the domain of **f** has no holes). We are now looking for a similar criterion for Shahshahani gradients.

If the vector field $\hat{\mathbf{f}}$ given by (19.17) is defined on a neighbourhood of int S_n (in $\mathbb{R}^n$), we may compute

$$\frac{\partial \hat{f}_i}{\partial x_j} = \delta_{ij}(f_i - \bar{f}) + x_i \left(\frac{\partial f_i}{\partial x_j} - \frac{\partial \bar{f}}{\partial x_j} \right) .$$

Since only the action of the Jacobian $D_{\mathbf{x}}\hat{\mathbf{f}}$ on vectors from the tangent space $T_{\mathbf{x}}S_n = \mathbb{R}_0^n$ is of relevance, we study the bilinear form $H_{\mathbf{x}}\hat{\mathbf{f}}$ given by

$$H_{\mathbf{x}}\hat{\mathbf{f}}(\xi, \eta) = \langle \xi, D_{\mathbf{x}}\hat{\mathbf{f}}(\eta) \rangle_{\mathbf{x}} \tag{19.20}$$

for $\xi, \eta \in \mathbb{R}_0^n$. Clearly

$$
\begin{aligned}
H_{\mathbf{x}}\hat{\mathbf{f}}(\xi, \eta) &= \textstyle\sum_{i,j=1}^{n} \frac{1}{x_i} \frac{\partial \hat{f}_i}{\partial x_j}(\mathbf{x}) \xi_i \eta_j \\
&= \textstyle\sum_{i=1}^{n} \frac{1}{x_i} (f_i - \bar{f}) \xi_i \eta_i + \sum_{i,j=1}^{n} \frac{\partial f_i}{\partial x_j}(\mathbf{x}) \xi_i \eta_i .
\end{aligned}
\tag{19.21}
$$

At the interior rest points, the first term of (19.21) vanishes and the bilinear form $H_{\mathbf{x}}\hat{\mathbf{f}}$ reduces to the Jacobian $D_{\mathbf{x}}\mathbf{f}$.

Theorem 19.5.4 *For a vector field $\hat{\mathbf{f}}$ defined by (19.17) in a neighbourhood U of int S_n, the following conditions are equivalent:*

(a) $\hat{\mathbf{f}}$ is a Shahshahani gradient on int S_n;
(b) There exist functions $V, G : U \to \mathbb{R}$ such that

$$f_i(\mathbf{x}) = \frac{\partial V}{\partial x_i} + G(\mathbf{x}) \tag{19.22}$$

holds on int S_n;
(c) The Jacobian bilinear form $H_{\mathbf{x}}\hat{\mathbf{f}}$ is symmetric at every $\mathbf{x} \in$ int S_n;
(d) The relation

$$\frac{\partial f_i}{\partial x_j} + \frac{\partial f_j}{\partial x_k} + \frac{\partial f_k}{\partial x_i} = \frac{\partial f_i}{\partial x_k} + \frac{\partial f_k}{\partial x_j} + \frac{\partial f_j}{\partial x_i} \tag{19.23}$$

holds on int S_n for all i, j, k.

A function V satisfying (19.22) is a Shahshahani potential for $\hat{\mathbf{f}}$.

Proof (a) $\Rightarrow$ (b) If $\hat{\mathbf{f}} = \text{Grad } V$, then (19.15) implies

$$\sum_{i=1}^{n} \frac{1}{x_i} \hat{f}_i(\mathbf{x}) \xi_i = \sum_{i=1}^{n} f_i(\mathbf{x}) \xi_i = \sum_{i=1}^{n} \frac{\partial V}{\partial x_i} \xi_i$$

for all $\boldsymbol{\xi} \in \mathbb{R}_0^n$ and $\mathbf{x} \in \text{int } S_n$. With $\xi_i = 1$, $\xi_n = -1$ and $\xi_j = 0$ for all $j \neq i, n$ this yields

$$f_i(\mathbf{x}) - f_n(\mathbf{x}) = \frac{\partial V}{\partial x_i}(\mathbf{x}) - \frac{\partial V}{\partial x_n}(\mathbf{x}) .$$

It follows that $\frac{\partial V}{\partial x_i}(\mathbf{x}) - f_i(\mathbf{x})$ is independent of i, which proves (b).

(b) $\Rightarrow$ (c) Since the f_i are of the form

$$f_i(\mathbf{x}) = \frac{\partial V}{\partial x_i}(\mathbf{x}) + G(\mathbf{x}) + \left(\sum_{j=1}^{n} x_j - 1\right) \varphi_i(\mathbf{x})$$

for $\mathbf{x} \in U$, the φ_i being suitable functions on U, the partial derivatives are given by

$$\frac{\partial f_i}{\partial x_j} = \frac{\partial^2 V}{\partial x_i \partial x_j} + \frac{\partial G}{\partial x_j} + \varphi_i(\mathbf{x})$$

for $\mathbf{x} \in \text{int } S_n$. Inserting this into (19.21), we see that the terms with G and φ_i disappear because $\sum \xi_i = \sum \eta_j = 0$. What remains is a symmetric bilinear form.

(c) $\Rightarrow$ (d) The symmetry of $H_{\mathbf{x}}\hat{\mathbf{f}}(\boldsymbol{\xi}, \boldsymbol{\eta})$ implies

$$\sum_{i,j=1}^{n} \left(\frac{\partial f_i}{\partial x_j} - \frac{\partial f_j}{\partial x_i}\right) \xi_i \eta_j = 0$$

for all $\boldsymbol{\xi}, \boldsymbol{\eta} \in \mathbb{R}_0^n$. With $\boldsymbol{\xi} = \mathbf{e}_i - \mathbf{e}_k$ and $\boldsymbol{\eta} = \mathbf{e}_j - \mathbf{e}_k$ we obtain (d).

(d) $\Rightarrow$ (b) Define for $\mathbf{x} \in \text{int } S_n$

$$g_i(x_1, \ldots, x_{n-1}) = f_i(x_1, \ldots, x_{n-1}, 1 - x_1 - \cdots - x_{n-1}) . \tag{19.24}$$

Then g_i coincides with f_i on S_n and $\frac{\partial g_i}{\partial x_j} = \frac{\partial f_i}{\partial x_j} - \frac{\partial f_i}{\partial x_n}$ by the chain rule. Thus (d) implies (with $k = n$)

$$\frac{\partial g_i}{\partial x_j} - \frac{\partial g_n}{\partial x_j} = \frac{\partial g_j}{\partial x_i} - \frac{\partial g_n}{\partial x_i} .$$

These are just the Euclidean integrability conditions for $g_i - g_n$ ($1 \leq i \leq n-1$)

on $\mathbb{R}^{n-1}$. Thus we find a potential $V = V(x_1, \ldots, x_{n-1})$ with

$$g_i - g_n = \frac{\partial V}{\partial x_i} \qquad i = 1, \ldots, n-1 \ .$$

By (19.24) this implies (b) with $G = g_n$.

(b) $\Rightarrow$ (a) is theorem 19.5.1. $\qquad\qquad\qquad\qquad\qquad\qquad\qquad\quad$ □

The *triangular integrability condition* (19.23) is very useful for finding out whether a given system is a Shahshahani gradient. Some applications will be given soon.

Exercise 19.5.5 Show that an ecological differential equation $\dot{x}_i = x_i f_i(\mathbf{x})$ with $\mathbf{x} \in \mathbb{R}^n_+$ is a gradient with respect to the Riemannian metric

$$\langle \boldsymbol{\xi}, \boldsymbol{\eta} \rangle_{\mathbf{x}} = \sum_{i=1}^{n} \frac{1}{x_i} \xi_i \eta_i \qquad\qquad (19.25)$$

in int $\mathbb{R}^n_+$ if and only if $\dot{x}_i = f_i(\mathbf{x})$ is an Euclidean gradient.

Exercise 19.5.6 Show that the Lotka–Volterra equation (5.1) is a gradient with respect to (19.25) if and only if the interaction matrix is symmetric. Compute its potential.

19.6 Mixed strategists and gradient systems

Let us consider a game with N pure strategies R_1 to R_N and a population consisting of n types of players E_1 to E_n, where type E_i plays strategy R_j with probability p^i_j and hence is characterized by a vector $\mathbf{p}^i \in S_N$. If we denote the frequencies of E_i in the population by x_i, then

$$p_k = \sum_{i=1}^{n} x_i p^i_k \qquad\qquad (19.26)$$

is the frequency of the strategy R_k in the population. Thus $\mathbf{x} \in S_n$ denotes the state of the population and $\mathbf{p} = \mathbf{p}(\mathbf{x}) \in S_N$ the distribution of the pure strategies. We shall assume that the payoff F_k for strategy R_k depends on $\mathbf{x}$ only through the strategy mix $\mathbf{p}$ in the population. Then the payoff for the phenotype E_i is given by

$$f_i(\mathbf{x}) = \mathbf{p}^i \cdot \mathbf{F}(\mathbf{p}) = \sum_{j=1}^{N} p^i_j F_j(\mathbf{p}) \qquad\qquad (19.27)$$

while the average payoff in the population is

$$\bar{f}(\mathbf{x}) = \sum x_i (\mathbf{p}^i \cdot \mathbf{F}(\mathbf{p})) = \mathbf{p} \cdot \mathbf{F}(\mathbf{p}) \; . \tag{19.28}$$

The game-dynamical ansatz leading to (7.1) now yields

$$\dot{x}_i = x_i [(\mathbf{p}^i - \mathbf{p}) \cdot \mathbf{F}(\mathbf{p})] \tag{19.29}$$

which we shall call the *mixed strategist dynamics*.

We may also consider a *pure strategist dynamics* for a fictitious population whose phenotypes correspond to the pure strategies. It is given by

$$\dot{y}_i = y_i (F_i(\mathbf{y}) - \bar{F}(\mathbf{y})) \tag{19.30}$$

on S_N.

Theorem 19.6.1 *If the pure strategist game (19.30) is a Shahshahani gradient with potential $V(\mathbf{y})$, then the mixed strategist game (19.29) is a Shahshahani gradient with potential $V(\mathbf{p}(\mathbf{x}))$.*

Proof Indeed, if

$$F_i(\mathbf{y}) = \frac{\partial V}{\partial y_i}(\mathbf{y}) + G(\mathbf{y})$$

(cf. (19.22)), then

$$f_i(\mathbf{x}) = \mathbf{p}^i \cdot \mathbf{F}(\mathbf{p}) = \sum_j p_j^i (F_j(\mathbf{p}) - G(\mathbf{p})) + G(\mathbf{p}) = \frac{\partial V(\mathbf{p})}{\partial x_i} + G(\mathbf{p}) \; .$$

$$\square$$

Obviously, the replicator dynamics of every 2×2 game is a Shahshahani gradient, since the triangular integrability condition (19.23) is trivially satisfied. Hence every mixed strategist dynamics based on two pure strategies is a Shahshahani gradient. Denoting by $p_1, \ldots, p_n$ and p the first component of $\mathbf{p}^1, \ldots, \mathbf{p}^n$ resp. $\mathbf{p}$, we obtain from (19.29)

$$\dot{x}_i = x_i (p_i - p) [F_1(p) - F_2(p)] \tag{19.31}$$

where $p = \sum_{i=1}^n p_i x_i$. The Shahshahani potential is given by

$$V(p) = \int [F_1(p) - F_2(p)] \, dp \; . \tag{19.32}$$

If the 2×2 game is linear and described by the matrix A, then we obtain in the generic case where $a = a_{11} - a_{21} + a_{22} - a_{12} \neq 0$:

$$F_1(p) - F_2(p) = a(p - \hat{p}) \tag{19.33}$$

(where $\hat{p} = a^{-1}(a_{22} - a_{12})$) and hence

$$V(p) = \frac{1}{2}a(p - \hat{p})^2 . \tag{19.34}$$

For the sex ratio game (see section 6.2) we have

$$F_1(m) - F_2(m) = \frac{1}{m} - \frac{1}{1-m} \tag{19.35}$$

(we write now m, the frequency of males, instead of p). Hence

$$V(m) = \log m(1 - m) . \tag{19.36}$$

It follows that the product $m(1 - m)$ of the frequencies of males and females increases to its maximal value, obtained for $m = \frac{1}{2}$: furthermore, the population evolves in such a way that the product increases at a maximal rate.

Exercise 19.6.2 When does the frequency p in (19.31) converge to an evolutionarily stable strategy? (Hint: start out with the linear case (19.33).)

Exercise 19.6.3 For any three phenotypes p_1, p_2 and p_3, (19.31) admits a constant of motion:

$$Q(\mathbf{x}) = (p_2 - p_3) \log x_1 + (p_3 - p_1) \log x_2 + (p_1 - p_2) \log x_3 . \tag{19.37}$$

Show that

$$\dot{p} = [F_1(p) - F_2(p)] \text{ Variance } \tilde{P} \tag{19.38}$$

where $\tilde{P}$ is the random variable taking the value p_i with probability x_i. Determine the rest points and the phase portrait of (19.31).

Exercise 19.6.4 Find invariants of motion for (19.29) if $n > N$. (Hint: there is one for every nontrivial linear relation $\sum c_i \mathbf{p}^i = \mathbf{0}$; note that $\sum c_i = 0$.) Find an analogue of (19.38) for $\dot{\mathbf{p}}$ (cf. (9.39) and section 22.3.)

Exercise 19.6.5 Set up a discrete equivalent for the evolution of the sex ratio, and verify that m converges to $\frac{1}{2}$.

Exercise 19.6.6 Consider the case of $N > 2$ mating types: every 'sex' R_i can mate with every other R_j ($j \neq i$). Show that the payoff for an R_i-individual is

$$F_i(\mathbf{m}) = \frac{1 - m_i}{\sum_k m_k(1 - m_k)}$$

and that

$$V(\mathbf{m}) = \frac{1}{2} \log \sum_{k=1}^{N} m_k(1 - m_k) = \frac{1}{2} \log \left[1 - \frac{1}{N} - \sum_k \left(m_k - \frac{1}{N} \right)^2 \right]$$

is a Shahshahani potential for the corresponding equation (19.29). Hence the 'sex ratio' converges to $(\frac{1}{N}, \ldots, \frac{1}{N})$. Why are there only two mating types?

19.7 Notes

The continuous time selection equation (19.2) goes back to the pioneers of population genetics, R. A. Fisher, J. B. S. Haldane and S. Wright. They used it at least implicitly as an approximation for the discrete time model (18.1) when selective differences are small, to make the mathematics simpler. Some population geneticists refuse to use (19.2) since its derivation and in particular the Hardy–Weinberg assumption is not correct in general (see section 21.5). Our proof of the convergence theorem 19.2.1 is from Akin and Hofbauer (1982). For the discrete case, we refer to Lyubich (1992, theorem 9.4.4) and Losert and Akin (1983). For the connection between cliques of graphs and stable rest points and patterns of ESS see Vickers and Cannings (1988a,b), Broom *et al.* (1993) and Cannings *et al.* (1993). Selection with density dependent fitness was studied by Roughgarden (1979). The Riemannian metric which makes the selection equation the gradient of mean fitness was introduced by Shahshahani (1979), and even earlier by Svirezhev, see Svirezhev and Passekov (1990), thereby making precise the ideas of Kimura (1958). This was further investigated by Akin (1979). Theorem 19.5.4 combines results from Akin (1979), Sigmund (1984), and Hofbauer (1985). These gradient vector fields have recently also been used as 'interior point flows' in optimization theory, see Helmke and Moore (1994). Section 19.6 follows Sigmund (1987a).

20

Mutation and recombination

In this chapter we investigate the effects of mutation and recombination terms. We characterize mutation equations which are Shahshahani gradients and show that periodic orbits occur for the selection–mutation equation. We also show that average fitness will not increase, in general, for the selection–recombination dynamics.

20.1 The selection–mutation model

In this section we extend the model introduced in section 18.6. Consider a gene locus with n alleles $A_1, \ldots, A_n$ in a population with discrete generations and let $x_1, \ldots, x_n$ be the relative frequencies of the alleles at the time of mating. Assuming random mating, the relative frequency of new gene pairs (A_i, A_j) will be $x_i x_j$. Due to natural selection only a proportion $w_{ij} x_i x_j$ will survive to mature age ($w_{ij} = w_{ji} \geq 0$ being the fitness parameters). Hence the number of newly produced genes A_j is proportional to $\sum_k w_{jk} x_j x_k = x_j (W\mathbf{x})_j$. Now let ε_{ij} be the *mutation rate* from A_j to A_i (note the order of the indices), so that

$$\varepsilon_{ij} \geq 0 \quad \text{and} \quad \sum_{i=1}^{n} \varepsilon_{ij} = 1 \tag{20.1}$$

for all $j = 1, \ldots, n$ (where the ε_{ii} are suitably defined). Then the frequency x_i' of genes A_i in the gene pool of the new generation is proportional to $\sum_j \varepsilon_{ij} x_j (W\mathbf{x})_j$. More precisely it is given by

$$x_i' = \frac{1}{\bar{w}(\mathbf{x})} \sum_{j=1}^{n} \varepsilon_{ij} x_j (W\mathbf{x})_j \tag{20.2}$$

265

with $\bar{w}(\mathbf{x}) = \mathbf{x} \cdot W\mathbf{x}$ (the mean fitness of the population) as the usual normalization factor. This is the *discrete time selection–mutation equation*. We shall always assume $w_{ii} > 0$ for all i in order to guarantee that $\bar{w}(\mathbf{x}) > 0$ for all $\mathbf{x} \in S_n$.

In order to get a differential equation we replace as usual $x_i' - x_i$ by $\dot{x}_i$ and obtain

$$\dot{x}_i = \bar{w}(\mathbf{x})^{-1} \sum_{j,k} \varepsilon_{ij} x_j w_{jk} x_k - x_i \ . \tag{20.3}$$

A biologically more satisfactory way to derive a *continuous time model* might be the following. Consider a population with overlapping generations and let selection act in the way described in section 19.1 with Malthusian fitness parameters m_{ij}. Mutation effects, being small in general, will change the gene frequencies in a linear way. A simultaneous action of selection and mutation in a small time interval Δt will be of smaller order $(\Delta t)^2$, since the two effects are independent. Thus we arrive at the following continuous time model with separate selection and mutation terms:

$$\dot{x}_i = x_i \left[(M\mathbf{x})_i - \mathbf{x} \cdot M\mathbf{x} \right] + \sum_{j=1}^{n} \left(\varepsilon_{ij} x_j - \varepsilon_{ji} x_i \right) \ . \tag{20.4}$$

All three equations (20.2–4) describe dynamical systems on S_n.

Exercise 20.1.1 To model weak selection and weak mutation, replace ε_{ij} by $\delta \varepsilon_{ij}$ (for $i \neq j$) in (20.2) and set $w_{ij} = 1 + \delta m_{ij}$, with δ a small parameter. Show that both the discrete time model (20.2) and its continuous time counterpart (20.3) reduce to the differential equation (20.4) as $\delta \to 0$.

Exercise 20.1.2 Analyse (20.3) and (20.4) for $n = 2$ alleles as in section 18.6.

20.2 Mutation and additive selection

In the absence of selection, i.e. if all w_{ij} (and m_{ij}) are equal, equations (20.3) and (20.4) reduce to the *continuous time mutation equation*

$$\dot{x}_i = \sum_{j=1}^{n} \varepsilon_{ij} x_j - x_i \tag{20.5}$$

and (20.2) turns into the *discrete time mutation equation*

$$x_i' = \sum_{j=1}^{n} \varepsilon_{ij} x_j \ . \tag{20.6}$$

These equations are linear and they are easily analysed using Perron–Frobenius theory.

Theorem 20.2.1 *If the mutation matrix (ε_{ij}) is primitive then the mutation equation (20.6) has a unique rest point $\mathbf{p} \in \text{int } S_n$. This point is globally stable.*

Proof This follows directly from the theory of finite Markov chains. We only have to remark that the *columns* of the matrix $M = (\varepsilon_{ij})$ sum to 1. The left eigenvector of the dominant eigenvalue 1 is $\mathbf{1}$; denote the uniquely defined right eigenvector belonging to $\text{int } S_n$ by $\mathbf{u}$. The elements of the i-th row of M^k all converge to u_i. Hence $M^k \mathbf{x} \to \mathbf{u}$ for all $\mathbf{x} \in S_n$. □

Exercise 20.2.2 Prove in a similar way that if the mutation matrix (ε_{ij}) is irreducible, then the mutation equation (20.5) has a unique equilibrium $\mathbf{p}$ which is globally stable. Furthermore, the functions

$$V(\mathbf{x}) = \sum_{i=1}^{n} \frac{1}{p_i}(x_i - p_i)^2 \quad \text{and} \quad P(\mathbf{x}) = \prod_{i=1}^{n} x_i^{p_i}$$

are global Lyapunov functions for (20.5). (Hint: $(\delta_{ij} - \varepsilon_{ij})$ is a singular M-matrix. Proceed as in exercise 15.3.9.)

Exercise 20.2.3 Show that the function P from the previous exercise is also a Lyapunov function for the discrete time model (20.6).

This simple dynamics carries over to the case where 'dominance free' selection is allowed. Indeed, suppose in (20.2) or (20.3) that selection is multiplicative (i.e. $w_{ij} = w_i w_j$). The corresponding assumption for (20.4) is that Malthusian fitness is additive (i.e. $m_{ij} = m_i + m_j$): one has only to take the limit as in exercise 20.1.1. Equations (20.2–4) then simplify to

$$\bar{w} x_i' = \sum_{j=1}^{n} \varepsilon_{ij} w_j x_j \tag{20.7}$$

$$\bar{w} \dot{x}_i = \sum_{j=1}^{n} \varepsilon_{ij} w_j x_j - x_i \bar{w} \tag{20.8}$$

$$\dot{x}_i = x_i(m_i - \bar{m}) + \sum_{j=1}^{n} \varepsilon_{ij} x_j - x_i \tag{20.9}$$

with $\bar{w} = \sum_{i=1}^{n} x_i w_i$ and $\bar{m} = \sum_{j=1}^{n} x_i m_i$.

Exercise 20.2.4 Under the assumptions of the previous theorem, show that the models (20.7–9) give rise to a unique rest point $\mathbf{p} \in \text{int}\, S_n$. This point is globally stable.

Exercise 20.2.5 Let A be the matrix $(\varepsilon_{ij} - \delta_{ij})$. If A is symmetric and $\mathbf{p}$ is the interior fixed point then show that each of the following functions is a Lyapunov function for (20.5):

$$(\mathbf{x} - \mathbf{p})^2, \quad \mathbf{x} \cdot A\mathbf{x}, \quad \frac{\mathbf{x} \cdot A\mathbf{x}}{\mathbf{x} \cdot \mathbf{x}}, \quad (\mathbf{x} - \mathbf{p}) \cdot A(\mathbf{x} - \mathbf{p}) .$$

Exercise 20.2.6 Define for $\mathbf{x}, \mathbf{y} \in \text{int}\, S_n$

$$g_1(\mathbf{x}, \mathbf{y}) = \max_i \frac{x_i}{y_i} \qquad g_2(\mathbf{x}, \mathbf{y}) = \min_i \frac{x_i}{y_i} .$$

Show that for $\varepsilon_{ij} > 0$ the mutation equation (20.6) and its generalization (20.7) are contractions for the metric $d(\mathbf{x}, \mathbf{y}) = \log\left(g_1(\mathbf{x}, \mathbf{y})/g_2(\mathbf{x}, \mathbf{y})\right)$.

20.3 Special mutation rates

An important question is to what extent the fact that the selection equation admits a Lyapunov function, and even a potential, carries over to the selection–mutation model. Under the assumption of exercise 20.2.2 there will exist some Lyapunov function for (20.7–9). No biologically satisfactory candidate has been found so far.

We cannot hope to obtain a gradient for general mutation rates ε_{ij}. While ε_{ij}, being a stochastic matrix, has leading eigenvalue 1 and therefore all eigenvalues of the restriction to S_n are of absolute value less than 1 (which implies stability), the possibility of complex eigenvalues excludes gradient behaviour.

Exercise 20.3.1 Find 3×3 mutation matrices, with ε_{ij} $(i \neq j)$ arbitrarily small, which have complex eigenvalues.

This suggests singling out those mutation matrices which are *Shahshahani gradients*. For this we write the mutation equation (20.5) in the form

$$\dot{x}_i = x_i(f_i(\mathbf{x}) - \bar{f}). \tag{20.10}$$

This implies on $\text{int}\, S_n$:

$$f_i(\mathbf{x}) = \sum_{j=1}^{n} \varepsilon_{ij} \frac{x_j}{x_i} .$$

and hence $\frac{\partial f_i}{\partial x_j} = \frac{\varepsilon_{ij}}{x_i}$, $i \neq j$. The triangular integrability condition (19.23) then reads (for pairwise different i, j, k)

$$\frac{\varepsilon_{ij}}{x_i} + \frac{\varepsilon_{jk}}{x_j} + \frac{\varepsilon_{ki}}{x_k} = \frac{\varepsilon_{ik}}{x_i} + \frac{\varepsilon_{kj}}{x_k} + \frac{\varepsilon_{ji}}{x_j} \qquad (20.11)$$

for all $\mathbf{x} \in \operatorname{int} S_n$. Taking the limit $x_i \to 0$, we find that $\varepsilon_{ij} = \varepsilon_{ik}$ for all $j \neq i \neq k$. Thus *the mutation rates must depend on the target gene only*: they can be written as

$$\varepsilon_{ij} = \varepsilon_i \qquad (i \neq j) . \qquad (20.12)$$

(20.1) implies for these special mutation rates

$$\varepsilon_{ii} = 1 + \varepsilon_i - \varepsilon \quad \text{with} \quad \varepsilon = \varepsilon_1 + \cdots + \varepsilon_n .$$

(20.3) reduces to

$$
\begin{aligned}
\bar{w}(\mathbf{x})\dot{x}_i &= x_i\big[(W\mathbf{x})_i - \bar{w}(\mathbf{x})\big] + \varepsilon_i \bar{w}(\mathbf{x}) - \varepsilon x_i (W\mathbf{x})_i \\
&= x_i\big[(1-\varepsilon)(W\mathbf{x})_i - \bar{w}(\mathbf{x})\big] + \varepsilon_i \bar{w}(\mathbf{x}) .
\end{aligned} \qquad (20.13)
$$

This gives a replicator equation (20.10) with

$$f_i(\mathbf{x}) = (1-\varepsilon)\frac{(W\mathbf{x})_i}{\bar{w}(\mathbf{x})} + \frac{\varepsilon_i}{x_i} \quad \text{and} \quad \bar{f}(\mathbf{x}) = 1 . \qquad (20.14)$$

Obviously the functions $f_i(\mathbf{x})$ fulfil the integrability conditions and thus, by theorem 19.5.4, (20.3) is a Shahshahani gradient if (20.12) is satisfied. The *potential* is easily computed to be

$$V(\mathbf{x}) = \frac{1-\varepsilon}{2} \log \bar{w}(\mathbf{x}) + \sum_{i=1}^{n} \varepsilon_i \log x_i . \qquad (20.15)$$

A more appealing Lyapunov function is obtained by exponentiating $2V$:

$$W(\mathbf{x}) = \bar{w}(\mathbf{x})^{1-\varepsilon} \prod_{i=1}^{n} x_i^{2\varepsilon_i} . \qquad (20.16)$$

For $\varepsilon = 0$ (i.e. no mutation), $W(\mathbf{x})$ reduces to the mean fitness $\bar{w}(\mathbf{x})$. (20.13) then implies

$$\dot{V}(\mathbf{x}) = \frac{1}{2}\frac{\dot{W}(\mathbf{x})}{W(\mathbf{x})} = \sum_{i=1}^{n} x_i(f_i - \bar{f})^2 \geq 0 . \qquad (20.17)$$

Just as with the selection equation, *the rate of growth of the modified mean fitness function $W(\mathbf{x})$ is proportional to the variance of the terms $f_i(\mathbf{x})$.*

It is easy to derive a similar result for the uncoupled version of the continuous time equation (20.4). One finds

$$f_i(\mathbf{x}) = (M\mathbf{x})_i + \frac{\varepsilon_i}{x_i} \quad \text{and} \quad \bar{f}(\mathbf{x}) = \mathbf{x} \cdot M\mathbf{x} + \varepsilon \qquad (20.18)$$

and the potential is given by

$$V(\mathbf{x}) = \frac{1}{2}\mathbf{x} \cdot M\mathbf{x} + \sum_{i=1}^{n} \varepsilon_i \log x_i \, . \tag{20.19}$$

Exercise 20.3.2 Work this out in detail (either directly or using the weak selection limit argument of exercise 20.1.1).

Summarizing, we have shown

Theorem 20.3.3 *The mutation equation* (20.5) *is a Shahshahani gradient if and only if the mutation rates take the special form* (20.12). *In this case, for arbitrary selection terms, both continuous time selection–mutation equations* (20.3) *and* (20.4) *are Shahshahani gradients too.*

A nice application of this result is

Theorem 20.3.4 *Suppose that the model without mutation (i.e.* $\varepsilon_{ij} = 0$ *for* $i \neq j$*) admits a stable interior rest point. Then for every choice of mutation rates satisfying* (20.12) *with* $\varepsilon \leq 1$*, the equations* (20.3) *and* (20.4) *have a unique rest point in* S_n*, which is globally attracting.*

Exercise 20.3.5 Prove this. (Hint: show that $\bar{w}(\mathbf{x})$ and $V(\mathbf{x})$ are strictly concave and therefore attain a unique maximum.)

Exercise 20.3.6 Show that the conclusion need not hold if selection (in the absence of mutation) leads to a globally stable rest point on the boundary. (Hint: take a two-allele model where the fitter allele is recessive.)

Exercise 20.3.7 Show that for $n = 2$, the selection–mutation equation is always a Shahshahani gradient.

Exercise 20.3.8 Generalize the results of this section to the discrete time model (20.2). (Hint: use the result in exercise 18.4.3.)

20.4 Limit cycles for the selection–mutation equation

What happens if mutation rates are not of the special form (20.12)? The mutation equation (20.5) is then not a Shahshahani gradient, and the following theorem shows that for some selection matrices (m_{ij}) the continuous time selection–mutation equation (20.4) has periodic orbits. Since (20.4) is a limiting case of (20.2) and (20.3), this result carries over to these more general equations.

Theorem 20.4.1 *Consider an equation of the form*

$$\dot{x}_i = x_i((M\mathbf{x})_i - \mathbf{x}\cdot M\mathbf{x}) + \hat{f}_i(\mathbf{x}) \tag{20.20}$$

where $\hat{\mathbf{f}}(\mathbf{x})$ is again a vector field on S_n which we may write in replicator form (20.10). Whenever $\hat{\mathbf{f}}(\mathbf{x})$ is not a Shahshahani gradient, there exist symmetric matrices M such that (20.20) has periodic orbits.

Proof We prove first the following statement: given a point $\mathbf{x} \in \text{int } S_n$ and a linear map $S : \mathbb{R}_0^n \to \mathbb{R}_0^n$ which is symmetric with respect to the Shahshahani metric, there exists a symmetric matrix M such that $\mathbf{x}$ is a rest point of (20.20) and S is the symmetric part of its Jacobian at $\mathbf{x}$.
Indeed, for given $\mathbf{x}$, we can split up any fitness matrix M as

$$m_{ij} = \bar{m} + (m_i - \bar{m}) + (m_j - \bar{m}) + \theta_{ij} \tag{20.21}$$

where $m_i = (M\mathbf{x})_i$, $\bar{m} = \sum x_i m_i = \mathbf{x}\cdot M\mathbf{x}$, and the dominance terms θ_{ij} satisfy

$$\sum_{j=1}^{n} \theta_{ij} x_j = 0 . \tag{20.22}$$

Conversely, arbitrary numbers m_i and $\theta_{ij} = \theta_{ji}$ obeying (20.22) determine a fitness matrix M. Now choose m_i such that

$$x_i m_i + \hat{f}_i(\mathbf{x}) = 0 . \tag{20.23}$$

Since $\hat{\mathbf{f}}(\mathbf{x}) \in \mathbb{R}_0^n$ one then has $\sum x_i m_i = \bar{m} = 0$. Thus $\mathbf{x}$ is a rest point of (20.20). Obviously, after specifying m_i and $\bar{m}$ by (20.23), every symmetric operator S in the $(n-1)$-dimensional space $\mathbb{R}_0^n$ can be realized by suitably chosen θ_{ij} obeying (20.22).
We may now prove the theorem by viewing the Jacobian of (20.20) at $\mathbf{x}$ as a linear map on $T_{\mathbf{x}} S_n = \mathbb{R}_0^n$ which we split into its symmetric and antisymmetric part with respect to the Shahshahani inner product: $S + A$. By assumption (and using theorem 19.5.4(c)) there exists a point $\mathbf{x} \in \text{int } S_n$ where $A \neq 0$. We choose an orthogonal basis $\xi^1, \ldots, \xi^{n-1}$ of $\mathbb{R}_0^n$ such that $A\xi^j = \omega_j \xi^{n-j}$ and $A\xi^{n-j} = -\omega_j \xi^j$ for $j = 1, \ldots, k$ where $\pm i\omega_j$ $(1 \leq j \leq k)$ are the pairs of nonzero imaginary eigenvalues of A $(k \geq 1)$ and $A\xi^j = \mathbf{0}$ for $k < j < n - k$. Define the symmetric operators S_λ by $S_\lambda \xi^j = \lambda \xi^j$ for $j = 1$ and $j = n - 1$ and $S_\lambda \xi^j = -\xi^j$ for $1 < j < n-1$. Then $S_\lambda + A$ has eigenvalues $\lambda \pm i\omega_1$, $-1 \pm i\omega_j$ $(2 \leq j \leq k)$ and possibly -1. By the above statement we can realize S_λ by symmetric matrices M_λ. Then (20.20) with M_λ undergoes a

Hopf bifurcation at the rest point $\mathbf{x}$, as λ varies from -1 to 1. Hence there exist nontrivial periodic solutions.　　　　　　　　　　　□

The following example even displays *stable* limit cycles. Assume that all homozygotes A_iA_i have the same fitness and all heterozygotes too. When working with (20.4) this means $m_{ij} = s\delta_{ij}$ where s measures the selective advantage of the homozygotes. Let us assume that the mutation rates are *cyclically symmetric*, that is,

$$\varepsilon_{ij} = \varepsilon_{j-i} \tag{20.24}$$

where the index i is considered modulo n. Then $\sum_{i=0}^{n-1} \varepsilon_i = 1$. Now (20.4) reads

$$\dot{x}_i = sx_i(x_i - Q(\mathbf{x})) + \sum_{j=1}^{n} \varepsilon_{j-i}x_j - x_i \tag{20.25}$$

with $Q(\mathbf{x}) = \sum_{i=1}^{n} x_i^2$. Obviously the barycentre $\mathbf{m} = \frac{1}{n}\mathbf{1}$ is a rest point of (20.25). The Jacobian of (20.25) at a point $\mathbf{x}$ is given by

$$\frac{\partial \dot{x}_i}{\partial x_j} = s\delta_{ij}(x_i - Q(\mathbf{x})) + sx_i(\delta_{ij} - 2x_j) + \varepsilon_{j-i} - \delta_{ij} .$$

The divergence is then

$$\mathrm{div} = \sum_{i=1}^{n} \frac{\partial \dot{x}_i}{\partial x_i} = s(2 - (n+2)Q(\mathbf{x})) + n(\varepsilon_0 - 1) .$$

After subtracting the eigenvalue orthogonal to S_n, which is given by $-\bar{f}(\mathbf{x})$, i.e. by $-sQ(\mathbf{x})$, we obtain the divergence within S_n,

$$\mathrm{div}_0 = s(2 - (n+1)Q(\mathbf{x})) + n(\varepsilon_0 - 1) . \tag{20.26}$$

Since $Q(\mathbf{x}) = \sum x_i^2 \geq \frac{1}{n}\left(\sum x_i\right)^2 = \frac{1}{n}$, we have for positive values of s

$$\mathrm{div}_0 \leq s\left(1 - \frac{1}{n}\right) + n(\varepsilon_0 - 1) . \tag{20.27}$$

Thus the divergence is strictly negative on $S_n\backslash\{\mathbf{m}\}$ whenever

$$s \leq \frac{n^2}{n-1}(1 - \varepsilon_0) . \tag{20.28}$$

Now we specialize to $n = 3$ alleles. Then we can compute the eigenvalues $\gamma, \bar{\gamma}$ at $\mathbf{m}$ within S_3, using formula (5.20), as

$$\gamma = \frac{s}{3} - 1 + \varepsilon_0 + \varepsilon_1\lambda + \varepsilon_2\bar{\lambda}$$

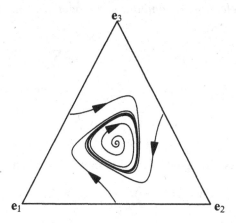

Fig. 20.1.

with $\lambda = \exp(2\pi i/3)$. They are complex if $\varepsilon_1 \neq \varepsilon_2$, and their real part is

$$\mathrm{Re}\,\gamma = \frac{s}{3} - \frac{3}{2}(\varepsilon_1 + \varepsilon_2).$$

For $s = \frac{9}{2}(\varepsilon_1 + \varepsilon_2)$ the eigenvalues are purely imaginary and a Hopf bifurcation occurs, with s as parameter. Since for all $s \leq \frac{9}{2}(\varepsilon_1 + \varepsilon_2)$, div$_0 < 0$ holds on $S_3 \setminus \{\mathbf{m}\}$ by (20.27) and (20.28), the Dulac criterion (see section 4.1) shows that there are no periodic orbits in this case, and hence that $\mathbf{m}$ is globally asymptotically stable. As this holds even in the critical case $s = \frac{9}{2}(\varepsilon_1 + \varepsilon_2)$, *stable* limit cycles appear for s slightly larger than $\frac{9}{2}(\varepsilon_1 + \varepsilon_2)$ (see fig. 20.1).

20.5 Selection at two loci

In this section we extend the selection–recombination model described in section 18.7. to an arbitrary number of alleles at the two loci. Let A_i $(1 \leq i \leq n)$ and B_j $(1 \leq j \leq m)$ be the possible alleles at the A- and the B-locus. Then there are nm different types of gametes $A_i B_j$, the frequencies of which we denote by x_{ij}. Assuming random mating and discrete generations, the proportion of $A_i B_j / A_k B_l$ individuals will change by natural selection from $x_{ij} x_{kl}$ in zygotes to $w_{ij,kl} x_{ij} x_{kl}$ in the adult stage. When haploid gametes are produced during meiosis, not only the parental combinations $A_i B_j$ and $A_k B_l$ but also the *recombinants* $A_i B_l$ and $A_k B_j$ will appear. Such crossovers happen with a certain probability r depending on the distance between the two loci. (Remember that this recombination fraction r takes its maximal value $\frac{1}{2}$

if the two loci are on different chromosomes.) This leads to the following *discrete two-locus selection–recombination model* for the gamete frequencies x'_{ij} in the next generation:

$$\bar{w}x'_{ij} = (1-r)x_{ij}\sum_{k,l} w_{ij,kl}x_{kl} + r\sum_{k,l} w_{il,kj}x_{il}x_{kj} \qquad (20.29)$$

with

$$\bar{w} = \sum_{i,j,k,l} w_{ij,kl}x_{ij}x_{kl} \qquad (20.30)$$

as the average fitness. Alternatively, this can be written as

$$\bar{w}x'_{ij} = x_{ij}\sum_{k,l} w_{ij,kl}x_{kl} - rD_{ij} \qquad (20.31)$$

with

$$D_{ij} = \sum_{k,l}(w_{ij,kl}x_{ij}x_{kl} - w_{il,kj}x_{il}x_{kj}) . \qquad (20.32)$$

The expressions D_{ij} are called *linkage disequilibrium functions*.

If we consider the allele frequencies p_i at locus A at the zygote stage, we obtain

$$p_i = \sum_{j=1}^{m} x_{ij} \qquad (20.33)$$

and hence for the next generation

$$p'_i = \sum_{j} x'_{ij} = \bar{w}^{-1}\sum_{k}\sum_{j,l} x_{ij}x_{kl}w_{ij,kl} . \qquad (20.34)$$

This last expression is just the allele frequency in the parent generation after selection. It does not depend on the recombination rate r.

The continuous time version of the two-locus selection equation takes the form

$$\dot{x}_{ij} = x_{ij}\left(\sum_{k,l} m_{ij,kl}x_{kl} - \bar{m}\right) - rD_{ij} \qquad (20.35)$$

with the disequilibrium functions now given by

$$D_{ij} = \sum_{k,l}(b_{ij,kl}x_{ij}x_{kl} - b_{il,kj}x_{il}x_{kj}) . \qquad (20.36)$$

These equations are obtained similarly to (19.2) or (20.4). Changes in the frequencies x_{ij} of A_iB_j gametes in a small time interval Δt are due to either births or deaths. With $b_{ij,kl}$, $d_{ij,kl}$ and $m_{ij,kl} = b_{ij,kl} - d_{ij,kl}$ denoting birthrates, deathrates and fitnesses of A_iB_j/A_kB_l individuals, this leads to the usual selection equation. A fraction r of the newborn has undergone a crossover, however, which gives the *recombination terms* $-rD_{ij}$ in (20.35). Note that the continuous time model has more parameters than the discrete time model.

Usually $w_{ij,kl} = w_{il,kj}$ holds since both A_iB_j/A_kB_l individuals and A_iB_l/A_kB_j individuals have A_iA_k on their A-locus and B_jB_l on the B-locus. Given the composition of the genotype, the arrangement of the alleles is irrelevant. With this assumption of *no position effects*, $D_{ij} = 0$ holds if the frequencies x_{ij} can be written as a product of the gene frequencies $p_i = \sum_j x_{ij}$ of A_i and $q_j = \sum_i x_{ij}$ of B_j. The set where this linkage equilibrium holds is the *Wright manifold* (or *linkage equilibrium manifold*)

$$\mathbf{W} = \{\mathbf{x} \in S_{nm} : x_{ij} = p_iq_j \text{ for all } i, j\} .$$

If there is no selection, i.e. if all birth- and deathrates are equal to 1 and d respectively, then (20.35) reduces to

$$\dot{x}_{ij} = -rD_{ij} \tag{20.37}$$

with

$$D_{ij} = x_{ij} - p_iq_j . \tag{20.38}$$

Then

$$\dot{p}_i = \sum_j \dot{x}_{ij} = -r\sum_j D_{ij} = 0$$

and similarly $\dot{q}_j = 0$. Thus recombination does not alter the gene frequencies (as is biologically obvious). Furthermore,

$$\dot{D}_{ij} = -rD_{ij} .$$

Therefore the linkage disequilibrium terms all tend to zero and the population state converges to the Wright manifold, where genes at the two loci are independently distributed.

Exercise 20.5.1 Show that the entropy $H(\mathbf{x}) = -\sum_{i,j} x_{ij} \log x_{ij}$ is also a Lyapunov function for the recombination equation (20.37).

Exercise 20.5.2 Analyse in a similar way the discrete time version of (20.37), showing that $x'_{ij} = x_{ij} - rD_{ij}$ and $D'_{ij} = (1 - r)D_{ij}$.

Since the convergence to the Wright manifold is at a uniform exponential rate, namely $-r$, general theorems show that for weak selection, all orbits converge to a manifold near $\mathbf{W}$, where the state is in *quasi linkage equilibrium*.

Furthermore, on this invariant manifold, the dynamics is a small perturbation of a gradient system.

At the other extreme, if $r = 0$ (which means no recombination or very tight linkage), then (20.35) can be interpreted as a one-locus selection equation for nm 'alleles' $A_i B_j$. Thus the selection part of (20.35) is again a gradient system with respect to the Shahshahani metric on S_{nm}.

Exercise 20.5.3 Assume that in (20.35) fitnesses are *additive* among the loci, i.e. that $m_{ij,kl} = a_{ik} + b_{jl}$. Show that the gene frequencies satisfy

$$\dot{p}_i = p_i \left(\sum_j a_{ij} p_j - \bar{m} \right) + \sum_j (B\mathbf{q})_j x_{ij}$$

$$\dot{q}_j = q_j \left(\sum_i b_{ij} q_i - \bar{m} \right) + \sum_i (A\mathbf{p})_i x_{ij}$$

with $\bar{m} = \bar{a} + \bar{b}$. The Wright manifold is invariant. Observe that the mean fitness depends only on the gene frequencies. Show by an argument like that in section 18.9 that the mean fitness increases. Show that on the invariant sets of constant mean fitness, entropy increases. Conclude that each solution converges to some rest point where both the selection and the recombination terms vanish.

Exercise 20.5.4 Assume that fitness is *multiplicative* in (20.31): $w_{ij,kl} = a_{ik} b_{jl}$. Show that the Wright manifold $\mathbf{W}$ is an invariant manifold, i.e. linkage equilibrium is preserved. On $\mathbf{W}$, the gene frequencies evolve independently on the two loci, and mean fitness increases. Is $\mathbf{W}$ necessarily attracting?

In the case of only two alleles at each of the two loci, the difference equation (20.31) reduces to (18.21) and the differential equation (20.35) to

$$\dot{x}_i = x_i \left(\sum_{j=1}^{4} m_{ij} x_j - \bar{m} \right) + \varepsilon_i b D \qquad i = 1, \dots, 4 \qquad (20.39)$$

where we adopt the notation of section 18.7: the birthrate of the double heterozygote is denoted by b and $\varepsilon_1 = -\varepsilon_2 = -\varepsilon_3 = \varepsilon_4 = -1$, $D = x_1 x_4 - x_2 x_3$.

Exercise 20.5.5 Consider the recombination equation $\dot{x}_i = \varepsilon_i D$. Show that the triangular integrability conditions (19.23) are satisfied on the Wright manifold $\mathbf{W}$, but not on the whole simplex S_4.

This implies that the recombination equation is not a Shahshahani gradient. Theorem 20.4.1 shows then that there exists a 4×4 selection matrix such that (20.39) has periodic orbits. Thus the two-locus two-allele system (20.39) is not gradient-like in general.

Exercise 20.5.6 Add a recombination term of the form $\varepsilon_i \rho D$ (for $i = 1, 2, 3, 4$) to the replicator equation (10.28). Show that in this case, the Wright manifold $\mathbf{W} = W_1$ attracts all orbits in int S_4, while the other W_K are no longer invariant. (Hint: show that $Z = (x_2 x_3)^{-1} x_1 x_4$ converges to 1.) The dynamics on W_1 is the same as without recombination, and can be written as

$$
\begin{aligned}
\dot{x} &= x(1 - x)(S - (R + S)y) \\
\dot{y} &= y(1 - y)(s - (r + s)x),
\end{aligned}
$$

where $x = x_1 + x_3$ and $y = x_1 + x_2$. This is just (10.15–16).

20.6 Notes

Sections 20.3 and 20.4 are based on Hofbauer (1985). Theorem 20.4.1 is due to Akin (1979). Theorem 20.3.4 is essentially due to Hadeler (1981). For further results on selection–mutation models we recommend the recent survey of Bürger (1997). Baake and Wiehe (1996) consider mutation in sequence space models. For game dynamics with mutation see Bomze and Bürger (1995) and Boylan (1994). Section 20.5 only scratches the vast and complex area of multilocus systems. We refer to Lyubich (1992), Nagylaki (1992), and Svirezhev and Passekov (1990). Akin (1982, 1983, 1987) and Hastings (1981) demonstrated that even in the simplest case of two alleles at two loci, stable limit cycles can occur. Brunovský *et al.* (1997) show that this cannot occur if selection is weak.

21

Fertility selection

If we allow for sex-dependent viabilities and different fertilities of mating pairs, one-locus genetics becomes rather complicated. We investigate special cases (two alleles, multiplicative or additive fertilities etc.), and show that the classical selection equation based on the Hardy–Weinberg equilibrium holds only under restricted conditions.

21.1 The fertility equation

All selection models treated so far have relied on two basic assumptions: random mating (or rather random union of genes) resulting in Hardy–Weinberg proportions for the zygote population, and selection by means of viability differences in the genotypes. With these assumptions we could express the selection equations in terms of gene frequencies rather than genotype frequencies.

However, natural selection generally works in a more complex way. Random mating is not the rule in nature (tall women, for example, usually prefer tall men), and some types of pairing may be more fertile than others. In this chapter we will consider selection models taking into account nonrandom mating, different fertilities of mating pairs and different viabilities of genotypes.

Let us study the case of one locus with n alleles $A_1, \ldots, A_n$. We consider a life cycle from zygote to adult age (with sex-dependent selection), followed by (possibly nonrandom) mating with different fertilities and production of the next zygote generation. Denoting the frequencies of $A_i A_i$ at the zygote stage by x_{ii} and those of $A_i A_j$ ($i \neq j$) by $2x_{ij}$, one obtains as the frequency of allele A_i

$$x_i = \sum_{j=1}^{n} x_{ij} \ . \tag{21.1}$$

We let $h(ij, rs)$ be the probability for a mating of an A_iA_j male with an A_rA_s female, multiplied with the fecundity (i.e. number of progeny) of such a pairing. Furthermore we denote by $m(ij)$ (resp. $f(ij)$) the viability (i.e. the probability of surviving up to maturity) of an A_iA_j male (resp. female). An A_iA_j zygote results from either an $A_iA_r \times A_jA_s$ or an $A_jA_s \times A_iA_r$ mating, with arbitrary r, s. This gives the frequencies of the genotypes A_iA_j in the next zygote generation:

$$x'_{ij} = \bar{F}^{-1} \sum_{r,s=1}^{n} \frac{1}{2} \left[h(ir, js)m(ir)f(js) + h(js, ir)m(js)f(ir) \right] x_{ir} x_{js} \ . \tag{21.2}$$

With

$$F(ir, js) = h(ir, js)m(ir)f(js) \tag{21.3}$$

and

$$f(ir, js) = \frac{1}{2} \left[F(ir, js) + F(js, ir) \right] \ , \tag{21.4}$$

(21.2) can be interpreted as a *fertility selection equation*

$$x'_{ij} = \bar{F}^{-1} \sum_{r,s=1}^{n} f(ir, js)x_{ir} x_{js} \ . \tag{21.5}$$

The normalization factor

$$\bar{F} = \sum_{i,j,k,l} f(ij, kl)x_{ij}x_{kl} \tag{21.6}$$

is the *mean fertility* of the population.

The corresponding *continuous time fertility equation* is obtained as usual by replacing $x'_i - x_i$ by $\dot{x}_i$ and multiplying by the common positive factor $\bar{F}$:

$$\dot{x}_{ij} = \sum_{r,s=1}^{n} f(ir, js)x_{ir} x_{js} - x_{ij}\bar{F} \ . \tag{21.7}$$

The state space of these equations is the simplex S_m with $m = n(n + 1)/2$, which is forward invariant for both (21.5) and (21.7).

21.2 Two alleles

If we denote the frequencies of the genotypes A_1A_1, A_1A_2 and A_2A_2 by x, y, z respectively, the two-allelic fertility equation (21.5) transforms into

$$x' = \bar{F}^{-1}\left[F_{11}x^2 + \frac{1}{2}(F_{12} + F_{21})xy + \frac{1}{4}F_{22}y^2\right]$$

$$y' = \bar{F}^{-1}\left[\frac{1}{2}F_{22}y^2 + \frac{1}{2}(F_{12} + F_{21})xy + \frac{1}{2}(F_{23} + F_{32})yz + (F_{31} + F_{13})xz\right]$$

$$z' = \bar{F}^{-1}\left[F_{33}z^2 + \frac{1}{2}(F_{23} + F_{32})yz + \frac{1}{4}F_{22}y^2\right] \tag{21.8}$$

and for continuous time into

$$
\begin{aligned}
\dot{x} &= f_{11}x^2 + f_{12}xy + \tfrac{1}{4}f_{22}y^2 - x\bar{F} \\
\dot{y} &= \tfrac{1}{2}f_{22}y^2 + f_{12}xy + f_{23}yz + 2f_{13}xz - y\bar{F} \\
\dot{z} &= f_{33}z^2 + f_{32}yz + \tfrac{1}{4}f_{22}y^2 - z\bar{F}
\end{aligned} \tag{21.9}
$$

with symmetrized fertility coefficients $f_{ij} = \frac{1}{2}(F_{ij} + F_{ji})$ and

$$\bar{F}(x, y, z) = f_{11}x^2 + 2f_{12}xy + f_{22}y^2 + 2f_{13}xz + 2f_{23}yz + f_{33}z^2 .$$

Since $x + y + z = 1$, this is a two-dimensional dynamical system. The best way to analyse it is to use the variables

$$u = \frac{x}{y} \qquad v = \frac{z}{y} . \tag{21.10}$$

Then (21.9) transforms — after dividing both equations by y — into

$$
\begin{aligned}
\dot{u} &= (f_{11} - f_{12})u^2 + \left(f_{12} - \tfrac{1}{2}f_{22}\right)u + \tfrac{1}{4}f_{22} - f_{32}uv - 2f_{13}u^2v \\
\dot{v} &= (f_{33} - f_{32})v^2 + \left(f_{32} - \tfrac{1}{2}f_{22}\right)v + \tfrac{1}{4}f_{22} - f_{12}uv - 2f_{13}uv^2 .
\end{aligned} \tag{21.11}
$$

This system is *competitive*: the off-diagonal terms of the Jacobian

$$\frac{\partial \dot{u}}{\partial v} = -f_{32}u - 2f_{13}u^2 \qquad \frac{\partial \dot{v}}{\partial u} = -f_{12}v - 2f_{13}v^2$$

are negative on the whole state space $\mathrm{int}\,\mathbb{R}_+^2$. This excludes by theorem 4.1.1 the possibility of periodic orbits.

In fact (21.11) is even a gradient system. Indeed, let us write it in the concise form

$$
\begin{aligned}
\dot{u} &= a(u) - vb(u) \\
\dot{v} &= c(v) - ud(v)
\end{aligned} \tag{21.12}
$$

with $b(u), d(v) > 0$ for $u, v > 0$. We define a Riemannian metric on $\mathrm{int}\,\mathbb{R}_+^2$ by

$$\langle(\xi_1, \eta_1), (\xi_2, \eta_2)\rangle_{(u,v)} = \frac{1}{b(u)}\xi_1\xi_2 + \frac{1}{d(v)}\eta_1\eta_2 . \tag{21.13}$$

Then

$$\langle(\dot{u},\dot{v}),(\xi,\eta)\rangle_{(u,v)} = \left(\frac{a(u)}{b(u)}-v\right)\xi + \left(\frac{c(v)}{d(v)}-u\right)\eta$$

$$= \frac{\partial V}{\partial u}\xi + \frac{\partial V}{\partial v}\eta$$

with

$$V(u,v) = \int \frac{a(u)}{b(u)}du - uv + \int \frac{c(v)}{d(v)}dv . \qquad (21.14)$$

Thus (21.12) is the gradient of the potential V with respect to the Riemannian metric (21.13). Unfortunately, the explicit expression for V is rather complicated, and does not seem to allow a meaningful biological interpretation.

Exercise 21.2.1 Compute $V(u,v)$ explicitly.

Exercise 21.2.2 Determine the behaviour of the fertility equations (21.8) and (21.9), if there are no fertility differences (i.e. all $F_{ij}=1$).

Exercise 21.2.3 For which fertility coefficients f_{ij} is the Hardy–Weinberg manifold $y^2 = 4xz$ invariant under (21.8) resp. (21.9)?

Exercise 21.2.4 Study the *symmetric fertility equation* (21.9) with $f_{11}=f_{33}=a$ and $f_{12}=f_{21}=f_{23}=f_{32}=b$, $f_{13}=f_{31}=c$, $f_{22}=4d$.

(a) Show that $x = z$ is an invariant line which contains up to three fixed points. Determine their stability properties.
(b) Show that a pair of *asymmetric* fixed points (with $x \neq z > 0$) exists if and only if $0 < [2d(a+c)-ab]^{-1}\left|2(a-b)\sqrt{ad}\right| < 1$.
(c) Show that at an asymmetric fixed point, the Jacobian has negative trace and determinant $a(b-a)(x-z)^2$.
(d) If $a > b$ then the corners corresponding to pure homozygote populations are stable, and the asymmetric fixed points are saddles (whenever they exist).
(e) If $a < b$ then the corners are saddle points and the asymmetric fixed points (if they exist) are stable. Show that uv^{-1} is a Dulac function for (21.11) in this case.
(f) Sketch all possible (robust) phase portraits.

Exercise 21.2.5 Show that the symmetric difference equation (21.8) gives rise to a period-doubling bifurcation on the line of symmetry $x = z$. Find parameter values a, b, c, d for which there is a stable orbit of period two, while none of the fixed points in S_3 is stable.

Exercise 21.2.6 Investigate the general two-allele equation (21.9).

(a) Show that there may be up to 5 interior fixed points.
(b) If one of the homozygotes is stable and the other unstable, then there are at most 4 interior fixed points. In generic situations their number is even. (Hint: adapt the index theorem 13.4.1 to this situation.)

21.3 Multiplicative fertility

For more than two alleles the fertility equations (21.5) and (21.7) have not yet been analysed in their general form. We will restrict our attention to the special case when the fertility of a mating type can be split up into the contributions of the male and the female part. So let us assume that in (21.3)

$$F(ij, kl) = m_{ij} f_{kl} . \tag{21.15}$$

We can then assign an average fertility to each allele A_i, namely

$$M(i) = \sum_{j=1}^{n} m_{ij} x_{ij} \tag{21.16}$$

in the male population, and

$$F(i) = \sum_{j=1}^{n} f_{ij} x_{ij} \tag{21.17}$$

in the female population. The fertility equations (21.5) and (21.7) take the form

$$x'_{ij} = \frac{M(i)F(j) + M(j)F(i)}{2\bar{F}} \tag{21.18}$$

and

$$\dot{x}_{ij} = \frac{1}{2}\left[M(i)F(j) + M(j)F(i)\right] - x_{ij}\bar{F} \tag{21.19}$$

with

$$\bar{F} = \sum_{i,j=1}^{n} M(i)F(j) = MF \tag{21.20}$$

where

$$M = \sum_{i=1}^{n} M(i) \quad \text{and} \quad F = \sum_{i=1}^{n} F(i) . \tag{21.21}$$

If male and female contributions are equal, i.e. $m_{ij} = f_{ij}$, then we have $M(i) = F(i)$, $M = F$, and (21.18) and (21.19) reduce to

$$x'_{ij} = M^{-2}M(i)M(j) \tag{21.22}$$

$$\dot{x}_{ij} = M(i)M(j) - x_{ij}M^2 . \tag{21.23}$$

In the discrete time model, genotype frequencies are in Hardy–Weinberg equilibrium after one generation. Indeed, the gene frequencies $x_i = \sum_{j=1}^n x_{ij}$ satisfy

$$x'_i = \sum_{j=1}^n x'_{ij} = \frac{M(i)}{M} \sum_{j=1}^n \frac{M(j)}{M} = \frac{M(i)}{M} , \tag{21.24}$$

so that

$$x'_{ij} = x'_i x'_j .$$

As soon as Hardy–Weinberg holds, (21.24) simplifies to

$$x'_i = x_i \frac{\sum m_{ij} x_j}{\sum m_{kl} x_k x_l} . \tag{21.25}$$

The dynamics of the fertility equation (21.5) with sex-independent multiplicative fertilities is easy to describe: *after one generation the state of the population reaches the Hardy–Weinberg manifold $x_{ij} = x_i x_j$ and follows subsequently the classical selection equation* (18.5) *with viability parameters m_{ij}. In particular the mean fitness $M = \sum m_{ij} x_{ij}$ increases monotonically after the first step.*

Exercise 21.3.1 Show that the differential equation (21.23) has all fixed points on the Hardy–Weinberg manifold, but that this manifold is in general not invariant. However, (21.23) is still equivalent to (19.2). (Hint: use the variables $X_i = M^{-1}M(i)$ instead of x_i.)

In the general case of different male and female contributions, (21.18) cannot be reduced to the classical equation on S_n. But still, the equations can be considerably simplified by introducing the variables

$$X_i = \frac{M(i)}{M} \quad \text{and} \quad Y_i = \frac{F(i)}{F} . \tag{21.26}$$

Indeed, X_i' is proportional to

$$\sum_j m_{ij} x_{ij} = \sum_j m_{ij} \frac{M(i)F(j) + M(j)F(i)}{2MF} = \frac{1}{2}\left(X_i \sum_j m_{ij} Y_j + Y_i \sum_j m_{ij} X_j \right) .$$

Hence (21.18) is equivalent to the following difference equation on $S_n \times S_n$:

$$
\begin{aligned}
X_i' &= (2\bar{m})^{-1}(X_i \textstyle\sum_{j=1}^n m_{ij}Y_j + Y_i \sum_{j=1}^n m_{ij}X_j) \\
Y_i' &= (2\bar{f})^{-1}(X_i \textstyle\sum_{j=1}^n f_{ij}Y_j + Y_i \sum_{j=1}^n f_{ij}X_j)
\end{aligned}
\tag{21.27}
$$

with

$$
\bar{m} = \sum_{i,j=1}^n m_{ij}X_iY_j \quad \text{and} \quad \bar{f} = \sum_{i,j=1}^n f_{ij}X_iY_j \; .
$$

In fact we do not need these computations to see that the fertility equation reduces to (21.27) for multiplicative fertility rates. Comparing with (21.3), we see that the case of multiplicative fertilities (21.15) is biologically equivalent to a pure viability selection model (viability meaning survival probability), without any fertility differences but with sex-dependent viability coefficients. Proceeding as in the derivation of the classical selection model (18.5), considering the gene frequencies X_i and Y_i of A_i in the adult gene pool one arrives immediately at (21.27). Since the x_{ij} in the fertility equation count genotype frequencies at the *zygote* stage, the *adult* gene frequencies X_i and Y_i are related to them by (21.26), (21.16), (21.17) and (21.21).

Unfortunately not much is known about the *two-sex selection equation* (if $n \geq 3$), nor about its continuous counterpart

$$
\begin{aligned}
\dot{X}_i &= \tfrac{1}{2}(X_i \textstyle\sum_j m_{ij}Y_j + Y_i \sum_j m_{ij}X_j) - X_i\bar{m} \\
\dot{Y}_i &= \tfrac{1}{2}(Y_i \textstyle\sum_j f_{ij}X_j + X_i \sum_j f_{ij}Y_j) - Y_i\bar{f} \; .
\end{aligned}
\tag{21.28}
$$

A particularly important open problem is whether (21.28) is a gradient (or gradient-like) system.

Exercise 21.3.2 Analyse (21.18), (21.19) or the two-sex equations (21.27) and (21.28) for $n = 2$ and compare the results with the one-sex case (see section 18.5).

(a) The homozygote A_1A_1 is stable if and only if $\frac{m_{12}}{m_{11}} + \frac{f_{12}}{f_{11}} \leq 2$.
(b) Show that there are at most 3 interior fixed points.
(c) Sketch all possible phase portraits of the differential equation.

Exercise 21.3.3 In the two-sex equations, with $m_{ij} = f_{ij}$, the subspace where $\mathbf{x} = \mathbf{y}$ is invariant and globally attracting.

Exercise 21.3.4 $f_{ij} = 1$ means that selection acts only in one sex, whereas $f_{ij} = am_{ij} + b$ (with $a > 0$) means that selection acts in the same way in both sexes but on a different scale. Show that in these cases $\mathbf{x} = \mathbf{y}$ holds at all rest points and that all eigenvalues of the linearization are real. Equation (21.28) seems a good candidate for a gradient system.

Exercise 21.3.5 Simplify the differential equation (21.19) using (21.26) as new variables. Does this lead to (21.28)?

21.4 Additive fertility

The assumption of multiplicative fertilities works out much better for the difference equation than for the differential equation. It turns out that an additive splitting is a more appropriate simplification for the continuous time case. So let us assume that in (21.3)

$$F(ij, kl) = m_{ij} + f_{kl} . \tag{21.29}$$

An important special case of (21.29) is when one of the two sexes has no influence on the fertility at all. (21.29) implies that the symmetrized fertility coefficients are of the form

$$f(ij, kl) = F_{ij} + F_{kl} \tag{21.30}$$

with

$$F_{ij} = \frac{m_{ij} + f_{ij}}{2} . \tag{21.31}$$

The differential equation (21.7) then simplifies to $\dot{x}_{ij} = \sum (F_{ir} + F_{js})x_{ir}x_{js} - x_{ij}\bar{F}$, i.e. to

$$\dot{x}_{ij} = x_i F(j) + x_j F(i) - 2x_{ij}\bar{F} . \tag{21.32}$$

Here x_i is again the frequency (21.1) of the allele A_i, while

$$F(i) = \sum_{k=1}^{n} F_{ik} x_{ik} \tag{21.33}$$

is its average fertility in the population, and

$$\bar{F} = 2F = 2\sum_{i=1}^{n} F(i) = 2\sum_{i,j=1}^{n} F_{ij}x_{ij} \tag{21.34}$$

is the mean fertility of the whole population. For the gene frequency x_i we obtain

$$\dot{x}_i = F(i) - x_i F . \tag{21.35}$$

Then

$$(x_{ij} - x_i x_j)^{\cdot} = x_i F(j) + x_j F(i) - 2x_{ij}F - x_j F(i) + x_i x_j F - x_i F(j) + x_i x_j F$$
$$= -(x_{ij} - x_i x_j)2F .$$

Since $F > 0$, the expression $x_{ij} - x_i x_j$ converges to zero and in the limit $t \to +\infty$ the Hardy–Weinberg relations hold:

$$x_{ij} = x_i x_j . \tag{21.36}$$

On the Hardy–Weinberg manifold we obtain

$$\dot{x}_i = F(i) - x_i F = x_i \left(\sum_k F_{ik} x_k - \sum_{r,s} F_{rs} x_r x_s \right) . \tag{21.37}$$

This is the continuous time selection equation (19.2) with selection parameters F_{ij}. Hence *the mean fertility $\bar{F}$ increases monotonically*.

Exercise 21.4.1 Show that the difference equation (21.5) with additive fertilities (21.30) is equivalent to a two-sex selection equation with no selection in one sex. The Hardy–Weinberg manifold is not invariant, however.

21.5 The fertility–mortality equation

The discrete time model in (21.5) was derived quite carefully, but the differential equation (21.7) was obtained in a rather cavalier way. A biologically more satisfactory way would be to consider a continuous time model for overlapping generations with $x_{ij}(t)$ as frequencies of the $A_i A_j$ genotypes at time t. During a small time interval Δt, the frequency $x_{ij}(t)$ will increase due to births by $\sum_{r,s} f(ir, js)x_{ir}x_{js}\Delta t$, with $f(ir, js)$ again measuring the fertility of an $A_i A_r \times A_j A_s$ mating, and it will decrease due to deaths by $d_{ij}x_{ij}\Delta t$, with d_{ij} as the death rate of the genotype $A_i A_j$. This leads to

$$\dot{x}_{ij} = \sum_{r,s=1}^{n} f(ir, js)x_{ir}x_{js} - d_{ij}x_{ij} - x_{ij}\bar{F} . \tag{21.38}$$

Here

$$\bar{F} = \sum_{i,j,k,l} f(ij, kl)x_{ij}x_{kl} - \sum_{i,j} d_{ij}x_{ij}$$

ensures again that (21.38) maintains the relation $\sum_{i,j} x_{ij} = 1$. Thus the natural continuous time counterpart corresponding to the discrete time model (21.5) involves both different fertility and mortality rates, the latter corresponding to the viabilities of the genotypes.

Exercise 21.5.1 Show that adding a constant to the death rates d_{ij} does not change the equation (21.38). In particular, if all death rates are equal, $d_{ij} = d$, then (21.38) reduces to the pure fertility equation (21.7).

Exercise 21.5.2 Analyse the pure mortality equation with all $f(ij, kl) = 1$.

(a) Show that the Hardy–Weinberg manifold is invariant if and only if the death rates are additive, i.e. $d_{ij} = d_i + d_j$.
(b) If there exists an interior equilibrium with Hardy–Weinberg proportions then all death rates are equal: $d_{ij} = d$.
(c) Analyse the case of two alleles and compare with the classical selection equation (19.2).

Exercise 21.5.3 If (21.38) leaves the Hardy–Weinberg manifold invariant, then the fertilities are additive (cf. (21.30)) and the death rates are also additive.

Exercise 21.5.4 Show the following

(a) If the fertilities are additive and there exists an interior equilibrium in Hardy–Weinberg proportions, then the death rates are additive as well.
(b) If both fertilities and mortalities are additive, and fertilities are positive, then each rest point of (21.38) is in Hardy–Weinberg proportions.
(c) The conclusion in (b) need not hold if some fertilities are zero. (Hint: choose $n = 2$, $f_{12} = f_{21} = 0$.)

Assume now that both fertility and mortality rates are additive, i.e. that

$$f(ij, kl) = f_{ij} + f_{kl} \quad \text{and} \quad d_{ij} = d_i + d_j . \qquad (21.39)$$

Then a similar computation to that in section 21.4 shows that the Hardy–Weinberg relations $x_{ij} = x_i x_j$ are invariant, and that the flow on the Hardy–Weinberg manifold is given by

$$\dot{x}_i = x_i \left[\sum f_{ik} x_k - d_i + \bar{d} - \bar{f} \right] \qquad (21.40)$$

with

$$\bar{d} = \sum d_k x_k \quad \text{and} \quad \bar{f} = \sum f_{ij} x_i x_j .$$

(21.40) is equivalent to the selection equation (19.2) with the Malthusian parameters $m_{ij} = f_{ij} - d_i - d_j$, with f_{ij} corresponding to the birth rate and $d_i + d_j$ to the death rate of $A_i A_j$.

Now remember that the derivation of the continuous selection equation (19.2) relied on the assumption of a Hardy–Weinberg equilibrium. Strictly speaking this assumption was not justified at all. A more satisfactory approach is to start with genotype frequencies and *prove* that Hardy–Weinberg holds. The above derivation (in particular, exercise 21.5.3) shows exactly when this approach works:

(a) the fertilities of a mating have to split up into additive contributions of each genotype; and (more seriously)

(b) the genes A_i and A_j have to contribute additively to the death rate of the genotype A_iA_j.

The moral is that the classical selection equation (19.2) is valid only under the additional assumption (b). Of course, this does not alter its value as the basic continuous time model in population genetics, in particular as there is no understanding of the dynamic behaviour of the fertility equation (21.7) or the fertility–mortality equation (21.38) for three or more alleles.

In contrast to the result in section 21.2, the fertility–mortality equation gives rise to periodic orbits even for two alleles:

Exercise 21.5.5 Show that after adding to the right hand side of (21.9) the mortality terms $-d_1x$, $-d_2y$ and $-d_3z$, a Hopf bifurcation can occur. (Hint: try for example $f_{11} = 14.55$, $f_{12} = 0.25$, $f_{22} = 14.5$, $f_{13} = 0.02$, $f_{23} = 3.1$, $f_{33} = 2.89$, $d_1 = 0.8$, $d_2 = 0.9$, $d_3 = 0.05$. The corners $(1, 0, 0)$ and $(0, 0, 1)$ are asymptotically stable. There are two saddles in S_3 and a centre $(\frac{1}{8}, \frac{1}{4}, \frac{5}{8})$. Use f_{22} as the bifurcation parameter.)

21.6 Notes

Bodmer (1965) has analysed the case of multiplicative fertilities and its relation to the two-sex equation, and studied the two-allele model. An excellent survey on discrete time selection models was given by Pollak (1979). The symmetric two-allele case (exercises 21.2.4–5) was analysed by Hadeler and Libermann (1975) and Butler *et al.* (1982). A detailed analysis of the two-sex equation, in particular the case $m_{ij} + f_{ij} = 1$ which arises in sex ratio models, is given in Karlin and Lessard (1986). The general continuous time model (21.38) is derived in Nagylaki (1992). Hadeler and Glas (1983) showed that the fertility equation (21.9) is gradient-like. Szucs (1993) proved convergence for the fertility–mortality equation (21.38) in the case of two alleles for additive fertilities, and Koth and Kemler (1986) found the stable limit cycles in exercise 21.5.5. Exercise 21.5.4 is from Akin and Szucs (1994), who give also sufficient conditions for the stability of the Hardy–Weinberg manifold.

22

Game dynamics for Mendelian populations

Game theory provides a tool for studying frequency-dependent selection. In particular, whenever the replicator dynamics is a Shahshahani gradient, then so is the corresponding selection equation. For games with two pure strategies, game theory and population genetics agree both for the continuous time model and for the discrete time model. For more than two strategies and more than three alleles, open problems remain. But under appropriate conditions, an ESS corresponds to an attracting set of rest points in the gene space, and long-term stability implies evolutionary stability.

22.1 Strategy and genetics

So far, we have not specified the connection between game-theoretical modelling and population genetics. The underlying idea is, of course, that genotypes specify strategies as behavioural phenotypes. But the replicator dynamics studied in part II of this book did not take the intricacies of Mendelian inheritance into account. It was firmly based on the implicit assumption of asexual reproduction. This simplified the analysis considerably; moreover, in the absence of detailed knowledge on the genetic background of a behavioural trait, all corresponding assumptions are bound to be arbitrary, and may obfuscate the essential aspects of the model. On the other hand, the neglect of Mendelian inheritance entails a serious loss of realism. For the Battle of the Sexes or the sex ratio game, an asexual model may seem paradoxical.

It is obvious that a genetic mechanism may prevent the establishment of an ESS. For example, if the corresponding phenotype is realized by the heterozygote, but not by the homozygotes, then it can never make up the whole population. But this objection can be levelled against every kind of adaptationist argument, not only against game-theoretical thinking.

We shall presently see that in many cases, strategic and genetic models agree quite well, and that there are close connections between population genetics and game-theoretical modelling.

Let us consider a game with N pure strategies R_1 and R_N, where the payoff F_i for strategy R_i depends only on the average frequencies p_k of the strategies R_k in the population.

Let us furthermore assume that the gene pool contains the alleles A_1 to A_n with frequencies x_1 to x_n, that the frequency of the gene pair (A_i, A_j) is given by $x_i x_j$ according to the Hardy–Weinberg relation, and that such a gene pair yields the phenotype $\mathbf{p}(ij) \in S_N$. The mean strategy mix $\mathbf{p} \in S_N$ in the population is given by

$$\mathbf{p} = \mathbf{p}(\mathbf{x}) = \sum_{ij} x_i x_j \mathbf{p}(ij) \ . \tag{22.1}$$

The frequency of the strategies 'played' by allele A_i is

$$\mathbf{p}^i(\mathbf{x}) = \sum_j x_j \mathbf{p}(ij) = \frac{1}{2} \frac{\partial \mathbf{p}}{\partial x_i} \tag{22.2}$$

(the allele A_i belongs with probability x_j to an (A_i, A_j)- or an (A_j, A_i)-individual playing strategy $\mathbf{p}(ij)$). The frequency-dependent payoff for (A_i, A_j)-individuals is

$$w_{ij} = \mathbf{p}(ij) \cdot \mathbf{F}(\mathbf{p}) \ , \tag{22.3}$$

and the average payoff for allele A_i is accordingly

$$f_i(\mathbf{x}) = \mathbf{p}^i \cdot \mathbf{F}(\mathbf{p}) \ . \tag{22.4}$$

The evolution of the gene pool is given by the selection equation (19.2), which now reads

$$\dot{x}_i = x_i \left[(\mathbf{p}^i - \mathbf{p}) \cdot \mathbf{F}(\mathbf{p}) \right] = x_i \left[f_i(\mathbf{x}) - \bar{f} \right] = \hat{f}_i(\mathbf{x}) \tag{22.5}$$

and is a replicator equation on S_n. This equation is closely related to the replicator equation for the pure strategists,

$$\dot{y}_i = y_i \left[F_i(\mathbf{y}) - \bar{F} \right] = \hat{F}_i(\mathbf{y}) \tag{22.6}$$

on S_N. Indeed, one has the following

Theorem 22.1.1 *If the pure strategist equation* (22.6) *is a Shahshahani gradient with potential* $V(\mathbf{y})$, *then the frequency-dependent selection equation* (22.5) *is a Shahshahani gradient with potential* $\frac{1}{2} V(\mathbf{p}(\mathbf{x}))$.

Proof If $\hat{\mathbf{F}}$ has the Shahshahani potential V, then by section 19.5 there exists a function $G : S_N \to \mathbb{R}$ such that

$$\mathbf{F}(\mathbf{y}) = \text{grad } V(\mathbf{y}) + G(\mathbf{y})\mathbf{1} \qquad (22.7)$$

for all $\mathbf{y} \in \text{int } S_N$. Then

$$f_i(\mathbf{x}) = \mathbf{p}^i \cdot \mathbf{F}(\mathbf{p}) = \frac{1}{2} \cdot 2\mathbf{p}^i \cdot \text{grad } V(\mathbf{p}) + G(\mathbf{p}) .$$

This implies by (22.2)

$$\mathbf{f}(\mathbf{x}) = \frac{1}{2} \text{grad } V(\mathbf{p}(\mathbf{x})) + G(\mathbf{p}(\mathbf{x}))\mathbf{1} . \qquad (22.8)$$

Hence $\hat{\mathbf{f}}$ has the Shahshahani potential $\frac{1}{2}V(\mathbf{p}(\mathbf{x}))$. $\qquad\square$

In particular, if $N = 2$ (or, of course, if $n = 2$), then (22.5) is a Shahshahani gradient.

Exercise 22.1.2 Prove directly that if (22.6) satisfies the triangular integrability condition (19.23), then so does (22.5).

Exercise 22.1.3 If $\mathbf{F}$ is linear and (22.6) is a Shahshahani gradient, i.e. if $\mathbf{F}(\mathbf{x}) = (B + S)\mathbf{x}$, where B is a symmetric $N \times N$ matrix and S a matrix with equal rows, then show that $\mathbf{x} \to \frac{1}{4}\mathbf{p} \cdot B\mathbf{p}$ is a Shahshahani potential for (22.5).

Exercise 22.1.4 If $N = 2$ and $\mathbf{F}$ is linear, i.e. given by a 2×2 matrix A, then a Shahshahani potential is given by

$$V(\mathbf{x}) = \frac{\alpha}{2} \left(\sum_{r,s} x_r x_s p_1(rs) - \frac{a_{22} - a_{12}}{\alpha} \right)^2 \qquad (22.9)$$

(provided $\alpha := a_{11} - a_{21} + a_{22} - a_{12} \neq 0$). If in particular the game admits a mixed ESS, i.e. if $a_{11} - a_{21}$ and $a_{22} - a_{12}$ are both negative, then the strategy mix in the population converges to the ESS 'whenever possible'. Describe the states in the gene pool for which this ESS is obtained.

In the case of two pure strategies and a linear payoff function, this yields a transparent picture of evolution for the continuous-time model (22.5). The population strategy approaches an ESS $\hat{\mathbf{p}}$ as closely as possible under the genetic constraints. Subsequently, mutants can only enter the population if they allow the population strategy to come closer to $\hat{\mathbf{p}}$. Once the population adopts an ESS strategy, only mutants can enter that do not alter the mean strategy of the population. One of the basic questions of ESS theory is to find out how much of this remains valid under more general circumstances.

(In the next section, we show that the result remains valid in the discrete time case too.)

Exercise 22.1.5 Discuss all possible phase portraits if, in addition to the assumptions of exercise 22.1.4, one has $n = 2$. (Hint: draw $p_1(\mathbf{x})$ as a function of x_1. The rest points of (22.5) in $]0,1[$ are the critical points of p_1, and the solutions of $p_1(x_1) = (a_{22} - a_{12})/\alpha$.)

Exercise 22.1.6 Find a common extension of theorems 19.6.1 and 22.1.1.

Exercise 22.1.7 If F is linear and the genetics is dominance free in the sense that $\mathbf{p}(ij) = \frac{1}{2}(\mathbf{p}(ii) + \mathbf{p}(jj))$, then show that (22.5) reduces to the haploid model $\dot{x}_i = x_i(a_i - \bar{f})$ with constant a_i.

22.2 The discrete model for two strategies

The discrete counterpart of (22.5) is, as usual, considerably more difficult to analyse, and we shall only discuss the case of two strategies R_1 and R_2 and linear payoff given by the 2×2 matrix A. We denote by $p(ij)$ (resp. p) the first component of $\mathbf{p}(ij)$ (resp. $\mathbf{p}(\mathbf{x})$), i.e. the frequency of R_1 among the A_iA_j-phenotype (resp. the whole population). Then

$$a_1 = a_{11}p + a_{12}(1 - p) \quad \text{and} \quad a_2 = a_{21}p + a_{22}(1 - p) \qquad (22.10)$$

are the payoffs for R_1 and R_2, while the average payoff for A_iA_j is

$$w_{ij}(\mathbf{x}) = p(ij)a_1 + (1 - p(ij))a_2 . \qquad (22.11)$$

Since the probability that allele A_i 'plays' R_1 (i.e. that it belongs to an R_1-strategist) is

$$p_i = \sum_{j=1}^{n} p(ij)x_j \qquad (22.12)$$

we obtain as the average payoff for allele A_i

$$p_i a_1 + (1 - p_i)a_2 = \sum_{j=1}^{n} x_j w_{ij}(\mathbf{x}) . \qquad (22.13)$$

If we denote by x_i' the frequency of A_i in the next generation, then the assumption that the increase x_i'/x_i of allele A_i is proportional to its payoff leads to the difference equation

$$x_i' = x_i \frac{p_i a_1 + (1 - p_i)a_2}{\bar{p}} \qquad (22.14)$$

where

$$\bar{p} = pa_1 + (1-p)a_2 \tag{22.15}$$

is the average payoff. Equivalently,

$$x_i' = x_i \frac{\sum_j w_{ij}(\mathbf{x})x_j}{\sum_{r,s} w_{rs}(\mathbf{x})x_r x_s}. \tag{22.16}$$

This is the discrete time analogue of (22.5) for $N = 2$.
   If $\mathbf{x}$ is a fixed point of (22.14) then

$$(p_i - p)(a_1 - a_2) = 0 \tag{22.17}$$

whenever $x_i > 0$. Thus we may distinguish two types of fixed points:

(a) *strategic fixed points* for which $a_1 = a_2$, i.e. where the payoffs for the two strategies are equal;
(b) *genetic fixed points* for which $p_i = p$ whenever $x_i > 0$. For those fixed points, all alleles which are present 'play' the same strategy mixture. Genetic fixed points are exactly the fixed points of the classical selection equation (18.5) with matrix $p(ij)$.

In the following discussion, we shall only consider the generic case $\alpha :=$ $a_{11} - a_{21} + a_{22} - a_{12} \neq 0$. From

$$a_1 - a_2 = \alpha p - (a_{22} - a_{12}) \tag{22.18}$$

we can easily deduce that $(a_1 - a_2)^2$ is a Lyapunov function for the continuous case (cf. (22.9)). This suggests the following theorem.

Theorem 22.2.1 $(a_1 - a_2)^2$ *is a strict Lyapunov function for* (22.14). *Furthermore,* $a_1 - a_2$ *does not change sign.*

Proof We shall first show that (22.14) is *locally adaptive*, in the sense that the frequency p of strategy R_1 increases if and only if its payoff a_1 is larger than a_2. Indeed, since $w_{ij}(\mathbf{x}) = w_{ji}(\mathbf{x})$, we can apply theorem 18.4.1 to get

$$\sum_{i,j} x_i' x_j' w_{ij}(\mathbf{x}) \geq \sum_{i,j} x_i x_j w_{ij}(\mathbf{x}) \tag{22.19}$$

which means by (22.11) that

$$a_2 + (a_1 - a_2)\sum_{i,j} x_i' x_j' p(ij) \geq a_2 + (a_1 - a_2)\sum_{i,j} x_i x_j p(ij). \tag{22.20}$$

Since

$$p = \sum_{i,j} x_i x_j p(ij) \qquad (22.21)$$

(22.20) yields

$$(a_1 - a_2)(p' - p) \ge 0 \qquad (22.22)$$

which corresponds to local adaptivity.

Case 1: $\alpha > 0$. Since $a_1 - a_2$ has the same sign as $p' - p$, (22.18) implies that $a_1' - a_2' \ge a_1 - a_2$ if $a_1 - a_2 \ge 0$, and $a_1' - a_2' \le a_1 - a_2$ if $a_1 - a_2 \le 0$. Hence $(a_1 - a_2)^2$ is an increasing Lyapunov function.

Case 2: $\alpha < 0$. Let us assume (without loss of generality) that $a_1 - a_2 > 0$. Then $a_1' - a_2' \le a_1 - a_2$ follows from local adaptivity as before, but we also want to show that

$$a_1' - a_2' \ge 0 \qquad (22.23)$$

and this takes some more work. (22.21) and (22.14) imply

$$p \;\doteq\; \bar{p}^{-2} \sum_{i,j} p(ij) x_i x_j [p_i a_1 + (1 - p_i) a_2][p_j a_1 + (1 - p_j) a_2] \quad (22.24)$$

$$= \bar{p}^{-2} \Big[(a_1 - a_2)^2 \sum p(ij) x_i x_j p_i p_j +$$
$$+ (a_1 - a_2) a_2 \sum p(ij) x_i x_j (p_i + p_j) + p a_2^2 \Big]. \qquad (22.25)$$

(22.18), when applied to $a_1' - a_2'$, therefore yields

$$\bar{p}^2 (a_1' - a_2') = \alpha p' \bar{p}^2 + (a_{12} - a_{22})[p^2(a_1 - a_2)^2 + 2p(a_1 - a_2) a_2 + a_2^2]$$
$$= (a_1 - a_2) R \qquad (22.26)$$

where

$$R = \alpha \big[(a_1 - a_2) \sum p(ij) x_i x_j p_i p_j + a_2 \sum p(ij) x_i x_j (p_i + p_j) \big] + a_2^2$$
$$+ (a_{12} - a_{22}) [p^2(a_1 - a_2) + 2p a_2]$$
$$= \alpha \big[a_1 \sum p(ij) x_i x_j p_i p_j - a_2 \sum p(ij) x_i x_j (-p_i - p_j + p_i p_j) \big]$$
$$+ (a_1 - a_2 - \alpha p) [p^2(a_1 - a_2) + 2p a_2] + a_2^2$$
$$= \alpha \big[a_1 \sum p(ij) x_i x_j p_i p_j + a_2 \sum (1 - p(ij)) x_i x_j (1 - p_i)(1 - p_j) \big]$$
$$- \alpha [p^3 a_1 + (1 - p)^3 a_2] + \bar{p}^2. \qquad (22.27)$$

Since $0 \le p(ij) \le 1$, we have

$$\sum p(ij) x_i x_j p_i p_j \le \Big(\sum p_i x_i \Big)^2 \le p^2 \qquad (22.28)$$

and the analogous estimate by $(1-p)^2$ for the second sum in (22.27). Thus

$$
\begin{aligned}
R &\geq \alpha\big[\left(p^2-p^3\right)a_1 + \left((1-p)^2-(1-p)^3\right)a_2\big]\bar{p}^2 \\
&= \alpha p(1-p)[pa_1+(1-p)a_2]\bar{p}^2 \\
&= \bar{p}[\bar{p}+\alpha p(1-p)] \qquad\qquad\qquad\qquad\qquad (22.29) \\
&= \bar{p}[a_{11}p+a_{22}(1-p)] \geq 0 . \qquad\qquad\qquad (22.30)
\end{aligned}
$$

Hence (22.26) proves (22.23). Thus $a_1 - a_2$ converges monotonically to 0, without changing sign. Since every point with $a_1 = a_2$ is a fixed point, $(a_1-a_2)^2$ is a strict Lyapunov function. $\qquad\qquad\qquad\square$

Exercise 22.2.2 Discuss the nongeneric case $\alpha = a_{11} - a_{21} + a_{22} - a_{12} = 0$.

Exercise 22.2.3 Under what conditions will the strategy mix of the population converge to an ESS?

Exercise 22.2.4 Discuss the case of $n = 2$ alleles completely.

22.3 Genetics and ESS

Now consider again the genetic model (22.5) for an evolutionary game with linear payoff

$$
\dot{x}_i = x_i(\mathbf{p}^i - \mathbf{p})\cdot A\mathbf{p} . \qquad\qquad\qquad (22.31)
$$

For the frequencies $\mathbf{p}$ of the strategies we obtain

$$
\dot{\mathbf{p}} = 2\sum \mathbf{p}^i \dot{x}_i = 2\sum x_i \mathbf{p}^i[(\mathbf{p}^i-\mathbf{p})\cdot A\mathbf{p}] = 2C(\mathbf{x})A\mathbf{p} \qquad (22.32)
$$

where the component $c_{kl}(\mathbf{x})$ of the matrix $C(\mathbf{x})$ is given by

$$
\begin{aligned}
c_{kl}(\mathbf{x}) &= \sum_i x_i p_k^i(\mathbf{x})(p_l^i(\mathbf{x}) - p_l(\mathbf{x})) \\
&= \sum x_i\left(p_k^i(\mathbf{x}) - p_k(\mathbf{x})\right)\left(p_l^i(\mathbf{x}) - p_l(\mathbf{x})\right) .
\end{aligned}
$$

Hence $C(\mathbf{x})$ is the covariance matrix of the mean strategies $\mathbf{p}^i(\mathbf{x})$ of the allele A_i. Since

$$
\boldsymbol{\xi}\cdot C(\mathbf{x})\boldsymbol{\xi} = \sum_{i=1}^n x_i[(\mathbf{p}^i-\mathbf{p})\cdot\boldsymbol{\xi}]^2 \geq 0 \qquad\qquad (22.33)
$$

the matrix $C(\mathbf{x})$ is positive semidefinite and its kernel consists of all vectors $\boldsymbol{\xi}$ orthogonal to all $\mathbf{p}^i - \mathbf{p}$, $i = 1,\ldots,n$.

We can now show that if at some point $\mathbf{x} \in \text{int } S_n$, the strategy distribution

$\mathbf{p}$ does not change, i.e. $\dot{\mathbf{p}} = \mathbf{0}$, then also $\dot{\mathbf{x}} = \mathbf{0}$, so that $\mathbf{x}$ is a rest point on the genetic level as well.

Indeed, $\dot{\mathbf{p}} = \mathbf{0}$ in (22.32) means that $A\mathbf{p}$ lies in the kernel of the covariance matrix $C(\mathbf{x})$. By (22.33) $A\mathbf{p}$ is orthogonal to all $\mathbf{p}^i - \mathbf{p}$, and hence $\dot{x}_i = 0$ in (22.31).

Suppose now that $\hat{\mathbf{p}} \in \text{int } S_N$ is an ESS for the payoff matrix A and that there exists at least one genetic state $\hat{\mathbf{x}}$ which displays this strategy, that is $\mathbf{p}(\hat{\mathbf{x}}) = \hat{\mathbf{p}}$. Let

$$S(\hat{\mathbf{p}}) = \{\mathbf{x} \in S_n : \mathbf{p}(\mathbf{x}) = \hat{\mathbf{p}}\} \qquad (22.34)$$

be the set of all such genetic states. $S(\hat{\mathbf{p}})$ is the intersection of $N - 1$ conic sections, and it consists of rest points by the above remark.

Let $\hat{\mathbf{x}} \in S_n$ be a rest point of (22.5) and $\hat{\mathbf{p}} = \mathbf{p}(\hat{\mathbf{x}})$. $\hat{\mathbf{x}}$ is said to be *strategically stable* if, for every neighbourhood U of $\hat{\mathbf{x}}$, there is some neighbourhood V of $\hat{\mathbf{x}}$ such that $\mathbf{x}(t)$ remains in U and $\mathbf{p}(\mathbf{x}(t))$ converges to $\hat{\mathbf{p}}$ for all $\mathbf{x} \in V$. (This notion is obviously between Lyapunov stability and local asymptotic stability.) The *strategy* $\hat{\mathbf{p}}$ is said to be strategically stable if every $\mathbf{x} \in S(\hat{\mathbf{p}})$ is strategically stable.

Theorem 22.3.1 *Let $\hat{\mathbf{p}} \in \text{int } S_N$ be an ESS for the payoff matrix A and assume that for all $\hat{\mathbf{x}} \in S(\hat{\mathbf{p}})$, the covariance matrix $C(\hat{\mathbf{x}})$ has full rank (the minimum of $N - 1$ and $n - 1$). Then $\hat{\mathbf{p}}$ is strategically stable.*

Proof Let us consider the more interesting case $n > N$ first. By assumption the $N \times N$ matrix $C(\hat{\mathbf{x}})$ is then positive definite, when restricted to the invariant subspace $\mathbb{R}_0^N$. According to exercise 9.6.2, $C(\hat{\mathbf{x}})A$ is stable on $\mathbb{R}_0^N$. Thus all $N - 1$ eigenvalues of the linearization

$$\dot{\mathbf{v}} = C(\hat{\mathbf{x}})A\mathbf{v} \qquad (22.35)$$

of (22.32), with

$$\mathbf{v} = (D_{\hat{\mathbf{x}}}\mathbf{p})(\mathbf{x} - \hat{\mathbf{x}}) = 2\sum \hat{x}_i \mathbf{p}^i (\mathbf{x} - \hat{\mathbf{x}}),$$

have negative real part. The remaining $n - N$ eigenvalues of (22.31), corresponding to the tangent space of the fixed manifold $S(\hat{\mathbf{p}})$, are zero. Hence $S(\hat{\mathbf{p}})$ is attracting, and each $\hat{\mathbf{x}} \in S(\hat{\mathbf{p}})$ is stable.

If $n \leq N, C(\hat{\mathbf{x}})$ has rank $n - 1$ and hence $S(\hat{\mathbf{p}})$ consists of one point $\hat{\mathbf{x}}$ only. By (22.35) and exercise 9.6.2, all $n - 1$ eigenvalues at $\hat{\mathbf{x}}$ have negative real part and hence $\hat{\mathbf{x}}$ is asymptotically stable. $\qquad\square$

Exercise 22.3.2 Consider the function $P(\mathbf{x}) = \prod x_i^{\hat{x}_i}$. Show that

$$\dot{P}/P = \left(\sum \hat{x}_i \mathbf{p}^i - \mathbf{p}\right) \cdot A\mathbf{p} \, .$$

Show that P is a Lyapunov function under the assumption of the theorem if $n \leq N$, but in general not for $n > N$.

Exercise 22.3.3 If $C(\hat{\mathbf{x}}) = 0$ (i.e. $\hat{\mathbf{x}}$ is a genetic rest point) and $\mathbf{p}(\hat{\mathbf{x}}) = \hat{\mathbf{p}}$ is an ESS for the matrix A, show that $\hat{\mathbf{p}}$ is strategically stable. (Hint: $P = \prod x_i^{\hat{x}_i}$ is a Lyapunov function in this case.)

Theorem 22.3.1 and exercise 22.3.3 together show that $\hat{\mathbf{p}}$ is strategically stable if $C(\hat{\mathbf{x}})$ has either rank 0 or maximal rank. For the case of $N = 2$ strategies this covers all cases and gives (another) complete proof of the fact that an ESS is strategically stable. For $N \geq 3$ open problems remain. However, the case of $n = 2$ and $n = 3$ alleles is completely solved. For $n = 2$, this is a simple exercise. For $n = 3$, the proof relies on centre manifold theory and cannot be reproduced here.

Theorem 22.3.4 *For $n = 3$ alleles, if $\hat{\mathbf{p}}$ is an ESS and $S(\hat{\mathbf{p}})$ is nonempty, then $\hat{\mathbf{p}}$ is strategically stable.*

Exercise 22.3.5 Consider a semidominance of the genetic pattern:

$$\mathbf{p}(ij) = \lambda_{ij}\mathbf{p}(ii) + (1 - \lambda_{ij})\,\mathbf{p}(jj) \quad 0 \leq \lambda_{ij} \leq 1 \, .$$

Assume $\hat{\mathbf{p}}$ is an ESS and lies in the interior of the convex hull of $\{\mathbf{p}(ii) : i = 1, \ldots, n\}$. Show that $S(\hat{\mathbf{p}}) \neq \emptyset$, that all orbits $\mathbf{x}$ near $S(\hat{\mathbf{p}})$ converge to $S(\hat{\mathbf{p}})$ and that there are no other interior fixed points. (Hint: note that the kernel of the covariance matrix $C(\hat{\mathbf{x}})$ coincides with the space spanned by $\{\mathbf{p}(\mathbf{x}) - \hat{\mathbf{p}} : \mathbf{x} \in S_n\}$.)

Exercise 22.3.6 Analyse a *genetic rock–scissors–paper game*. Use the payoff matrix A from (13.8) and assume $n = 3$ with $\mathbf{p}(ii) = \mathbf{e}_i$ and $\mathbf{p}(ij) = p\mathbf{e}_i + q\mathbf{e}_j + r\mathbf{e}_k$ (with $p + q + r = 1$, $p \geq 0$, $q \geq 0$, $r \geq 0$) if (i, j, k) is a cyclic permutation of $(1, 2, 3)$. Recall that $\hat{\mathbf{p}} = (\frac{1}{3}, \frac{1}{3}, \frac{1}{3})$ is an ESS for A if and only if $\varepsilon > 0$.
(a) Show that $\hat{\mathbf{x}} = (\frac{1}{3}, \frac{1}{3}, \frac{1}{3})$ is a rest point and compute the linearization.
(b) Show that $\hat{\mathbf{x}}$ is a sink if $r < 1$ and $\varepsilon > 0$.
(c) Show that in the critical case $\varepsilon = 0$, $\hat{\mathbf{x}}$ is asymptotically stable if $p < q$ (i.e. if the heterozygote prefers the 'better' of the two strategies) and $\hat{\mathbf{x}}$ is unstable if $p > q$.
(d) Deduce the occurrence of super- and subcritical Hopf bifurcations. This shows that even in the case of semidominance ($r = 0$) the state $\hat{\mathbf{x}}$

corresponding to the ESS $\hat{\mathbf{p}}$ need not be globally stable for small $\varepsilon < 0$ and can have a small basin of attraction (bounded by the unstable limit cycle).
(e) Using exercise (17.2.2) determine the stability conditions for bd S_3 (it is a heteroclinic cycle for small r and — in the reverse direction — for r close to 1).
(f) For large r ($r > \frac{2}{3}$), additional rest points arise in int S_3. Find phase portraits with seven interior rest points and four limit cycles.

22.4 ESS and long-term evolution

In long-term evolution, we have to investigate the notion of stability against invasion attempts by *any* conceivable allele. In particular, we have even to allow mutants at other gene loci to introduce new strategies. Let us consider a discrete time selection–recombination model involving two loci (the extension to more loci is straightforward). If $w_{ij,kl}$ is the viability of the genotype $A_i B_j / A_k B_l$, then we obtain (20.29) for the evolution of the frequencies x_{ij} of the gametes $A_i B_j$, and we can write (20.34) for the frequencies x_i of the alleles A_i as

$$x_i' = \frac{x_i w_i}{\bar{w}} \tag{22.36}$$

where w_i is the *marginal fitness* of allele A_i,

$$w_i = \frac{1}{x_i} \sum_{j,k,l} x_{ij} x_{kl} w_{ij,kl}. \tag{22.37}$$

Obviously, allele A_i will increase in frequency if and only if its marginal fitness w_i is larger than the average fitness $\bar{w}$ of the population.

Let us now assume that the genotype $A_i B_j / A_k B_l$ plays the mixed strategy $\mathbf{p}(ij, kl)$. Its payoff, then, is given by

$$\mathbf{p}(ij, kl) \cdot A\mathbf{p} \tag{22.38}$$

where $\mathbf{p} = \sum x_{ij} x_{kl} \mathbf{p}(ij, kl)$ is the average strategy in the population and A is the payoff matrix. If we define the *marginal strategy* $\mathbf{p}_i$ of allele A_i by

$$\mathbf{p}_i = \frac{1}{x_i} \sum_{j,k,l} x_{ij} x_{kl} \mathbf{p}(ij, kl) \tag{22.39}$$

then the *marginal fitness* of A_i is given by

$$w_i = \mathbf{p}_i \cdot A\mathbf{p} \tag{22.40}$$

and the mean fitness in the population can be written as

$$\bar{w} = \mathbf{p} \cdot A\mathbf{p}. \tag{22.41}$$

Obviously the frequency of A_i increases if and only if $\mathbf{p}_i \cdot A\mathbf{p} > \mathbf{p} \cdot A\mathbf{p}$.

We consider now genetic rest points which are resistant against the introduction of *any* new allele A_s. Such a new allele can occur in many genetic combinations $A_s B_j / A_k B_l$ (and in the homozygote type $A_s B_j / A_s B_l$) where the A_k, B_j and B_l belong to the resident allele pool. Any such combination determines a mixed strategy $\mathbf{p}(sj, kl)$ which is a convex combination of the available pure strategies (we do not allow new pure strategies to be introduced: the game-theoretical framework is fixed in advance, and only the genetic opportunities can be expanded.)

Exercise 22.4.1 Show that if a rest point of the genetic frequencies cannot be invaded by any conceivable mutant allele, then the corresponding population strategy $\hat{\mathbf{p}}$ must be a Nash equilibrium. (Hint: if $\hat{\mathbf{p}}$ is not a Nash equilibrium, there exists an alternative strategy $\mathbf{q}$ such that $\mathbf{q}\cdot A\hat{\mathbf{p}} > \hat{\mathbf{p}}\cdot A\hat{\mathbf{p}}$. Consider invasion by a mutant allele which induces strategy $\mathbf{q}$ in any organism carrying it, irrespective of the genetic background.)

We note that a stable equilibrium with a mixed population strategy $\mathbf{p}$ can be invaded by a mutant allele with marginal strategy $\mathbf{p}$ — there is no selective force to drive this mutant from the allele pool. Thus we cannot expect local asymptotic stability, in general, but we can hope for strategic stability. A rest point of the genetic frequencies is said to be *long-term stable* if it is strategically stable against any invasion attempt.

Exercise 22.4.2 If we consider a game with two pure strategies and a genetic mechanism with a single locus only, then show that a genetic rest point is long-term stable if and only if its population strategy is an ESS. (Hint: essentially, this was shown in section 22.2.)

Theorem 22.4.3 *If all resident genotypes adopt the same strategy $\hat{\mathbf{p}}$ at a genetic rest point, and if $\hat{\mathbf{p}}$ is long-term stable, then it is an ESS.*

Proof Suppose that $\hat{\mathbf{p}}$ is not a local ESS. There exists by (6.9) an alternative strategy $\mathbf{q}$ such that $\mathbf{q}\cdot A\mathbf{p}_\varepsilon \geq \mathbf{p}_\varepsilon \cdot A\mathbf{p}_\varepsilon$ holds for small values of ε, where $\mathbf{p}_\varepsilon$ is the strategy mixture $(1-\varepsilon)\hat{\mathbf{p}} + \varepsilon\mathbf{q}$. Consider a mutant allele A_s which induces strategy $\mathbf{q}$ irrespective of its genetic background. Then this allele will not be selected against, provided its frequency is sufficiently small. Accordingly, the population strategy will not be driven back to $\hat{\mathbf{p}}$, which contradicts the assumption. $\qquad\square$

It is not clear whether the converse is always valid.

Theorem 22.4.4 *A genetic monomorphism with strategy $\hat{\mathbf{p}}$ is long-term stable if and only if it is an ESS.*

Proof At a monomorphism, only one allele is present at each locus. This leads to a one-locus model with resident allele A_1 and new mutant allele A_2. The mean population strategy, then, is

$$\mathbf{p} = (1 - x_2)^2\hat{\mathbf{p}} + (1 - x_2)x_2\mathbf{p}(12) + x_2^2\mathbf{p}(22).$$

Clearly, if $\mathbf{p}(12) = \mathbf{p}(22) = \hat{\mathbf{p}}$, the frequency x_2 of A_2 will not change. If $\mathbf{p}(12) = \hat{\mathbf{p}}$ but $\mathbf{p}(22) \neq \hat{\mathbf{p}}$, then the marginal strategy $\mathbf{p}_2$ of A_2 is given by $(1 - x_2)\hat{\mathbf{p}} + x_2\mathbf{p}(22)$ and $\mathbf{p} = (1 - x_2)\hat{\mathbf{p}} + x_2\mathbf{p}_2$. For small x_2, the ESS condition (6.9), which now reads $\mathbf{p}{\cdot}A\mathbf{p} < \hat{\mathbf{p}}{\cdot}A\mathbf{p}$, implies that $\mathbf{p}_2{\cdot}A\mathbf{p} < \mathbf{p}{\cdot}A\mathbf{p}$ and hence the frequency of A_2 declines. There remains the case $\mathbf{p}(12) \neq \hat{\mathbf{p}}$. The marginal strategy $\mathbf{p}_2$ is given by $\mathbf{p}(12) + x_2(\mathbf{p}(22) - \mathbf{p}(12))$. Again, if the initial frequency x_2 is sufficiently small, $\mathbf{p}_2$ remains in a compact neighbourhood of $\mathbf{p}(12)$ and the ESS condition implies that $\mathbf{p}_2$ fares less well than average, and hence decreases monotonically to 0. $\square$

We stress in this context that if the newly arising homozygotes and heterozygotes also play $\mathbf{p}$, the population will not remain genetically monomorphic.

22.5 Notes

The first papers that combine strategy and genetics are Stewart (1971), Maynard Smith (1981) and Hofbauer *et al.* (1982). The case of two strategies was studied by Eshel (1982) and by Lessard (1984), whose treatment is the basis for section 22.2. The compatibility with Shahshahani gradients (theorem 22.1.1) is from Sigmund (1987b). An equivalent version of theorem (22.3.1) is given in Cressman (1988). Thomas (1985) attacks the problem on the genetic level (see exercise 22.3.2). Exercise 22.3.6 is from Hofbauer *et al.* (1982). For a discussion of the relation between game theoretic and population genetic models, we refer to Cressman (1992). Theorem 22.3.4 is from Cressman *et al.* (1996). The relation between ESS and long-term evolution has been studied in Eshel (1982, 1991, 1996) and the 'streetcar theory' of Hammerstein and Selten (1994) and Hammerstein (1996). We also refer to Libermann (1988) and Matessi and Di Pasquale (1996). For a general treatment with nonlinear payoff functions, and a critical assessment of the streetcar theory, we refer to Weissing (1996).

References

Akin, E. (1979): *The Geometry of Population Genetics.* Lecture Notes in Biomathematics **31**. Berlin-Heidelberg-New York: Springer-Verlag.

Akin, E. (1980): Domination or equilibrium. *Math. Biosciences* **50**, 239–250.

Akin, E. (1982): Cycling in simple genetic systems. *J. Math. Biol.* **13**, 305–324.

Akin, E. (1983): Hopf bifurcation in the two locus genetic model. *Memoirs Amer. Math. Soc.* **284**. Providence, RI.

Akin, E. (1987): Cycling in simple genetic systems II: the symmetric case. In: A. Kurzhanski and K. Sigmund (eds), *'Dynamical Systems'*, Lecture Notes in Economics and Mathematical Systems **287**. Berlin: Springer.

Akin, E. (1990): The differential geometry of population genetics and evolutionary games. In: S. Lessard (ed), *Mathematical and Statistical Developments of Evolutionary Theory*, pp. 1–93. Dordrecht: Kluwer.

Akin, E. (1993): *The General Topology of Dynamical Systems.* Amer. Math. Soc., Providence.

Akin, E., and J. Hofbauer (1982): Recurrence of the unfit. *Math. Biosciences* **61**, 51–63.

Akin, E., and V. Losert (1984): Evolutionary dynamics of zero-sum games. *J. Math. Biol.* **20**, 231–258.

Akin, E. and J. M. Szucs (1994): Approaches to the Hardy–Weinberg manifold. *J. Math. Biol.* **32**, 633–643.

Amann, E. (1989): *Permanence of Catalytic Networks.* Dissertation, University of Vienna.

Amann, E. and J. Hofbauer (1985): Permanence in Lotka–Volterra and replicator equations. In: W. Ebeling and M. Peschel (eds), *Lotka–Volterra Approach to Cooperation and Competition in Dynamic Systems.* Berlin: Akademie–Verlag.

Anderson, H. M., V. Hutson and R. Law (1992): On the conditions for permanence in ecological communities. *Amer. Nat.* **139**, 663–668.

Anderson, R. M. and R. M. May (1991): *Infectious Diseases of Humans.* Oxford University Press.

Andronov, A., E. Leontovich, I. Gordon and A. Maier (1973): *Qualitative Theory of Second-Order Dynamic Systems.* New York: Halsted Press.

Arneodo, A., P. Coullet and C. Tresser (1980): Occurrence of strange attractors in three dimensional Volterra equations. *Physics Letters* **79A**, 259–263.

Arneodo, A., P. Coullet, J. Peyraud and C. Tresser (1982): Strange attractors in Volterra equations for species in competition. *J. Math. Biol.* **14**, 153–157.

Arrow, K. J. and F. H. Hahn (1971): *General Competitive Analysis.* San Francisco: Holden-Day.

Arrowsmith, D. K. and C. M. Place (1990): *An Introduction to Dynamical Systems.* Cambridge University Press.

Axelrod, R. (1984): *The Evolution of Cooperation.* New York: Basic Publ.

Baake, E. and T. Wiehe (1996): Bifurcations in haploid and diploid sequence space models. *J. Math. Biol.* **35**, 321–343.

Baum, L. E. and Eagon, J. A. (1967): An inequality with applications to statistical estimation for probabilistic functions of Markov processes and to a model for ecology. *Bull. Amer. Math. Soc.* **73**, 360–363.

Berger, U. (1997a): The best-response dynamics in role games. Preprint. University of Vienna.

Berger, U. (1997b): Projective geometry and the best-response dynamics. Preprint. University of Vienna.

Berman, A., and D. Hershkovitz (1983): Matrix diagonal stability and its applications. *SIAM J. Alg. Discr. Math.* **4**, 377–382.

Berman, A. and Plemmons, R. J. (1979): *Nonnegative Matrices in the Applied Mathematical Sciences.* New York: Academic Press.

Binmore, K. (1992): *Fun and Games: a Text on Game Theory.* Lexington, MA: Heath and Co.

Binmore, K. and L. Samuelson (1992): Evolutionary stability in repeated games played by finite automata. *J. Economic Theory* **57**, 278–305.

Bishop, T., and C. Cannings (1978): A generalized war of attrition. *J. Theor. Biol.* **70**, 85–124.

Björnerstedt, J. and J. W. Weibull (1996): Nash equilibrium and evolution by imitation. In: K. Arrow and E. Colombatto (eds), *The Rational Foundations of Economic Behaviour*, pp 155–171. New York: MacMillan.

Bodmer, W. (1965): Differential fertility in population genetic models. *Genetics* **51**, 411–424.

Bomze, I. (1983): Lotka–Volterra equations and replicator dynamics: A two dimensional classification. *Biol. Cybern.* **48**, 201–211.

Bomze, I. M. (1986): Non-cooperative two-person games in biology: a classification. *Int. J. Game Theory* **15**, 31–57.

Bomze, I. (1991): Cross entropy minimization in uninvadable states of complex populations. *J. Math. Biol.* **30**, 73–87.

Bomze, I. M. (1995): Lotka–Volterra equation and replicator dynamics: New issues in classification. *Biological Cybernetics* **72**, 447–453.

Bomze, I. and R. Bürger (1995): Stability by mutation in evolutionary games. *Games Econ. Behav.* **11**, 146–172.

Bomze, I. and B. Pötscher (1989): *Game Theoretical Foundations of Evolutionary Stability*. Berlin: Springer.

Bomze, I. and E. van Damme (1992): A dynamical characterization of evolutionarily stable states. *Annals of Operations Research* **37**, 229–244.

Börgers, T. and R. Sarin (1993): Learning through reinforcement and replicator dynamics. Preprint.

Boylan, R. T. (1994): Evolutionary equilibria resistant to mutation. *Games Econ. Behav.* **7**, 10–34.

Brannath, W. (1994): Heteroclinic cycles on the tetrahedron. *Nonlinearity* **7**, 1367–1384.

Broom, M., C. Cannings and G. T. Vickers (1993): On the number of local maxima of a constrained quadratic form. *Proc. Royal Soc. London A* **443**, 573–584.

Brown, G. W. (1951): Iterative solution of games by fictitious play. In: T. C. Koopmans (ed), *Activity Analysis of Production and Allocation*, pp. 374–376. New York, Wiley.

Brown, J. S. and Vincent, T. L. (1987): A theory for the evolutionary game. *Theor. Pop. Biol.* **31**, 140–166.

Brunovský, P., J. Hofbauer and T. Nagylaki (1997): Convergence of multilocus systems under weak epistasis or weak selection. Preprint.

Bürger, R. (1997): Mathematical properties of mutation–selection models. *Genetica.* To appear.

Busenberg, S., and P. van den Driessche (1993): A method for proving the non-existence of limit cycles. *J. Math. Anal. Appl.* **172**, 463–479.

Butler, G. J., H. I. Freedman and P. Waltman (1982): Global dynamics of a selection model for the growth of a population with genotype fertility differences. *J. Math. Biol.* **14**, 25–35.

Butler, G., H. I. Freedman and P. Waltman (1986): Uniformly persistent systems. *Proc. Amer. Math. Soc.* **96**, 425–430.

Butler, G., and P. Waltman (1986): Persistence in dynamical systems. *J. Diff. Equations* **63**, 255–263.

Cabrales, A (1993): Stochastic replicator dynamics. Preprint. San Diego.

Cabrales, A. and J. Sobel (1992): On the limit points of discrete selection dynamics. *J. Economic Theory* **57**, 407–419.

Cannings, C., J. P. Tyrer and G. T. Vickers (1993): Routes to polymorphism. *J. Theor. Biol.* **165**, 213–223.

Chawanya, T. (1995): A new type of irregular motion in a class of game dynamics systems. *Progress Theor. Phys.* **94**, 163–179.

Chawanya, T. (1996): Infinitely many attractors in game dynamics systems. *Progress Theor. Phys.* **95**, 679–684.

Chenciner, A. (1977): Comportement asymptotique de systèmes differentiels du type "compétition d'espèces". *Comptes Rendus Acad. Sc. Paris* **284**, 313–315.

Cheng, K. S. (1981): Uniqueness of a limit cycle for predator–prey systems. *SIAM J. Math. Anal.* **12**, 541–548.

Cheng, K. S., Hsu, S. B. and Lin, S. S. (1981): Some results on global stability of predator–prey system. *J. Math. Biol.* **12**, 115–126.

Christiansen, F. B. (1991): On conditions for evolutionary stability for a continuously varying character. *Am. Nat.* **138**, 37–50.

Clark, C. E. and T. G. Hallam (1982): The community matrix in three species community models. *J. Math. Biol.* **16**, 25–31.

Cohen, J. E. (1988): Untangling an 'entangled bank': recent facts and theories about community food webs. In A. Hastings (ed), *Community Ecology*. New York: Springer.

Coste, J., J. Peyraud, and P. Coullet (1979): Asymptotic behavior in the dynamics of competing species. *SIAM J. Appl. Math.* **36**, 516–542.

Cottle, R. W. (1980): Completely Q-matrices. *Math. Programming* **19**, 347–351.

Cressman, R. (1988): Frequency-dependent viability selection (a single-locus, multi-phenotype model). *J. Theor. Biol.* **130**, 147–165.

Cressman, R. (1990): Strong stability and density-dependent evolutionarily stable strategies. *J. Theor. Biol.* **145**, 319–330.

Cressman, R. (1992): *The Stability Concept of Evolutionary Game Theory.* Berlin: Springer.

Cressman, R. (1996): Frequency dependent stability for two-species interactions. *Theor. Pop. Biol.* **49**, 189–210.

Cressman, R., Dash, A. T. and Akin, E. (1986): Evolutionary games and two species population dynamics. *J. Math. Biol.* **23**, 221–230.

Cressman, R., A. Gaunersdorfer and J. F. Wen (1996): Dynamic stability in two-market competition models. Preprint.

Cressman, R., J. Hofbauer and W. G. S. Hines (1996): Evolutionary stability in strategic models of single-locus frequency-dependent viability selection. *J. Math. Biol.* **34**, 707–733.

Cross, G. W. (1978): Three types of matrix stability. *Linear Algebra and Appl.* **20**, 253–263.

Czárán, T. (1997): *Spatiotemporal Models of Population and Community Dynamics.* Chapman and Hall.

Dawid, H. (1996): *Adaptive Learning by Genetic Algorithms. Analytical Results and Applications to Economical Models.* Lecture Notes in Economics and Mathematical Systems, **441**. Berlin: Springer.

Dawkins, R. (1989): *The Selfish Gene* (2nd edn). Oxford University Press.

Dekel, E. and S. Scotchmer (1992): On the evolution of optimizing behavior. *J. Economic Theory* **57**, 392–46.

de Melo, W. and S. van Strien (1993): *One-dimensional Dynamics.* New York: Springer.

Devaney, R. L. (1986): *An Introduction to Chaotic Dynamical Systems.* Menlo Park: Benjamin/Cummings.

Dieckmann, U. and R. Law (1996): The dynamical theory of coevolution: a derivation from stochastic ecological processes. *J. Math. Biol.* **34**, 579–612.

Edwards, A. W. F. (1977): *Foundations of Mathematical Genetics.* Cambridge University Press.

Edwards, A. W. F. (1994): The fundamental theorem of natural selection. *Biol. Rev.* **69**, 443–474.

Eigen, M., W. Gardiner, P. Schuster, and R. Winkler-Oswatitsch (1981): The origin of genetic information. *Scientific American* **244**, 78–95.

Eigen, M., and P. Schuster (1979): *The hypercycle: A principle of natural selforganization.* Berlin-Heidelberg: Springer.

Enquist, M. and O. Leimar (1983): Evolution of fighting behavior. Decision rules and assessment of relative strength. *J. Theor. Biol.* **102**, 387–410.

Eshel, I. (1982): Evolutionarily stable strategies and viability selection in Mendelian populations. *Theor. Pop. Biol.* **22**, 204–217.

Eshel, I. (1983): Evolutionary and continuous stability. *J. Theor. Biol.* **103**, 99–111.

Eshel, I. (1991): Game theory and population dynamics in complex genetical systems: The role of sex in short term and in long term evolution. In: R. Selten (ed) *Game equilibrium models I: Evolution and Game Dynamics*, pp. 6–28, Berlin: Springer.

Eshel, I. (1996): On the changing concept of evolutionary population stability as a reflection of a changing point of view in the quantitative theory of evolution. *J. Math. Biol.* **34**, 485–510.

Eshel, I. and E. Akin (1983): Coevolutionary instability of mixed Nash solutions. *J. Math. Biol.* **18**, 123–133.

Eshel, I. and A. Motro (1981): Kin selection and strong evolutionary stability of mutual help. *Theor. Pop. Biol.* **19**, 420–433.

Ewald, P. W. (1993): The evolution of virulence. *Scientific American* (4), 56–62.

Ewens, W. J. (1969): A generalized fundamental theorem of natural selection. *Genetics* **63**, 531–537.

Ewens, W. J. (1989): An interpretation and proof of the fundamental theorem of natural selection. *Theor. Pop. Biol.* **36**, 167–180.

Ewens, W. J. (1992): An optimizing principle of natural selection in evolutionary population genetics. *Theor. Pop. Biol.* **42**, 333–346.

Farkas, M. (1994): *Periodic Motions.* Applied Mathematical Sciences **104**. New York: Springer.

Ferrière, R. and G. A. Fox (1995): Chaos and Evolution. *Trends Ecol. Evol.* **10**, 480–485.

Ferrière, R. and M. Gatto (1995): Lyapunov exponents and the mathematics of invasion in oscillatory or chaotic populations. *Theor. Pop. Biol.* **48**, 126–171.

Fiedler, M. and V. Ptak (1962): On matrices with non-positive off-diagonal elements and positive minors. *Czech. Math. J.* **12**, 382–400.

Field, M. and Swift, J. W. (1991): Stationary bifurcations to limit cycles and heteroclinic cycles. *Nonlinearity* **4**, 1001–1043.

Fisher, R. A. (1930): *The Genetical Theory of Natural Selection.* Oxford: Clarendon Press.

Foster, D. and H. P. Young (1990): Stochastic evolutionary game dynamics. *Theor. Pop. Biol.* **38**, 219–232.

Foster, D. and H. P. Young (1995): On the nonconvergence of fictitious play in coordination games. IIASA working paper.

Freedman, H. (1980): *Deterministic Mathematical Models in Population Ecology.* Edmonton: HIFR Co.

Freedman, H. and P. Moson (1990): Persistence definitions and their connections. *Proc. Amer. Math. Soc.* **109**, 1025–1033.

Freedman, H. I. and P. Waltman (1977): Mathematical analysis of some three species food chain models. *Math. Biosci.* **33**, 257–273.

Freedman, H. I., and P. Waltman (1984): Persistence in models of three interacting predator–prey populations. *Math. Biosci.* **68**, 213–231.

Freedman, H. I., and P. Waltman (1985): Persistence in a model of three competitive populations. *Math. Biosci.* **73**, 89–101.

Friedman, D. (1991): Evolutionary games in economics. *Econometrica* **59** 637–666.

Fudenberg, D. and D. K. Levine (1998): *Theory of Learning in Games.* MIT Press, to appear.

Fudenberg, D. and J. Tirole (1991): *Game Theory.* MIT Press.

Gale, D. and H. Nikaido (1965): The Jacobian matrix and global univalence of mappings. *Math. Annalen* **159**, 81–93.

Garay, B. M. (1989): Uniform persistence and chain recurrence. *J. Math. Anal. Appl.* **139**, 372–381.

Gard, T. and T. Hallam (1979): Persistence in food webs I: Lotka–Volterra food chains. *Bull. Math. Biol.* **41**, 877–891.

Gardini, L., R. Lupini and M. G. Messia (1989): Hopf bifurcation and transition to chaos in Lotka Volterra equation. *J. Math. Biol.* **27**, 259–272.

Gaunersdorfer, A. (1992): Time averages for heteroclinic attractors. *SIAM J. Appl. Math* **52**, 1476–1489.

Gaunersdorfer, A. and J. Hofbauer (1995): Fictitious play, Shapley polygons and the replicator equation. *Games Econ. Behav.* **11**, 279–303.

Gaunersdorfer, A., J. Hofbauer and K. Sigmund (1991): On the dynamics of asymmetric games. *Theor. Pop. Biol.* **39**, 345–357.

Gilpin, M. E. (1979): Spiral chaos in a predator–prey model. *Amer. Nat.* **113**, 306–308.

Goh, B. S. (1979): Stability in models of mutualism. *Amer. Nat.* **113**, 261–275.

Gopalsamy, K. (1992): *Stability and Oscillations in Delay Differential Equations of Population Dynamics.* Dordrecht: Kluwer.

Gouzé, J.-L. (1993): Global behavior of n-dimensional Lotka–Volterra systems. *Math. Biosci.* **113**, 231–243.

Hadeler, K. P. (1981): Stable polymorphisms in a selection model with mutation. *SIAM J. Appl. Math.* **41**, 1–7.

Hadeler, K. P. (1983): On copositive matrices. *Linear Algebra and Appl.* **49**, 79–89.

Hadeler, K. P., and D. Glas (1983): Quasimonotone systems and convergence to equilibrium in a population genetic model. *J. Math. Anal. Appl.* **95**, 297–303.

Hadeler, K. P., and U. Liberman (1975): Selection models with fertility differences. *J. Math. Biol.* **2**, 19–32.

Hallam, T. G. and S. A. Levin (1986): *Mathematical Ecology: An Introduction.* Biomathematics **17**. Berlin: Springer.

Hallam, T., L. Svoboda and T. Gard (1979): Persistence and extinction in three species Lotka–Volterra competitive systems. *Math. Biosciences* **46**, 117–124.

Hamilton, W. D. (1967): Extraordinary sex ratios. *Science* **156**, 477–88.

Hamilton, W. D. (1996): *Narrow Roads of Gene Land.* New York: Freeman.

Hammerstein, P. (1996): Darwinian adaptation, population genetics and the streetcar theory of evolution. *J. Math. Biol.* **34**, 511–532.

Hammerstein, P., and G. Parker (1982): The asymmetric war of attrition. *J. Theor. Biol.* **96**, 647–682.

Hammerstein, P. and R. Selten (1994): Game theory and evolutionary biology. In: R. J. Aumann and S. Hart (eds), *Handbook of Game Theory II*, pp. 931–993. Amsterdam: North-Holland.

Harrison, G. W. (1978): Global stability of food chains. *Amer. Nat.* **114**, 455–457.

Hastings, A. (1981): Stable cycling in discrete-time genetic models. *Proc. Nat. Acad. Sci. USA* **11**, 7224–7225.

Helbing D. (1992): A mathematical model for behavioral changes by pair interactions and its relation to game theory. In: G. Haag, U. Mueller and K.G. Troitzsch (eds), *Economic Evolution and Demographic Change. Formal Models in Social Sciences*. Berlin: Springer.

Helmke, U. and J. B. Moore (1994): *Optimization and Dynamical Systems*. London: Springer.

Hines, W. G. S. (1980): Three characterizations of population strategy stability. *J. Appl. Prob.* **17**, 333–340.

Hines, W. G. S. (1987): Evolutionarily stable strategies: a review of basic theory. *Theor. Pop. Biol.* **31**, 195–272.

Hirsch, M. W. (1982): Systems of differential equations which are competitive or cooperative: I. Limit sets. *SIAM J. Math. Anal.* **13**, 167–179.

Hirsch, M. W. (1988): Systems of differential equations which are competitive or cooperative: III. Competing species. *Nonlinearity* **1**, 51–71.

Hirsch, M. W. and S. Smale (1974): *Differential Equations, Dynamical Systems, and Linear Algebra*. New York: Academic Press.

Hofbauer, J. (1981a): On the occurrence of limit cycles in the Volterra–Lotka equation. *Nonlinear Analysis* **5**, 1003–1007.

Hofbauer, J. (1981b): A general cooperation theorem for hypercycles. *Monatsh. Math.* **91**, 233–240.

Hofbauer, J. (1984): A difference equation model for the hypercycle. *SIAM J. Appl. Math.* **44**, 762–772.

Hofbauer, J. (1985): The selection mutation equation. *J. Math. Biol.* **23**, 41–53.

Hofbauer, J. (1987): Heteroclinic cycles on the simplex. In: M. Farkas, V. Kertész, G. Stépán (eds), *Proc. Int. Conf. Nonlinear Oscillations*. Budapest: Janos Bolyai Math. Society. 1987.

Hofbauer, J. (1988): Saturated equilibria, permanence, and stability for ecological systems. In: L. J. Gross, T. G. Hallam, S. A. Levin (eds), *Mathematical Ecology. Proc. Trieste 1986*. World Scientific.

Hofbauer, J. (1989): A unified approach to persistence. *Acta Appl. Math.* **14**, 11–22.

Hofbauer, J. (1990): An index theorem for dissipative semiflows. *Rocky Mountains J. Math.* **20**, 1017–1031.

Hofbauer, J. (1994): Heteroclinic cycles for ecological differential equations *Tatra Mount. Math. Publ.* **4**, 105–116.

Hofbauer, J. (1995a): Stability for the best response dynamics. Preprint.

Hofbauer, J. (1995b): Imitation dynamics for games. Preprint.

Hofbauer, J. (1996): Evolutionary dynamics for bimatrix games: a Hamiltonian system? *J. Math. Biol.* **34**, 675–688.

Hofbauer, J., V. Hutson and W. Jansen (1987): Coexistence for systems governed by difference equations of Lotka–Volterra type. *J. Math. Biol.* **25**, 553–570.

Hofbauer, J., J. Mallet-Paret and H. L. Smith (1991): Stable periodic solutions for the hypercycle system. *J. Dynamics and Diff. Equs.* **3**, 423–436.

Hofbauer, J., P. Schuster, and K. Sigmund (1979): A note on evolutionarily stable strategies and game dynamics. *J. Theor. Biol.* **81**, 609–612.

Hofbauer, J., P. Schuster, and K. Sigmund (1981): Competition and cooperation in catalytic selfreplication. *J. Math. Biol.* **11**, 155–168.

Hofbauer, J., P. Schuster, and K. Sigmund (1982): Game dynamics for Mendelian populations. *Biol. Cybern.* **43**, 51–57.

Hofbauer, J., P. Schuster, K. Sigmund, and R. Wolff (1980): Dynamical systems under constant organization. Part 2: Homogeneous growth functions of degree 2. *SIAM J. Appl. Math.* **38**, 282–304.

Hofbauer, J. and K. Sigmund (1988): *The Theory of Evolution and Dynamical Systems.* Cambridge University Press.

Hofbauer, J. and K. Sigmund (1989): On the stabilising effect of predators and competitors on ecological communities. *J. Math. Biol.* **27**, 537–548.

Hofbauer, J. and K. Sigmund (1990): Adaptive dynamics and evolutionary stability. *Appl. Math. Lett.* **3**, 75–79.

Hofbauer, J. and J. W.-H. So (1989): Uniform persistence and repellors for maps. *Proc. Amer. Math. Soc.* **107**, 1137–1142.

Hofbauer, J. and J. W.-H. So (1990): Multiple limit cycles for predator–prey models. *Math. Biosciences* **99**, 71–75.

Hofbauer, J. and J. W.-H. So (1994): Multiple limit cycles in three dimensional Lotka–Volterra equations. *Appl. Math. Letters* **7**, 65–70.

Hofbauer, J. and J. Swinkels (1998): A universal Shapley example. (in preparation)

Hofbauer, J. and J. W. Weibull (1996): Evolutionary selection against dominated strategies. *J. Economic Theory* **71**, 558–573.

Hopkins, E. (1995): Learning, matching and aggregation. Preprint.

Horn, R. and C. R. Johnson: (1991) *Topics in Matrix Analysis.* Cambridge University Press.

Houston, A. I. and J. M. McNamara (1988): Fighting for food: A dynamic version of the hawk–dove game. *Evolutionary Ecology* **2**, 51–64.

Hsu, S., S. Hubbell, and P. Waltman (1978): Competing predators. *SIAM J. Appl. Math.* **35**, 617–625.

Hutson, V. (1984): A theorem on average Ljapunov functions. *Monatsh. Math.* **98**, 267–275.

Hutson, V. (1990): The existence of an equilibrium for permanent systems. *Rocky Mountain J. Math.* **20**, 1033–1040.

Hutson, V. and R. Law (1985): Permanent coexistence in general models of three interacting species. *J. Math. Biol.* **21**, 285–298.

Hutson, V. and Moran, W. (1982): Persistence of species obeying difference equations. *J. Math. Biol.* **15**, 203–213.

Hutson, V. and K. Schmitt (1992): Permanence and the dynamics of biological systems. *Math. Biosci.* **111**, 1–71.

Hutson, V. and G. T. Vickers (1983): A criterion for permanent coexistence of species with an application to a two-prey one-predator system. *Math. Biosci.* **63**, 253–269.

Jansen, V. A. A. (1995): Regulation of predator–prey systems through spatial interactions: a possible solution to the paradox of enrichment. *Oikos* **74**, 384–390.

Jansen, W. (1987): A permanence theorem for replicator and Lotka–Volterra systems. *J. Math. Biol.* **25**, 411–422.

Johnson, C. R. (1974): Sufficient conditions for *D*-stability. *J. Economic Theory* **9**, 53–62.

Kandori, M. (1996): Evolutionary game theory in economics. Preprint. Tokyo.

Kaniovski, Yu. M. and H. Peyton Young (1995): Learning dynamics in games with stochastic perturbations. *Games Econ. Behav.* **11**, 330–363.

Karlin, S. and S. Lessard (1986): *Sex Ratio Evolution.* Monographs in Population Biology **22**. Princeton University Press.

Khazin, L. G. and E. E. Shnol (1991): *Stability of Critical Equilibrium States.* Manchester University Press.

Kimura, M. (1958): On the change of population fitness by natural selection. *Heredity* **12**, 145–167.

Kingman, J. (1961): A matrix inequality. *Quarterly J. Math.* **12**, 78–80.

Kirk, V. and M. Silber (1994): A competition between heteroclinic cycles. *Nonlinearity* **7**, 1605–1621.

Kirlinger, G. (1986): Permanence in Lotka–Volterra equations: linked prey–predator systems. *Math. Biosci.* **82**, 165–191.

Kirlinger, G. (1988): Permanence of some ecological systems with several predator and one prey species. *J. Math. Biol.* **26**, 217–232.

Kirlinger, G. (1989): Two predators feeding on two prey species: a result on permanence. *Math. Biosci.* **96** 1–32.

Koth, M. and F. Kemler (1986): A one locus, two allele selection model admitting stable limit cycles. *J. Theor. Biol.* **122**, 263–267.

Krupa, M. (1996): Robust heteroclinic cycles. *J. Nonlinear Science.* **7**, 129–176.

Kuznetsov, Yu. A. (1995): *Elements of Applied Bifurcation Theory.* Applied Math. Sciences **112**. New York: Springer.

Law, R. and J. C. Blackford (1992): Self-assembling food webs: a global viewpoint of coexistence of species in Lotka–Volterra communities. *Ecology* **73**, 567–578.

Law, R. and R. D. Morton (1993): Alternative permanent states of ecological communities. *Ecology* **74**, 1347–1361.

Lessard, S. (1984): Evolutionary dynamics in frequency dependent two-phenotype models. *Theor. Pop. Biol.* **25**, 210–234.

Lessard, S. (1990): Evolutionary stability: one concept, several meanings. *Theor. Pop. Biol.* **37**, 159–170.

Lessard, S. (1996): Fisher's fundamental theorem of natural selection revisited. Preprint.

Liberman, U. (1988): External stability and ESS: Criteria for initial increase of new mutant allele. *J. Math. Biol.* **26**, 477–485.

Lindgren, K. (1991): Evolutionary phenomena in simple dynamics. In: C.G. Langton *et al.* (eds), *Artificial Life II.* Santa Fe Institute for Studies in the Sciences of Complexity, Vol. X, pp. 295–312.

Lipsitch, M., S. Siller and M. A. Nowak (1996): The evolution of virulence in pathogens with vertical and horizontal transmission. *Evolution* **50**, 1729–1741.

Logofet, D. O. (1993): *Matrices and Graphs. Stability Problems in Mathematical Ecology.* Boca Raton: CRC Press.

Losert, V., and E. Akin (1983): Dynamics of games and genes: Discrete versus continuous time. *J. Math. Biol.* **17**, 241–251.

Lu Zhengyi (1996): *Global Stability Analysis and Computer Aided Proof for Lotka–Volterra Systems.* Beijing: Academia Sinica.

Lyubich, Yu. I. (1992): *Mathematical Structures of Population Genetics Springer.* Biomathematics **22**. Berlin: Springer

MacArthur, R. (1970): Species packing and competitive equilibria for many species. *Theor. Pop. Biol.* **1**, 1–11.

Matsui, A. (1992): Best response dynamics and socially stable strategies. *J. Economic Theory* **57**, 343–362.

May, R. M. (1973): *Stability and Complexity in Model Ecosystems.* Princeton University Press.

May, R. M. (1976): Simple mathematical models with very complicated dynamics. *Nature* **261**, 459–467.

May, R. M., and W. Leonard (1975): Nonlinear aspects of competition between three species. *SIAM J. Appl. Math.* **29**, 243–252.

Maynard Smith, J. (1974): The theory of games and the evolution of animal conflicts. *J. Theor. Biol.* **47**, 209–221.

Maynard Smith, J. (1981): Will a sexual population evolve to an ESS? *Amer. Nat.* **177**, 1015–1018.

Maynard Smith, J. (1982): *Evolution and the Theory of Games.* Cambridge University Press.

Maynard Smith, J. and J. Hofbauer (1987): The "battle of the sexes": a genetic model with limit cycle behavior. *Theor. Pop. Biol.* **32**, 1–14.

Maynard Smith, J., and G. Price (1973): The logic of animal conflicts. *Nature* **246**, 15–18.

Mesterton-Gibbons, M. (1992): *An Introduction to Game-Theoretic Modelling.* Redwood City: Addison–Wesley.

Metz, J. A. J., S. A. H. Geritz, G. Meszéna, F. J. A Jacobs and J. S. van Heerwaarden (1996a): Adaptive dynamics: a geometrical study of the consequences of nearly faithful reproduction. In: S. J. Van Strien and S. M. Verduyn Lunel (eds), *Stochastic and Spatial Structures of Dynamical Systems*, pp. 183–231. Amsterdam: North Holland.

Metz, J. A. J., S. D. Mylius and O. Diekmann (1996b): When does evolution optimize? IIASA Working Paper 96-04.

Molander, P. (1985): The optimal level of generosity in a selfish, uncertain environment. *J. of Conflict Resolution* **29**, 611–618.

Molchanov, A. M. (1961): Stability in the case of a neutral linear approximation. *Doklady Akad. Nauk* **141**, 24–27.

Monderer, D. and L. Shapley (1996): Potential games. *Games Econ. Behavior* **14**. 124–143.

Moran, P. A. P. (1964): On the nonexistence of adaptive topographies. *Ann. Human Genetics* **27**, 283–293.

Nachbar, J. (1990): 'Evolutionary' selection dynamics in games: convergence and limit properties. *Intern. J. of Game Theory* **19**, 59–89.

Nagylaki, T. (1992): *Introduction to Theoretical Population Genetics* (Biomathematics **21**). Berlin: Springer.

Nash, J. (1996): *Essays on Game Theory.* Cheltenham, UK: Elgar.

Nee, S. (1990): Community Construction. *Trends in Ecology and Evolution* **5**, 337–340.

Nikaido, H. (1968): *Convex Structure and Economic Theory.* New York: Academic Press.

Nowak, M. A. (1990): An evolutionarily stable strategy may be inaccessible. *J. Theor. Biol.* **142**, 237–241.

Nowak, M. A. and R. M. May (1994): Superinfection and the evolution of virulence. *Proc. Royal Society London B* **255**, 81–89.

Nowak, M. A., R. M. May and K. Sigmund (1995a): Immune responses against multiple epitopes. *J. Theor. Biol.* **175**, 325–353.

Nowak, M. A., R. M. May and K. Sigmund (1995b): The arithmetics of mutual aid. *Scientific American* **272** (June), 76–81.

Nowak, M. A. and Sigmund, K. (1990): The evolution of reactive strategies in iterated games. *Acta Applicandae Math.* **20**, 247–265.

Nowak, M. A. and K. Sigmund (1994): The alternating Prisoner's Dilemma. *J. Theor. Biol.* **168**, 219–226.

Nowak, M. A. and K. Sigmund (1995): Automata, repeated games, and noise. *J. Math. Biol.* **33**, 703–722.

Oshime, Y. (1988): Global boundedness of cyclic predator–prey systems with self-limiting terms. *Japan J. Appl. Math.* **5**, 153–172.

Palm, G. (1984): Evolutionary stable strategies and game dynamics for *n*-person games. *J. Math. Biol.* **19**, 329–334.

Pimm, S. L., J. H. Lawton and J. E. Cohen (1991): Food web patterns and their consequences. *Nature* **350**, 669–674.

Plank, M. (1995): Hamiltonian structures for the *n*-dimensional Lotka–Volterra equations. *J. Math. Phys.* **36**, 3520–3534.

Pohley, H.-J. and B. Thomas (1983): Nonlinear ESS-models and frequency-dependent selection. *Biosystems* **16**, 87–100.

Pollak, E. (1979): Some models of genetic selection. *Biometrics* **35**, 119–137.

Posch, M. (1997): Cycling in a stochastic learning algorithm for normal form games. *J. Evolutionary Economics* **7**, 193–207.

Quirk, R. (1981): Qualitative stability of matrices and economic theory: a survey article. In: H. J. Greenberg and J. S. Maybee (eds), *Computer-Assisted Analysis and Model Simplification* (Proceedings of a symposium in Boulder, CO, 1980), New York: Academic Press.

Rand, D., Wilson, H. B. and McGlade, J. M. (1994): Dynamics and evolution: evolutionarily stable attractors, invasion exponents and phenotype dynamics. *Phil. Trans. Roy. Soc. London B* **24**, 261–283.

Redheffer, R. (1985): Volterra multipliers. *SIAM J. Discrete Math.* **6**, 592–623.

Redheffer, R. (1989): A new class of Volterra differential equations for which the solutions are globally asymptotically stable. *J. Differential Equations* **82**, 251–268.

Reyn, J. W. (1987): Phase portraits of a quadratic system of differential equations occurring frequently in applications. *Nieuw Archief voor Wiskunde* **5**, 107–155.

Riechert, S. and Hammerstein, P. (1983): Game theory in the ecological context, *Ann. Rev. Ecol. Syst.* **14**, 377–409.

Ritzberger, K. (1995): The theory of normal form games from the differentiable viewpoint. *Int. J. Game Theory*, **23**, 201–236.

Ritzberger, K. and J. W. Weibull (1995): Evolutionary selection in normal-form games. *Econometrica* **63**, 1371–1399.

Robinson, C. (1995): *Dynamical Systems: Stability, Symbolic Dynamics, and Chaos*. Boca Raton: CRC Press.

Robinson, J. (1951): An iterative method of solving a game. *Ann. Math.* **54**, 296–301.

Rosenzweig, M. L., and R. H. MacArthur (1963): Graphical representation and stability condition of predator–prey interaction. *Amer. Nat.* **97**, 209–223.

Roughgarden, J. (1979): *Theory of Population Genetics and Evolutionary Ecology*. New York: Macmillan.

Samuelson, L. (1997): Evolutionary Games and Equilibrium Selection. MIT Press.

Samuelson, L. and J. Zhang (1992): Evolutionary stability in asymmetric games. *J. Economic Theory* **57**, 363–391.

Scheuer, P., and S. Mandel (1959): An inequality in population genetics. *Heredity* **13**, 519–524.

Schlag, K. (1994): Why imitate, and if so, how? Preprint, Bonn.

Schreiber, S. J. (1997): Generalist and specialist predators that mediate permanence in ecological communities. *J. Math. Biology* **36**, 133–148.

Schuster, P., and K. Sigmund (1981): Coyness, philandering and stable strategies. *Anim. Behavior* **29**, 186–192.

Schuster, P., and K. Sigmund (1983): Replicator Dynamics. *J. Theor. Biol.* **100**, 533–538.

Schuster, P., K. Sigmund, J. Hofbauer, and R. Wolff (1981a): Selfregulation of behavior in animal societies. Part I: Symmetric contests. *Biol. Cybern.* **40**, 1–8.

Schuster, P., K. Sigmund, J. Hofbauer, and R. Wolff (1981b): Selfregulation of behavior in animal societies II: Games between two populations without self'-interaction. *Biol. Cybern.* **40**, 9–15.

Schuster, P., K. Sigmund, J. Hofbauer, R. Gottlieb, and P. Merz (1981c): Selfregulation of behavior in animal societies. Part III: Games between two populations with self'-interaction. *Biol. Cyb.* **40**, 17–25.

Schuster, P., K. Sigmund, and R. Wolff (1979a): Dynamical systems under constant organization. Part 3: Cooperative and competitive behavior of hypercycles. *J. Diff. Equations.* **32**, 357–368.

Schuster, P., K. Sigmund, and R. Wolff (1979b): On ω-limits for competition between three species. *SIAM J. Appl. Math.* **37**, 49–54.

Schuster, P., K. Sigmund, and R. Wolff (1980): Mass action kinetics of selfreplication in flow reactors. *J. Math. Anal. Appl.* **78**, 88–112.

Scudo, F., and J. Ziegler (1978): *The Golden Age of Theoretical Ecology, 1923–1940*. Lecture Notes in Biomathematics **22**. Berlin-Heidelberg-New York: Springer.

Selten, R. (1980): A note on evolutionarily stable strategies in asymmetrical animal conflicts. *J. Theor. Biol.* **84**, 93–101.

Selten, R. (1983): Evolutionary stability in extensive two-person games. *Math. Social Sciences* **5**, 269–363.

Shahshahani, S. (1979): A new mathematical framework for the study of linkage and selection. *Memoirs Amer. Math. Soc.* **211**. Providence, RI.

Shapley, L. (1964): Some topics in two-person games. *Ann. Math. Studies* **5**, 1–28.

Shigesada, N., K. Kawasaki and E. Teramoto (1984): The effects of interference competition on stability, structure and invasion of a multi-species system. *J. Math. Biol.* **21**, 97–133.

Shigesada, N., K. Kawasaki and E. Teramoto (1989): Direct and indirect effects of invasions of predators on a multiple-species community. *Theor. Pop. Biol.* **36**, 311–338.

Sigmund, K. (1984): The maximum principle for replicator equations. In: W. Ebeling and M. Peschel (eds), *Lotka–Volterra Approach to Dynamical Systems*. Berlin: Akademie Verlag.

Sigmund, K. (1987a): Game dynamics, mixed strategies and gradient systems. *Theor. Pop. Biol.* **32**, 114–126.

Sigmund, K. (1987b): A maximum principle for frequency dependent selection. *Math. Biosci.* **84**, 189–197.

Sigmund, K. (1993): *Games of Life*. Harmondsworth: Penguin.

Sigmund, K. (1995): Darwin's 'Circles of Complexity': Assembling Ecological Communities. *Complexity* **1**, 40–44.

Šiljak, D. D. (1978): *Large Scale Dynamical Systems: Stability and Structures*. Amsterdam: North-Holland.

Sikder, A. and A. B. Roy (1994): Persistence of a generalized Gause-type predator prey pair linked by competition. *Math. Biosci.* **122**, 1–23.

Smale, S. (1976): On the differential equations of species in competition. *J. Math. Biol.* **3**, 5–7.

Smith, H. L. (1986a): On the asymptotic behavior of a class of deterministic models of cooperating species. *SIAM J. Appl. Math.* **46**, 368–375.

Smith, H. L. (1986b): Competing subcommunities of mutualists and a generalized Kamke theorem. *SIAM J. Appl. Math.* **46**, 856–874.

Smith, H. L. (1995): *Monotone Dynamical Systems: An Introduction to the Theory of Competitive and Cooperative Systems*. Amer. Math. Soc. Math. Surveys and Monographs, Vol. **41**.

Smith, H. L. and P. Waltman (1995): *The Theory of the Chemostat: Dynamics of Microbial Competition*. Cambridge University Press.

So, J. (1979): A note on global stability and bifurcation phenomenon of a Lotka–Volterra food chain. *J. Theor. Biol.* **80**, 185–187.

Stadler, P. and R. Happel (1993): The probability of permanence. *Math. Biosci.* **113**, 25–60.

Stadler, P. F. and P. Schuster (1996): Permanence of sparse autocatalytic networks. *Math. Biosci.* **131**, 111–134.

Stadler, P. F., P. Schuster and A. S. Perelson (1994): Immune networks modeled by replicator equations. *J. Math. Biol.* **33**, 111–137.

Stewart, F. (1971): Evolution of dimorphism in a predator–prey model. *Theor. Pop. Biol.* **2**, 493–506.

Sugden, R. (1986): *The Economics of Rights, Co-operation, and Welfare.* Oxford: Blackwell.

Svirezhev, Yu. M. and Logofet, D. O. (1983): *Stability of Biological Communities.* Moscow: Mir.

Svirezhev, Yu. M. and V. P. Passekov. (1990): *Fundamentals of Mathematical Evolutionary Genetics.* Dordrecht: Kluwer.

Swinkels, J. (1993): Adjustment dynamics and rational play in games. *Games Econ. Behav.* **5**, 455–484.

Szucs, J. M. (1993): Equilibria and dynamics of selection at a diallelic autosomal locus in the Nagylaki–Crow continuous model of a monoecious population with random mating. II: The case of constant fertilities. *J. Math. Biol.* **31**, 601–609.

Takens, F. (1994): Heteroclinic attractors: Time averages and moduli of topological conjugacy. *Bol. Soc. Brasil. Mat.* **25**, 107–120.

Takeuchi, Y. and N. Adachi (1980): The existence of globally stable equilibria of ecosystems of the generalized Volterra type. *J. Math. Biol.* **10**, 401–415.

Takeuchi, Y. and N. Adachi (1983): Existence and bifurcation of stable equilibrium in two-prey, one-predator communities. *Bull. Math. Biol.* **45**, 877–900.

Takeuchi, Y. and N. Adachi (1984): Influence of predation on species coexistence in Volterra models. *Math. Biosci.* **70**, 65–90.

Takeuchi, Y., N. Adachi and H. Tokumaru (1978): Global stability of ecosystems of the generalized Volterra type. *Math. Biosci.* **42**, 119–136.

Taylor, P. D. (1979): Evolutionarily stable strategies with two types of players. *J. Appl. Prob.* **16**, 76–83.

Taylor, P. D. (1989): Evolutionary stability in one-parameter models under weak selection. *Theor. Pop. Biol.* **36**, 125–143.

Taylor, P. D., and L. Jonker (1978): Evolutionarily stable strategies and game dynamics. *Math. Biosci.* **40**, 145–156.

Taylor, P. J. (1988): The construction and turnover of complex community models having generalized Lotka–Volterra dynamics. *J. Theor. Biol.* **135**, 569–588.

Thomas, B. (1985): Genetical ESS-models I and II. *Theor. Pop. Biol.* **28**, 18–49.

Thomas, B. (1985): On evolutionarily stable sets. *J. Math. Biol.* **22**, 105–115.

Thomas, L. C. (1984): *Games, Theory and Applications.* New York: Wiley.

Troger, H. and A. Steindl (1991): *Nonlinear Stability and Bifurcation Theory. An Introduction for Engineers and Scientists.* Wien: Springer.

van Damme, E. (1991): *Stability and Perfection of Nash Equilibria* (2nd edn). Berlin: Springer.

van Damme, E. (1994): Evolutionary game theory. *European Economic Review* **38**, 847–858.

van den Driessche, P. and M. L. Zeeman (1997): Three-dimensional competitive Lotka–Volterra systems with no periodic orbits. *SIAM J. Appl. Math.*, to appear.

Vega-Redondo, F. (1996): *Evolution, Games, and Economic Theory.* Oxford University Press.

Verhulst, F. (1990): *Nonlinear Differential Equations and Dynamical Systems.* Berlin: Springer

Vickers, G. T. and C. Cannings (1988a): On the number of stable equilibria in a one–locus, multi–allelic system. *J. Theor. Biol.* **131**, 273–277.

Vickers, G. T. and C. Cannings (1988b): Patterns of ESS's. *J. Theor. Biol.* **132**, 387–408 and 409–420.

Vincent, T. L., Y. Cohen and J. S. Brown (1993): Evolution via strategy dynamics. *Theor. Pop. Biol.* **44**, 149–176.

Weibull, J. (1995): *Evolutionary Game Theory.* Cambridge, MA: MIT Press.

Weissing, F. (1991): Evolutionary stability and dynamic stability in a class of evolutionary normal form games. In: R. Selten (ed), *Game Equilibrium Models I.* Berlin: Springer.

Weissing, F. J. (1996): Genetic versus phenotypic models of selection: can genetics be neglected in a long-term perspective? *J. Math. Biol.* **34**, 533–555.

Yodzis, P. (1990): *Introduction to Theoretical. Ecology.* Harper and Row.

Young, H. P. (1993): Evolution of conventions. *Econometrica* **61** 57–84.

Zeeman, E. C. (1980): Population dynamics from game theory. In: *Global Theory of Dynamical Systems.* Lecture Notes in Mathematics **819**. New York: Springer.

Zeeman, E. C. and M. L. Zeeman (1994): On the convexity of carrying simplices in competitive Lotka–Volterra systems. In: K.D. Elworthy *et al.* (eds), *Differential Equations, Dynamical Systems, and Control Science.* pp. 353–364. Marcel Dekker

Zeeman, M. L. (1993): Hopf bifurcations in competitive three-dimensional Lotka–Volterra systems. *Dynamics and Stability of Systems* **8**, 189–216.

Zeeman, M. L. (1995): Extinction in competitive Lotka–Volterra systems. *Proc. AMS* **123**, 87–96.

Zeeman, M. L. (1996): On directed periodic orbits in three-dimensional competitive Lotka–Volterra systems. In: M. Martelli *et al.* (eds), *Proceedings of the International Conference on Differential Equations and Applications to Biology and to industry*, pp. 563–572. River Edge, NJ: World Scientific.

Index

Printed in the United States
By Bookmasters